The Trans-Alaska Pipeline Controversy

Technology, Conservation, and the Frontier

Peter A. Coates

T0136775

University of Alaska Press
1993

Library of Congress Cataloging-in-Publication Data

Coates, Peter A., 1957-
   The Trans-Alaska pipeline controversy : technology, conservation,
and the frontier / Peter A. Coates. -- 1993 pbk. ed.
      p.   cm.
   Originally published: Bethlehem, Pa. : Lehigh University Press ;
London ; Cranbury, NJ : Associated University Presses, c 1991. With
new pref.
   Includes bibliographical references and index.
   ISBN 0-912006-67-6 (pbk. : alk. paper) : $24.95
   1. Trans Alaska Pipeline System. 2. Petroleum pipelines-
-Environmental aspects--Alaska. 3. Trans Alaska Pipeline System-
-Public opinion. 4. Environmentalists--United States-- Attitudes.
5. Conservationists--United States--Attitudes. 6. Alaska-
-History--1867-1959. 7. Alaska--History--1959- I. Title.
TD195.P4C63 1993
333.8'2314'09798--dc20   93-32931
                                                                      CIP

© 1991 by Associated University Presses, Inc.
Original hardcover edition published in the United States by
         Lehigh University Press and in London and Toronto by
         Associated University Presses.

1993 paperback edition by the University of Alaska Press.
All rights reserved. First printing of paperback edition.

International Standard Book Number: 0-912006-67-6
Library of Congress Catalogue Number: 93-32931

Printed in the United States of America by Thomson-Shore, Inc.

This publication was printed on acid-free paper that meets the
minimum requirements for the American National Standard for
Information Science—Permanence of Paper for Printed Library
Materials ANSI Z39.48-1984.

Cover design by Dixon Jones, IMPACT/Graphics, Rasmuson Library.
Cover photograph by Kenneth R. Kollodge, Alaska Chromes, Inc.

For Graziella, Lotte, and Allan

and to the memory of Prince William Sound

Contents

Maps and Charts

Preface to the 1993 Edition

I finished this study of the Trans-Alaska Pipeline and earlier conflicts over massive construction projects in Alaska with a flourish, warning that frontier and environmental historians ignored Alaska at their peril. That may have struck some as melodramatic, though no reviewer has raised this. I certainly might have been more explicit about the relationship between Alaska and historians of the American West. This seems an appropriate opportunity to pursue the topic a bit further, especially since it's the centenary (almost to the day) of Frederick Jackson Turner's delivery of his frontier thesis in Chicago. Turner and the concept of the frontier are seriously out of favor these days. A recent survey of western history, which hopes to displace that hardy perennial, Billington and Ridge's *Westward Expansion*, as the basic undergraduate text, makes the point by avoiding the ethnocentric "F" word altogether. (If we ban or boycott undesirable and pejorative words, has the crucial part of the battle against entities like racism, sexism and imperialism been won? I am reminded of the purging of "nigger" from *Huckleberry Finn*.) Yet like it or not, frontier remains a preoccupation for Alaskans. An indispensable part of their vocabulary, the frontier as a cultural factor continues to influence perceptions. And no matter how much the New Western History grumbles about the sloppy, interchangeable use of the terms frontier and West, for most of the Alaskans and other Americans I deal with, the West is the frontier and the frontier is the West.

The conviction of boosters and conservationists locked in the struggles recounted here that Alaska belongs to the larger saga of the national frontier and the West has not been shared by most western historians. Even though Walter Prescott Webb liberated the study of the frontier from parochialism in the 1950s by locating western American history within the global dialectic between colonial periphery and European metropole, he, too, insisted that "the Great Frontier" had ended along with the specifically American frontier in 1890. He had nothing but scorn for claims regarding the survival of frontier regions. Alaska was for him a "rejected" land, one of those empty, unappropriated territories that were climatically unsuited for settlers from the temperate zone to recast in their own image. Whereas the protagonists in my story have sided with Turner in his abstract, rather vague, even metaphysical approach to the West as a state of mind and a process as well as a particular place, most western historians since the 1930s have gone along with Webb and the more tangible, if narrow, approach to the West as a specific region. And that regional conception has usually excluded Alaska. But here also, a strong case can be made for inclusion.

In 1971 in the *Western Historical Quarterly*, W. Eugene Hollon called for greater attention to Alaska, a point reiterated in the same journal by Howard Lamar in 1986.[1] With the exception of Orlando Miller, Morgan Sherwood and Melody Webb, few have approached Alaska within the broader context of western and frontier history.[2] David Wrobel in his study of the impact of perceived frontier closure on all aspects of American life recognizes the existence of Alaska as a self-styled last frontier but there are no more than a handful of brief references.[3] Historians on the cutting edge of the post-Turner, post-1890 New Western History (which always seems to be capitalized) have begun to pay lip service. As noted in my concluding remarks, Patricia Nelson Limerick, in her influential revisionist work, *The Legacy of Conquest* (1987), asserted that many of the patterns and processes of twentieth-century western history also applied to Alaska, regretting that she had neither time nor space to include it in her tale.[4] Part of the problem may simply be that not enough western, frontier or environmental historians have been to Alaska. (With great circle jet travel, Britain is closer to Alaska than are many other parts of the United States. Compare a top-down polar perspective on the globe with the traditional Mercator's projection.) When more do venture north, they might change their minds—as Walter Prescott Webb did. In 1962, a year before his death, he did some summer teaching at the University of Alaska Fairbanks, and was captivated by the frontier essentials of the new state (see my third chapter).

Something similar seems to have happened recently to prominent members of that growing breed of western-cum-environmental historians, the exposure of the environmental impact of white colonization being central in the endeavor to reveal the less flattering underside of western history. Donald Worster visited Alaska while working on *Under Western Skies: Nature and History in the American West* (1992) and, as he explains in the preface, "could not resist the temptation to write something about that extraordinary place and its history, so little known in the Lower 48, so resonant with the western past." The longest essay in his collection is a potted petroleum history of Alaska, focusing on the Trans-Alaska Pipeline. But Worster does not engage with the frontier analogy and western theme and his Alaskan case study appears to be mainly the product of personal whim: "Some readers may wonder why I include a piece on Alaska, which by my own definition is on the periphery of the region we call the West, indeed constitutes a whole region unto itself."[5] William Cronon, telling a story about an Alaskan ghost town and abandoned mining complex, is struck by "just how *western* it is…nearly everything about it evokes the West."[6] Better late than never, as they say. Clyde Milner's collection, *Major Problems in the History of the American West* (1989), also provides a little northern exposure, with an essay by Alfred Runte on the batch of Alaska national parks created in 1980.[7] Milner's maps of the U.S., incidentally, continue the tradition of including Alaska as a small inset in the bottom left-hand corner, reduced to the size of New Mexico.

Yet Alaska plainly falls within lately formulated criteria of westernness. Michael P. Malone has offered four "fundamental bonds" that define the West as a region: aridity, heavy dependence on the federal government, the fresh legacy of the frontier process, and the persisting importance of extractive industries.[8] Alaska, though included explicitly only in connection with the second point, fully satisfies three of these criteria. While some historians identify hydraulic society and pastoralism as the "modes of production" that give western history its special flavor, the historian of Alaska wishing to take a wider view is also likely to home in on extractive industries as the bedrock and leitmotif of the region's white history. Webb was conventional enough to share Turner's bias toward the agrarian West—even if his West was further west. But if the process of frontiering is treated more expansively as the exploitation of all unappropriated natural resources (and agriculture, after all, was usually a case of soil mining rather than land husbandry), then the rush to exploit the Prudhoe Bay oil fields underscored the vibrancy of the frontier process in North America.

As for the role of the federal government, World War II was an important divide in Alaska, too (see chapter 2), with the territory participating fully in the military West, though Gerald Nash omits it from his coverage in *The American West Transformed* (1985).[9] And let us not forget that some sixty percent of Alaska remains federally owned and that this proportion is unlikely to decrease.[10]

As well as meeting these three conditions, Alaska's candidacy also arguably satisfies the first: aridity. The southeastern panhandle may represent the northernmost thrust of the damp Pacific Northwest coastal bioregion but subarctic and arctic Alaska experience between twelve to twenty and eight to twelve inches of annual precipitation respectively. Historians (especially environmental) should not overlook that elementary geographical and meteorological truth: deserts can be cold as well as hot. (Don't be fooled by all that bog.) The effect of rigorous climate (if not water shortage, though water supplies are a real problem for residents of arctic communities like Barrow) has been much the same as in the American Southwest; a predominantly urban, "oasis" society.[11] Some may also need reminding of what Orlando Miller wrote in his epilogue to *The Frontier in Alaska and the Matanuska Colony* (1975).[12] Despite the sourdough mystique, "The Last Frontier" license plates and the "white Indians" dotting the bush who were popularized by John McPhee's best seller, *Coming Into the Country* (1977), the majority of Alaskans are urban folk and live in Anchorage, which, according to the wags in Fairbanks, is at least near Alaska. Add Fairbanks and Juneau and the urban proportion is overwhelming.

Gerald Thompson goes further than Malone, outlining ten unifying elements that, by consensus of his students (in Toledo, Ohio) define the contemporary West. To aridity, mining, and the federal government are added cowboys, Indians, mountains, Hispanic influence, space, long territorial experience, and urbanization.[13] If we discount the first and the fourth—the western garb of Texan and Oklahoman oil field workers and of truckers on the pipeline haul road, and forlorn Mexican restaurants like the one at Tok Junction (is it still there?) don't really count—we find that Alaska meets at least seven out of Thompson's ten. Mountains are no problem. Alaska has by far the biggest. But let me dwell briefly on space. Not the final frontier kind of Star Trek but the old-fashioned terrestrial sort that old fogey Turner was talking about when he defined the frontier according to the federal census's criterion of population density. Geographers have often been more perspicacious than historians and, as pointed out by the urban planner Frank J. Popper in 1986, measured by the classic yardstick of fewer than two people per square mile, ninety-six percent of Alaska qualified as frontier in 1980—the largest percentage of any state. Alaska is central to and a natural part of Popper's argument. There is no sense of needing to justify its inclusion.[14]

Other western historians refuse to be shackled by any strict regional definition of the West, stressing its infinite variety and the predominance of subregions—a catholic approach that easily accommodates Alaska as a subregion of its own or as part of the larger Pacific Northwest. Whether this book constitutes New Western History (Alaska Branch), however, is a tricky question. Perhaps it takes too seriously the popular need for and appeal of Old West, Turnerian mythology, as expressed in gung-ho, triumphalist boosterism *and* the thirst for that environmental purity apparently experienced by the first pioneers on the eve of the frontier process.

Alaska exemplifies many standard themes and fixations of the New Western History: the equivocal nature of territorial/state attitudes to Washington, D.C.; the dichotomy between bombastic, individualistic frontier rhetoric and the reality of big capital and corporate control; the urge to conform to the rest of the Union and follow the past, notwithstanding the ethos of conscious apartness and novelty found in frontier regions; the federal government's powerful say even after statehood; the significance of World War II as a watershed and the role of other external events as driving forces in western history; and, by no means least, what Limerick calls the "persistence of Natives."[15] As elsewhere in the West, Alaska's indigenous population is rising and may soon outstrip its precontact level.

Despite its geographical remoteness from the metropole and the frequent emphasis on the separate evolution of frontier zones, much of Alaska's Euro-American history has been fashioned within a wider world. Since the arrival in the 1740s of the first *promyshlenniki*, Alaska has been sucked into a growing global economy, functioning as a storehouse furnishing raw materials to various metropoli—especially in the form of energy. Alaska's role within what we might as well call the capitalist world system intensified when whale oil lit the streets of nineteenth-century American cities. (No wonder cartoons showed whales in partying mood when oil was struck in Pennsylvania in 1859.) The major protagonists in Alaska's Euro-American history have been a succession of outsiders conforming broadly to a single model—whether Yankee whalers, Kennecott copper kings, Seattle salmon trusts or London-based British Petroleum.

Another hallmark of the American West and its recent history is the clash between unmodified and reconstructed (postfrontier?) values over the best and proper allocation of natural resources (different uses often being preclusive)—as well as a deeper dispute over the very commodification of nature into natural resources. When British undergraduates embark on my U.S. environmental history course they are bemused when informed that issues like wolf control in Alaska make the front pages not only of local and Seattle newspapers but also the *New York Times* and get picked up and distributed worldwide by CNN. Emotive wildlife issues aside, the politics of nature in the American West since the 1950s have been dominated by energy issues. And wildlife and energy issues often become entangled, like dolphin in drift nets, as demonstrated in recent years by the *Exxon Valdez* oil spill and debate over the future of the Arctic National Wildlife Refuge (ANWR). As I wrapped up this manuscript in the spring of 1989, oil was gushing into Prince William Sound and I was struggling like a seal to keep ahead of the fast-moving slick of events and publications. At least two books on the spill appeared before my own broader history.

During the controversy prior to the pipeline's construction, diffuse public fears generated by the Alyeska project and general questionings of the autocratic thrust of technology and a destructive way of life based on excessive consumption of nonrenewable fossil fuels crystallized around terrestrial concerns: the fate of the Brooks Range and arctic caribou herds, perceived to be the last great American wilderness and wildlife spectacle. Only a handful of Alaskans, it seems, especially fishermen in Southeast, paid much attention to threats associated with the marine leg of the transportation system. And as it turned out, the charismatic sea otter (in the same higher bracket as the giant panda when it comes to the human hierarchy of animal value), became the focus of popular outrage as supreme representative of a once-pristine *watery* Alaskan wilderness.

Those who had fought the pipeline project felt massively deceived and betrayed. The promise to adopt double-bottomed tankers was never delivered nor enforced by federal regulatory agencies. Some would argue that the State of Alaska's casual attitude to regulation was as culpable as Big Oil's shoddy approach to safety and environmental protection. Alyeska's watchdog served instead (as watchdogs have a habit of doing) as Alyeska's poodle. Even at the height of confusion, drift and ineptitude, as the slick spread unchecked, there was never any thought that the State of Alaska would apply its ultimate sanction ("the Nuke") and shut down the pipeline. For a state dependent for some eighty percent of its revenue on oil taxes and royalties, that would have been a self-inflicted wound.

Happenings far away, especially war and world oil prices, continue to dictate Alaskan affairs. As I put down my manuscript in 1989, *the* environmental flash point in the U.S. was drilling in the Arctic National Wildlife Refuge. The Iraqi invasion of Kuwait in August 1990 ended the fifteen-month breathing space opponents of drilling on the coastal plain of ANWR gained from the Exxon spill. In November 1991, Senate opponents successfully filibustered an administration backed energy bill largely due to its provision for opening the refuge to drilling. An energy bill went through Congress in 1992 minus the ANWR drilling provision, adding credence to the view that a decisive corner has been turned. As one opponent of drilling declared in 1991, employing a rather insensitive image that brought to mind the Hickel Highway: "We have drawn a line on the tundra." Encouraged by the Democrats' recapture of the White House and Mr. Ozone's elevation to the vice-presidency, environmentalists feel the climate is receptive for a bid to secure permanent protection for all of ANWR through wilderness designation for the coastal plain.

Meanwhile, a thoroughly and proudly unreconstructed Walter Hickel is back in the governor's mansion. Elected in 1990 as an independent on the strength of urban votes and on a platform virtually indistinguishable from that of 1966, Hickel is again touting a medley of fabulous schemes. Most fanciful are his plans for an offshore pipeline to carry Alaskan water to parched southern California (reminiscent of Ralph Parsons's land-based NAWAPA project advanced in the 1960s to divert southward the flow of rivers like the Yukon and Tanana), and for a rail tunnel under the Bering Straits. (How long before I can buy a ticket from London Victoria through to Fairbanks?) Of more immediate relevance is the revived Yukon Pacific Corporation proposal for a gas pipeline paralleling TAPS to Valdez, a project in which the governor has considerable personal interests. Advocates believe the project will be economically viable early next century, when oil production on the North Slope declines and the gas is no longer required to enhance oil recovery. Atlantic Richfield has plans for a different line to carry natural gas from northwest Canada to a terminal on the Kenai Peninsula. Both proposed lines run through Fairbanks. Will the city be able to apply effectively the lessons learnt from the TAPS construction boom of the mid-1970s and emerge less scathed by the impacts of rapid growth on the human community?

In the interim before pipe is ordered for any gas line (the size of the U.S. trade deficit with Japan makes one wonder if the steel will come from across the Pacific this time), there will be plenty to occupy public debate over economic development and environmental protection in Alaska: among other things, the consequences of the Alaska Native Claims Settlement Act (ANCSA) of 1971, the wrangling over whether to spend the State of Alaska's share of the Exxon spill settlement funds on acquiring threatened lands, and the controversy over the subsistence hunting provisions of the Alaska National Interest Lands Conservation Act of 1980 which gave priority to "rural residents" but conflict with the natural resource provisions of the state constitution. If I were to rewrite and expand the book in hand, Native Alaskan interests and perspectives would move in from the periphery. After all, the Arctic as a whole is the theatre not only of some of the world's most tendentious environmental conflicts. It is a major arena for confrontations between the first and the fourth worlds (first world and first nations?) and some of the most powerfully articulated indigenous rights campaigns. ANCSA granted Native Alaskans an ambiguous autonomy and equivocal guarantee of cultural survival. Though granting sixteen percent of the Alaskan population almost twelve percent of the state's lands, ANCSA did so within a Euro-American framework and on corporate terms. The lands and the cash settlement were allocated to village and regional corporations in which individual Natives hold shares. According to ANCSA's original provisions, individual shares could be sold on the open market at the end of 1991. Amendments in 1987 maintained the inalienability of stock until a majority of shareholders vote to remove restrictions on sale and ensured exemption of undeveloped lands from tax and creditors.

Yet hundreds of thousands of Native-owned acres constitute inholdings within national parks, forests, and wildlife refuges. Native corporate priority (like corporate priority elsewhere) is to generate revenue for shareholders and many corporations limp along on shaky foundations (please excuse the mixed metaphor). Some village units are facing bankruptcy. If environmentalists cannot find the means to buy these inholdings—and the Hickel administration is ideologically hostile to public purchase of private lands—Native corporations will lease to the logging industry. Despite the revival of a tribalism skeptical of moves to alienate the common inheritance, Native Alaska, through internal and external pressures, is likely to play an increasingly prominent role in natural resource extraction. The joint exploitation of one of the world's most valuable lead and zinc deposits at Red Dog may be the first of a series of cooperative ventures between Natives and white multinational corporations.

At least one Native corporation is building a timber mill, and the destination for much of its output will be Asia. The post-1945 American West has been drawn increasingly into the orbit of the Pacific Rim. Alaska, which has always looked west to the far east as well as southeast to the rest of the U.S., may be reverting to its former role as part of the Pacific and Asiatic world. North Slope oil still cannot be exported but the advocates of gas pipelines also have Japan as well as Taiwan and Korea in mind. Old-growth Sitka spruce, sold by the U.S. Forest Service at heavily subsidized prices, are chewed up by the Japanese-owned Alaska Pulp Corporation, converted into rayon in Japan, then made into clothing in Korea and finally exported to the U.S. The Native economy, too, is becoming integrated into Asia's. Salmon roe from the Yukon, for example, is exported en bloc to satisfy Japanese demand. For all the rhetoric about the "owner state" (shades of the 1980s Sagebrush Rebellion), what are the chances of Alaska successfully transcending its colonial economy, even if that economy, like those of some subsaharan African nations, is geared to tourism's "non-consumptive" uses of nature?

A number of reviewers have noted that less than half of this book actually deals with the Trans-Alaska Pipeline itself. Don't judge a book by its cover (or title)! One of the major topics I covered in my attempt to show that conflict between grand technological enterprises and those with other visions of Alaska's future has a respectable history was the Atomic Energy Commission's Project Chariot (see chapter 4). The relationship of science and scientists to environmental debates and structures of economic and political power is an underdeveloped theme in the book and others are taking such matters far beyond what I was able to accomplish. Chariot became a public issue again in the summer of 1992 as a result of Dan O'Neill's archival investigations under the Freedom of Information Act. Documents revealed that in 1962 (just when it was officially abandoning Chariot) the then Atomic Energy Commission and U.S. Geological Survey conducted another kind of experiment. Some 15,000 pounds of contaminated soil, consisting of fallout from a Nevada test and by-products of reactors, was distributed at Cape Thompson. The aim was to observe the effects of erosion and leaching on radioactive material. After the experiments, the material was consigned in drums to a burial mound. O'Neill's findings prompted a flurry of investigations and a revival of the fears experienced by the Native community thirty years ago. This summer the Department of Energy plans to ship the material by barge to a nuclear waste site at Hanford, Washington. But one of Chariot's loose ends has been tied up. In May 1993, two of the dissident scientists, William Pruitt and Leslie Viereck, received honorary doctorates from the University of Alaska at Fairbanks, their employer at the time the controversy over the project's environmental impact studies broke into the open.

Though I write this preface no closer to Alaska than when I composed the original, I have shifted considerably further west, at least by modest British standards, and the Americanist in me derives satisfaction from facing the new world. Much of the Atlantic world was drawn into Bristol's commercial domain between the initial fifteenth-century forays of local fishermen into the rich cod banks off Newfoundland and the heyday of the triangular trade. But the city's arctic (let alone Alaskan) connections are practically nonexistent. Before I moved here, these links consisted of the local BBC Natural History Unit's Alaskan documentaries and an incident in October 1577. The arctic explorer, Martin Frobisher, a West Countryman like many of that era's adventurers, landed in Bristol after another futile voyage in search of the elusive Northwest Passage. He brought with him not only fool's gold but a family of three "Eskimaux," "savage people" from Baffin Island. Much to the delight of the locals, the male paddled around the harbor in an animal-skin kayak harpooning duck. All succumbed within a month, their deaths attributed variously to pneumonia, the mild and damp climate, and their inability to eat anything but raw flesh. The doctor's report on the man's death regretted that the Queen had been deprived of the chance to see him.[16]

I am grateful that, despite my Alaskan hankerings, I've proved more resilient and adaptable. This new edition is for the two Bristolians who have made my life more complicated since 1991. Giuliana is old enough to tell the difference between a moose and a caribou—though not between a caribou and a reindeer. Ivana would make a tasty morsel for a grizzly just emerged from winter and feeling peckish. (How hard it is not to perpetuate those bear stories!) They will get to that northern paradise soon. And the sooner the better, for the ecological systems of the world's far northern regions as a whole are embattled regardless of distance from the world's industrial centers. Chernobyl (1986) restated Project Chariot's lessons about the susceptibility of arctic food chains to contamination. The phenomenon of arctic haze and accumulations of auto and factory emissions from Europe in arctic ice packs underline for the early 1990s the one-ness and vulnerability of all planetary natural systems in the same way that concentrations of DDT in the fatty tissues of antarctic penguins symbolized the global nature of the environmental crisis in the early 1970s.

Building on the best of indigenous knowledge and awareness, Alaska remains one of the last best hopes for humankind to pioneer a gentler way forward in its relations with the rest of nature, a new softer path requiring harder decisions about daily lives and everyday environments. Much more is demanded than the old formula geared to the creation of special, prelapsarian places like parks and wilderness areas which are invested with higher value than ordinary places. For though this expresses a profound and paradoxical Euro-American cultural urge for people-free zones, for nature without cultural defilement, for safe havens where nobody belongs unless just a visitor, it tends to perpetuate the apartness of people and nature and reinforce the old culture-nature dichotomy. Nor is scrubbing every pebble to restore pristine, pre-slick glory sufficient.

At least there are still no Winnebago motor homes and Avis rental cars permitted north of Disaster Creek at Dietrich on the Dalton Highway. If I want to dip my feet in Prudhoe Bay again I may have to stick out my thumb once more.

—*Peter Coates*
Bristol, England
July 1993

NOTES

1. Hollon, "Adventures of a Western Historian," *Western Historical Quarterly* 2 (July 1971): 258; Lamar, "Much to Celebrate: The Western History Association's Twenty Fifth Birthday," Ibid., 17 (October 1986): 415.

2. Orlando W. Miller, *The Frontier in Alaska and the Matanuska Colony* (New Haven, Conn.: Yale University Press, 1975); Morgan B. Sherwood, *Big Game in Alaska: A History of Wildlife and People* (New Haven: Yale University Press, 1981); Melody Webb, *The Last Frontier: A History of the Yukon Basin of Canada and Alaska* (Albuquerque: University of New Mexico Press, 1985).

3. David M. Wrobel, *The End of American Exceptionalism: Frontier Anxiety from the Old West to the New Deal* (Lawrence: University Press of Kansas, 1993).

4. Patricia Nelson Limerick, *The Legacy of Conquest: The Unbroken Past of the American West* (New York: W. W. Norton, 1987), 26.

5. Donald Worster, *Under Western Skies: Nature and History in the American West* (New York: Oxford University Press, 1992), ix, viii.

6. William Cronon, "Kennecott Journey: The Paths out of Town," in *Under an Open Sky: Rethinking America's Western Past*, eds. William Cronon, George Miles and Jay Gitlin (New York: W. W. Norton, 1992): 32. Note how the titles of these proliferating collections of essays are practically indistinguishable.

7. Clyde A. Milner II, ed., *Major Problems in the History of the American West* (Lexington: D.C. Heath, 1989).

8. Michael P. Malone, "Beyond the Last Frontier: Toward a New Approach to Western American History." *Western Historical Quarterly* 20 (November 1989): 417.

9. Gerald D. Nash, *The American West Transformed: The Impact of World War Two* (Bloomington: Indiana University Press, 1985), has no index entry for Alaska but does nod tentatively toward the far north (ix).

10. Malone, "Beyond the Last Frontier," 418. Malone, despite the Alaska Native Claims Settlement Act and the provisions of statehood, declares that ninety-six percent of Alaska is federally owned.

11. For the term "oasis" society see Gerald D. Nash, *The American West in the Twentieth Century: A Short History of an Urban Oasis* (Albuquerque: University of New Mexico Press, 1977): 5, 299.

12. Miller, *The Frontier in Alaska*, 228-29.

13. Gerald Thompson, "Another Look at Frontier/Western Historiography," *Montana The Magazine of Western History* 40 (Summer 1990); as reprinted in *Trails: Toward a New Western History* eds. Patricia Nelson Limerick, Clyde A. Milner II, and Charles E. Rankin. (Lawrence: University Press of Kansas, 1991): 95.

14. Popper, "The Strange Case of the Contemporary American Frontier," *Yale Review* 76 (December 1986); as reprinted in *Major Problems in the History of the American West*, ed. Milner: 657. Popper is the chief theoretician of the "Buffalo Commons." Alaska, of course, has had this exotic, shaggy ingredient of the Old West grafted on at Big Delta.

15. Limerick, *Legacy of Conquest*, title of chapter 6.

16. Vilhjalmur Stefansson, ed., George Best, *The Three Voyages of Martin Frobisher in Search of a Passage to Cathay and India by the North-West, A.D. 1576-8* (1578) London: The Argonaut Press, 1938): 237-40.

Preface

One evening in mid-August a few years ago I stood beside the old Trans-Alaska Pipeline haul road, now called the Dalton Highway, just beyond Coldfoot in the southern Brooks Range. At the beginning of the century, in its heyday, Coldfoot consisted of a gambling den, two roadhouses, two stores, seven saloons (including ten prostitutes), and no churches. Now there's only a cemetery with a gas station, motel, and service area nearby. The last place to buy fuel, a cup of coffee, or a hamburger is 120 miles south at the Yukon River Bridge. There are no further facilities until Prudhoe Bay, 240 miles ahead at the end of the road. As I waited, another enormous rig rolled north to the oilfields. A tremendous cloud of dust in its wake left me coughing. The plume was visible long after the truck had passed out of sight and for a long time after it could no longer be heard. I had spent the afternoon in the "Furthest North Truckstop in the World"—probably the most expensive too. I couldn't make the seventy-five cent tea bag last forever and was getting nowhere soliciting rides. So I left for some cooler, fresher air, hoping for a change of luck. It did cross my mind that it was maybe time to return south to the security of the subterranean university archives in Fairbanks. After all, there had been no shortage of adventures during my excursion so far, and I was fortunate to have traveled this distance.

A short ride from two Fairbanksans out in search of huckleberries had taken me to Livengood, where the Dalton Highway begins. North of Livengood, I resumed my journey with Sam, a man with long black hair and a frightened look. He'd been to Fairbanks to have an outboard motor repaired and to resupply. He was heading back to his gold claim in the southern foothills of the Brooks Range. After we crossed the Yukon River, he turned off the road and drove down to the gravel beach where the ice bridge used to be. He refueled from spare cans. Gas cost twice as much there as it did in Fairbanks. So he always carried enough for the trip to his claim and to get him back to Fairbanks. Beyond the Yukon, crossing open country, his side window suddenly shattered. He slammed on the brakes, pulled a .44 magnum from under my seat and told me to get out and keep down. Having pushed me out, he slid out my side and crouched beside me. "Some sonofabitch

11

is shooting at us," he exclaimed. (Another analysis of the startling incident might point to a rock flung up by his nearside front tire.) Our next stop was the official sign announcing the Arctic Circle. Here we encountered a young Londoner on the first leg of a bicycle trip from Prudhoe Bay to Tierra del Fuego. Sam was again astounded to meet someone who wasn't armed. ("I'm surprised you didn't pick up a piece in Fairbanks," he had remarked to me.) The Cockney cyclist was running low on food. He hadn't realized there was no place to buy supplies in any quantity before the Yukon. He'd resorted to panhandling from crews at pump stations. My colleague gave the bemused adventurer a small cannister of mace to deploy in the event of a tricky encounter with a griz.

Feeling game for almost anything, I accepted an invitation to spend the night at the mining camp. The equipment and supplies Sam had brought were loaded into a boat under the bridge over the south fork of the Koyukuk near where we parked. I walked in the few miles upstream to the camp. My armed escort stumbled over every root and tensed up at the slightest rustle in the bushes. The men at camp were a mix of young and old, part hippy, part hillbilly. A few of them had worked on the pipeline. There were a lot of beards and ponytails and a small arsenal. Gun stories were the staple of conversation. Men were comparing weapons and blasting away at rocks on the other side of the river, despite the regulations prohibiting the discharge of guns within the pipeline corridor. Sam's pride and joy was a sinister looking Israeli automatic carbine. I was offered a shot but politely declined. The miners fired at anything that moved and I wondered if there was a creature left for miles.

A pale lad of about fifteen called Dan, who had run away from his family in Arkansas, spent his time fishing and doing camp chores. He had teamed up with Hank, an older man with a Jack Nicholson smile who cursed the international bankers who were sucking the wealth of the U. S. and denounced the federal government and big business for cutting out the small miner; "that's not the American way." Sam wanted to take his Cat across the river to excavate another stream. But it was too late in the year and he told me the EPA said it would interfere with spawning. "What about the beaver who pisses upstream?" he wanted to know. He believed the EPA should fine the beaver for polluting the water instead of threatening him and his crew. The men tossed their beer cans into the river with regularity. Despite the beer cans and the beaver urine, the grayling Dan hauled onto the bank (every minute it seemed) were sweet and tender, and I polished off a mess for supper and again for breakfast. Dan agreed they were "fine eating." Almost

as good as catfish, he opinioned, but not as tasty as burbot by a long
stretch. Still visible to the west in the half light was the gleaming, silver
Trans-Alaska Pipeline. To the east was a vast expanse of spruce forest.
"There's a lot of country out there," commented Hank laconically. It
tried to snow and there was a whir of chopper blades as pipeline security
flew overhead on its daily round.

A Frontier Transportation Company truck, air brakes wheezing, disturbed these reflections. Gazing down from his cab, the grinning driver
stressed that picking up riders was strictly against company regulations.
He stopped because I looked unarmed. He didn't want to read in the
paper when he got home to Anchorage about a hitchhiker who was
mauled by a grizzly in the Brooks Range. (From what I'd seen, the
humans in the Arctic were much more dangerous than any beast.) I
clambered up beside him and soon we were riding in convoy with three
other Frontier vehicles also delivering methane to SOHIO base camp
at Prudhoe. My driver, the Arctic Bandit, kept in close contact with
his colleagues over the CB.

"Who you got riding with you there?," asked a southbound driver.

"Picked up a pilgrim headed for Prudhoe. Wants to dip his feet in
the Arctic Ocean. He's a limey," he added, as if that explained everything.

Before long I spied the hamlet of Wiseman to the west across the
middle fork of the Koyukuk. The trucker told me that casual visitors
and curious strangers were not welcome there. You needed a good reason
to visit. I had planned to take a look at Wiseman, but it was hard
to turn down a ride all the way to the end of the road.

There was no trouble at the checkpoint manned by a couple of teenag-
ers for the State of Alaska Highway Department thirty-five miles north
of Coldfoot at Dietrich. Beyond this point my presence was officially
illegal. The tundra was already turning red, russet, and gold. The
U-shaped valleys and slopes reminded me of Scottish glens. The trees
had given out quite a way back down the road. My driver told me
there was once a sign marking the northernmost tree. Needless to say,
someone stole it (the tree, not the sign). It was drizzling and too misty
to see anything as we climbed over the continental divide at Atigun
Pass. Descending from the Brooks Range onto the North Slope during
the few hours of darkness in a late-summer arctic night the lights of
Prudhoe Bay could be clearly seen some forty miles away. Apart from
a fox that shot across the truck's beams, the only wildlife we saw were
some small black bears rooting in the bushes south of Atigun and a
few scattered caribou.

After a seven hour run, the Arctic Bandit set me down outside Deadhorse airport in the early hours. Understandably, he wasn't prepared

to take me into the oilfield complex. It was already light as I pitched my conspicuous orange tent on a relatively flat patch of ground near the shore of South Lake, just outside the restricted area. To my rear was an assortment of prefab huts, trailers, cranes, and trucks. Prudhoe could be any orderly industrial park anywhere, except that it happens to front onto the Arctic Ocean. The oilfield area is off limits to anyone without a permit. The only way to get to the ocean, a further seven miles, was to sneak a ride in. I stood at the fork in the road at the boundary of the oilfield complex. One road led to ARCO, the other to SOHIO-BP. The first pickup that came along took the road to ARCO. It stopped for me and we passed through the checkpoint without a hitch. As I crawled out at a safe distance from the main buildings, one of the workers gave me his lunch pack. Munching on a hard-boiled egg, I strolled off down the gravel road toward the ocean. Plenty of vehicles passed me, seemingly unconcerned. Some even returned my wave. Small mammals (lemmings?) scuttled across the road and disappeared into the flat yet bumpy, pond-pocked tundra.

I reached the sea near East Dock. The water was gray and smooth. The beach was a dark mixture of gravel and pebbles, backed by a low, muddy bank. The overcast sky blended almost imperceptibly with the waveless water. There was an abundance of driftwood and washed-up trash, including rotting oranges, but definitely no polar bear tracks. The only polar bear I was likely to see around Prudhoe was on the half-price jigsaw puzzle at the Deadhorse trading post. I removed boots and socks and waded in the disappointingly warm water. During my homeward trudge a helicopter passed low overhead. Soon afterwards, a truck came purposefully towards me. It drew up, and the security guard asked if I was aware of the gravity of my trespassing offense. The oil companies are naturally worried about industrial espionage, drug trafficking, and environmentalist sabotage. The guard tried to find out who brought me in. What did the driver look like? What type of truck was it? I was unable to provide any incriminating information. The eventual discovery of a Cambridge University Library card, for some reason the only piece of identification I had on me, seemed to convince him that I was just a harmless foreign eccentric. The absence of a camera also assisted my cause. In fact, he became quite chummy as he drove me off the property and deposited me at Deadhorse, convinced I was headed out on the next plane to Fairbanks.

The following morning dawned gray and wet. I stood beside the Dalton Highway again, wondering how far I would manage to get that day. The first southbound rig, having barely gathered momentum, slowed down for me. I climbed up and asked how far he was going. "All the way," he replied, meaning Anchorage, some 800 miles south. That afternoon I was back at Coldfoot Services drinking more tea, while my ride-giver, Bob, napped for a few hours in the sleeper at the back of his cab. All the regulars had their own coffee mugs hanging in rows on the wall behind the service counter. "Long Distance" came in. The wallpaper in the bathroom was decorated with trucks like a kid's bedroom and covered with anti-Texan and anti-Teamsters slogans. Probably the least crude was the exhortation, "Flush Twice, Jesse Carr lives in California" (Jesse Carr was boss of the powerful Alaskan Teamsters Local 959). I was grateful for a few quiet hours and the opportunity to stretch. Bob hailed from southern Missouri and, like many other Alaskans, came north during the pipeline construction boom. He was a great country-and-western fan. I was growing a little tired of endless songs about the teenage queen who worked at the candy store and left the boy next door for the bright lights, only to renounce her fancy home, swimming pool, and shiny cars to return to marry the boy next door who still worked at the candy store; despite her wealth and fame she was plain sad and lonesome and empty inside. The hard wooden box seat I had bounced on for seven hours was becoming unbearable and the handle of the cooler at my back was abrading my spine.

South of Coldfoot, while a tape I had already heard three times played again mercilessly, it started to rain heavily. One section of road was especially muddy and required all Bob's driving skills. We lurched past a Winnebago in distress but stopped to help a trucker put on chains. South of the Yukon, the leaves on the birch are still green. We had slipped back from fall into summer. A few hundred yards short of the weigh station at Fox, I jumped down and walked around to the other side through the willows, to avoid problems for Bob. I passed by a place where the pipeline dips underground following an elevated section. Fading graffito inquired, "Where will it all end?" Then Bob picked me up again for the loop around Fairbanks. After fifteen-and-a-half hours of my company, he deposited me virtually on the doorstep. He looked fresh and undaunted as he surged off at 10:30 P.M. down the Parks Highway for the final 360 miles to the end of his road. The song about the boy next door who worked at the candy store had begun to play again. Enough outdoor research. Tomorrow I would return to a softer seat and a more sedate pursuit of the Trans-Alaska Pipeline.

Acknowledgments

I have accumulated a range of more traditional debts in addition to those I owe to the people who delivered me safely to Prudhoe Bay and back. I want to thank Morgan Sherwood for his help on many occasions. I also gained from the comments and assistance of Rod Nash. I am grateful to a number of Americanists in Britain, notably John Thompson, who calmly guided the dissertation out of which this book grew. Charlotte Erickson, Rupert Wilkinson, and Steve Spackman provided useful assessments at a later stage. The Bancroft Library, the Suzzallo Library, and, especially, the Rasmuson Library, supplied ample research material. Renee Blahuta and the rest of the staff in the University of Alaska Archives and throughout the Rasmuson always offered cheerful and skilful asistance. Renee, et al. kindly countenanced my occupation of their vaults during three visits; two of them unreasonably long, the final sojourn far too short. Without the rich and eccentric collection of Alaskana at the Scott Polar Research Institute, Cambridge, England, my initial interest in Alaska could not have advanced very far. The institute provided a congenial local habitat and was ever a source of serendipity.

Bill Schneider, Ron Inouye, and Dan O'Neill provided unfailing and enthusiastic support and friendship and hospitality without limits. Dan, Gary Holthaus, and the Alaska Humanities Forum made possible a third visit to Fairbanks in the fall of 1987. Chapter four of this book has been enriched by discussions with Dan, many via trans-polar computer correspondence. (The Bitnet electronic mail service has been a blessing for someone based outside the United States. A request sent to Ron or Dan in Fairbanks in the evening usually resulted in the delivery of information on the screen next day.) Chapter four was also improved considerably by Jim Ducker's suggestions and vigilance. Michael Mc-Closkey, Bob Weeden, David Klein, Arthur Lachenbruch, Henry Coulter, James Gillett, Jim Payne, Geoff Larminie, Barbara Bodenhorn, Tony Thompson, Larry Sokoloff, Charly Stimpfig, Stephen Cutcliffe, and Lauren Lepow also assisted me in various ways. David Nelson Blair gave the manuscript the benefit of his editorial expertise and I wish to apologize again for the irritating one-and-a-half spacing. The British Academy funded three years of research. George Marshall graciously granted me access to his papers and those of his brother, Bob. Edgar Wayburn kindly allowed me to consult his papers. I also wish to thank Brock Evans, David Hales, Richard Presby, and Karyl Winn for granting permission to publish archival material. John Ratterman of Alyeska and Stephen Wofford of the Lawrence Livermore National Laboratory generously provided photographs. Chris Cromarty of the Geography Department at University College, London, contributed his cartographic skills and produced some black and white prints from color slides.

Parts of chapter 4 appeared in article form as "Project Chariot: Alaskan Roots of Environmentalism," *Alaska History* 4:2 (Fall 1989). The author wishes to thank *Alaska History* for permission to republish the material that first appeared there.

The final draft was pruned and polished on the rails during a daily commute. I gained little inspiration from the East Anglian landscape that fled by. This is a desolate region, much more deserving of the wasteland epithet than the Arctic. It has been smothered by humanity for centuries, and, more recently, laid waste by agribusiness. Nevertheless, I would like to thank the pinstripe brigade for not smothering me entirely with *The Daily Telegraph* and *The Times*. Graziella Mazza didn't type or proofread a single word. She had equally important, tropical things of her own to do these past few years. Still, she tolerated with only occasional ill humor my terminal obsession with the silver snake and a place to which she had never given a thought (let alone a positive one) before we met. Graziella and my parents, Lotte and Allan, have the tremendous misfortune never to have been to Alaska. This book is for these three impoverished souls.

Cambridge, England, 1989

Introduction

In 1977 oil began to flow south from the Arctic through the Trans-Alaska Pipeline System (TAPS)[1] to the ice-free port of Valdez for shipment to the lower forty-eight.[2] Before, during, and after construction, the project's magnitude prompted comparisons with heroic American engineering feats of the past such as the transcontinental railroads and the Panama Canal. There were also more exotic analogies. Speaking in Valdez shortly after Congress had authorized the project, Rogers C. B. Morton, the secretary of the interior, compared the pipeline to the Egyptian pyramids.[3] Elliott Growe, the chief engineer on the Canol Pipeline constructed by the U.S. Army during the Second World War to carry oil from Canada's Northwest Territories to military bases in Alaska,[4] and Frank Moolin, the senior project manager of Alyeska's Pipeline Department, compared the pipeline to the Great Wall of China.[5] Allusions of this kind fitted a well-established historical pattern. An Englishman traveling through the United States in the early 1820s reported that the Erie Canal (1817–25) was frequently compared to these ancient monuments.[6]

Journalists and other commentators hailed TAPS as the most ambitious construction scheme in American history and the most expensive private industrial undertaking in world history.[7] (Larger projects were government-financed and built, like Egypt's Aswan Dam.) Other aspects of this episode were also superlative. The pipeline proposal spawned a debate commensurate to the size of the project. Indeed, congressional approval in the summer of 1973 was preceded by a four-year controversy (1969–73) that was the most passionately fought conservation battle in American history since the controversy (1908–13) over the city of San Francisco's proposal to dam Hetch Hetchy Valley in Yosemite National Park to provide itself with a water supply from the Sierra Nevada.[8]

Plenty has been written in various categories about aspects of the multifaceted TAPS controversy. An award-winning series of articles, "The 1969 Oil Rush: Alaska's Challenge: Oil and Wilderness," by Tom Brown, a journalist for the *Anchorage Daily News,* was published in book form by the Sierra Club in 1971 under the title *Oil on Ice: Alaskan Wilderness at the Crossroads.*[9] Mary Clay Berry, another journalist who worked in Alaska, wrote *The Alaska Pipeline: The Politics of Oil and*

Native Land Claims (1975). Her book remains the best available discussion of the TAPS debate from the perspective of Alaskan Native issues.

Obviously slanted, semi-official accounts of the pipeline contest, written from the standpoints of the respective adversarial forces, were also soon available. For a version of events sympathetic to the Alyeska Pipeline Service Company (ALPS), one should consult Lawrence J. Allen, *The Trans-Alaska Pipeline, 1, The Beginning* (1975), and 2, *South to Valdez* (1976); also 3, *Emerging Alaska,* by Kristina Lindbergh and Barry L. Provorse (1977). Closest in tone and attitude is James P. Roscow, *800 Miles to Valdez: The Building of the Alaska Pipeline* (1977); a panegyrical paean to the unique technological challenges posed by the demanding Alaskan environment and triumphantly overcome by a formidable engineering achievement. Harvey Manning's *Cry Crisis! Rehearsal in Alaska (A Case Study Of What Government By Oil Did To Alaska And Does To The Earth)* (1974), represented one of the least compromising positions within the anti-TAPS alliance, that of Friends of the Earth.

The dramatic flavor and epic dimensions of the pipeline saga made it an attractive subject for popular literature. Sensationalist pulp novels, perhaps inevitably, portrayed conservationist opposition to the pipeline as extremist and subversive. Eco-terrorist plots in the style of Edward Abbey's famous fictional characters (*The Monkey Wrench Gang,* 1975), to sabotage a project that could not be stopped by other means appealed irresistibly to authors. A character in one novel is the rebellious son of a Texas oil tycoon. An archaeology student involved in the investigation of sites along the proposed pipeline route (mandated by Department of the Interior stipulations) becomes part of a plan to blow up the new bridge across the Yukon.[10] Another writer creates a more elaborate scenario. "Operation Cut-Off" is a Weathermen-style conspiracy masterminded by an oil magnate's daughter who is a drop-out from the University of California, Berkeley. The aim is to bomb all twelve pump stations along the pipeline and the operations center at Valdez. The ultimate target is the *Globtik Alamo,* the first North Slope crude-laden supertanker leaving the pipeline terminal for refineries in the lower forty-eight. An explosion in the ship's engine-room causes it to run aground in heavy fog on the notorious Middle Rock in Valdez Narrows.[11]

Academic specialists in various fields, though none of them historians, have already written on aspects of the TAPS controversy. As early as 1972, when the oil industry's application for a construction permit was embroiled in legal difficulties, Peter M. Hoffman, a law student, discussed the conflict to that date as a contribution to the development of new American legal standards concerning the environment.[12] In 1975, Henry R. Myers, a physicist, published a sound preliminary appraisal of events as a case study of the limitations of federal decision making in the absence of a national energy policy.[13] Yet to date, there has been no book-length treatment of the controversy to supersede these early accounts.[14]

At this point, it should be explained that this is not a comprehensive history of the TAPS controversy. I have not set out to analyze the nature of the political process leading to the authorization of TAPS. Moreover, unlike Hoffman and Myers, who are highly critical of the Department of the Interior's predisposition to issue the pipeline permit, it has not been my intention to judge the wisdom of the decisions by the Department of the Interior and Congress to approve the project. Furthermore, while Native Alaskan interests and concerns feature on occasion, I do not systematically explore them. It should also be emphasized that while dealing with the various interest groups involved in issues of conservation and development in Alaska, this is not a political scientist's study of special interest and regional pressure groups and their methods.

As an environmental and intellectual historian, I was most concerned with the opposition to TAPS from conservationists and, especially, with their ideologies. I believe the most interesting aspect of the pipeline controversy from a historical point of view—and certainly the least discussed—is its relationship to earlier engineering projects and technological innovations in Alaska and the debates that accompanied them. Accordingly, while inspired by the TAPS conflict, I also study events between the Second World War and 1968. In an introductory chapter, the book's scope has been extended back to the beginning of the American period of Alaskan history (1867). This is necessary to achieve a fuller understanding of the controversy's place in Alaskan history. This chapter also provides an overview and a synthesis for those unfamiliar with Alaskan history.

Tracing the evolution of conservationist concern regarding Alaska since American acquisition involves reviewing those developments that have aroused concern, investigating the changes in the nature and strength of this concern, and considering the character of the debates over these issues. This requires some attention to those who advocated such developments, especially those in Alaska itself, and particularly those who can be called boosters. I use the term *booster* in a catholic sense to describe someone (not necessarily resident in Alaska) imbued with the entrepreneurial spirit who believes that Alaska's natural resources have great material value and advocates their rapid and thorough development without government interference (preferably with government assistance). In Alaska, as elsewhere in the United States, *booster* was often a self-appellation. Though conservationists used the term pejoratively—and continue to do so—my frequent use of it implies no judgment.

Up to a point, the description of the controversy provoked by the TAPS proposal as Alaska's "first environmental crisis" was accurate enough.[15] It was the first major Alaskan dispute over natural resources since ecological issues rose to prominence in the American hierarchy of personal values and public priorities in the late 1960s. Likewise, there was a good deal of truth in an exasperated remark by Ted Stevens, a U.S. senator from Alaska, in the fall of 1969. He said that "suddenly out of the woodwork come thousands of people talking about ecology."[16] Many critics of TAPS had no previous knowledge of the Alaskan environment or established interest in Alaska. A number of the organizations to which they belonged had been only recently founded. Timing was crucial in explaining the dimensions and intensity of the conflict.

Many features of conservationist involvement in the TAPS controversy, like the sheer scale of the project and the controversy themselves, were unprecedented. This was partly a reflection of the discontinuity between two eras of American environmental history, conservation and environmentalism, which was itself related to broader, postwar socioeconomic changes. This break with the past was also associated with a new technological age and a new order of environmental threat, which, for many Americans, the pipeline epitomized. Nevertheless, the attention that these striking novel features deserve should not cause one to overlook some less obtrusive yet profound elements of continuity in terms of the conservationist community's role in Alaskan history. Though it was the first large-scale postwar engineering project to materialize there, TAPS was not the first substantial modern challenge to the environmental status quo in Alaska. Large numbers of American conservationists did "discover" Alaska in 1969. But this was not the first occasion that interest and serious concern had been aroused. The process of emerging awareness and growing involvement can be traced through the campaign to create the Arctic Wildlife Range and the opposition to abortive schemes like Project Chariot, a nuclear engineering proposal, and the Rampart Dam hydroelectric plan. Since these events have hitherto been largely neglected by historians, as full an account is provided of them here as of the pipeline controversy itself.

Comparison between Chariot and Rampart on the one hand and TAPS on the other reveals many obvious differences (not least that the pipeline was built). During the pipeline controversy the stakes were higher, the adversaries were more numerous and powerful, and the attentive audience was larger. However, there were plenty of common themes, and I have a particular interest in one recurring feature of these debates:

a preoccupation (shared by conservationists and boosters) with the significance of Alaska in connection with the frontier idea and the frontier's meaning for American history. For the intellectual historian of American conservation the TAPS debate is familiar and predictable in some respects. At this level, its historical distinctiveness did not lie in the production and articulation of new ideas about Alaska and the frontier. It resided chiefly in the large-scale acceptance and application of well-established ones, especially by conservationists.

This study considers continuity as much as change, viewing the TAPS proposal and controversy as an extension (even a culmination) of established processes, policies, and attitudes. I have no desire to downplay the regional distinctiveness of Alaskan history or to relegate it to a subcategory of western American history. However, to be properly understood and appreciated, Alaskan history must be approached within the framework of the history of the West and the frontier. This larger context highlights the features that Alaskan history shares with the rest of the American past and enhances our appreciation of what is truly distinctive about its course. Alaskan history, in turn, opens a large window on a fundamental aspect of national history; the relationship between people and the rest of nature. Alaska and various controversies there over the fate of natural resources have a central position in American environmental history and the history of conservation.

The Trans-Alaska
Pipeline Controversy

1
The Frontier Image and Environmental Reality, 1867–1940

> Other than the grassland, probably the arctic has been more
> maligned than any other region. . . . The arctic land is not
> a wasteland nor the arctic sea a barren sea.
> —James C. Malin, *The Grassland of North America*

The colonization of the New World drastically altered the face of the land, displacing indigenous flora, fauna, and peoples.[1] Widespread cultivation, ranching, mineral exploitation, and internal improvements such as roads, canals, and railroads accelerated the evolution of a cultural (human-shaped) landscape out of the substantially natural environment.[2] In Alaska, some of the agents of modification (though fewer and later) were the same as they were in the rest of the nation; natural resource exploitation, technological innovation, and internal improvements. Other agents, notably agriculture and the form of settlement associated with it, were virtually absent. Attitudes and responses to environmental changes induced by humans and their technology in Alaska were influenced by the earlier frontier experience in the United States, especially that of nineteenth-century West, and the mythology this process gave rise to. This study of environmental perception and environmental reality concentrates on the views and reactions of conservationists (of various kinds) and those who sought to boost Alaskan economic development.[3]

Luigi Barzini, the Italo-American writer and journalist, has commented: "Official mythologies are common to all countries. All countries cherish one or two particular periods of their histories, which they ennoble and embellish, to justify and give meaning to their present and to give a purpose to their future. This habit may be merely useful or ornamental to great, old, and solid nations. It is extremely important to recent and ramshackle ones."[4] The conquest of the West is probably the part of their history that Americans cherish most.[5] From the conquest of the West sprang "the Myth of the Frontier," which, as Richard Slotkin observes, "is arguably the longest-lived of American myths . . . and a powerful continuing presence in contemporary culture."[6] A notable feature of the TAPS controversy was frontier consciousness and this provides a striking example of how frontier myths, images, and symbols have operated in recent American history.

Many protagonists in the TAPS debate saw Alaska as the final extension of the American frontier, physically and in spirit. A *leitmotif* of Alaskan history has been—and continues to be—a tension between two visions of Alaska based on this shared frontier image. These visions are not necessarily absolute and irreconcilable. However, in their most extreme and ideal forms they do constitute radically divergent perceptions. Conservationists and boosters were united in admiration for the frontier and in agreement on its importance as an ingredient in American culture and history. However, they differed, often diametrically, in the ways they expressed this affection and how they formulated the best means to ensure the survival of their revered Alaskan frontier.[7] At this level, the TAPS controversy was a dispute over the lessons of frontier history. It was also a debate over the role and meaning of conservation.

For pipeline advocates, the ideal was the same as it had been for generations of Alaskans eager for swift economic advance; to allow Alaska, the last American frontier, to function unimpeded as another West. Booster rhetoric downplayed the difference between the pioneering act on the western frontier and its latter-day Alaskan equivalent. Seeking a usable past for the purposes of advocacy, they often cited both the national and the earlier Alaskan historical experiences as reliable and guilt-free guides to contemporary behavior. Approving the past in some ways, pipeline proponents wanted a break in others. For critics the error of national history was overexploitation of natural resources. For advocates the crime in Alaska was underexploitation, which they attributed in large part to conservation.

In contrast, some conservationists emphasized the limitations, technologically and ecologically, of the analogy between contemporary Alaska and the nineteenth-century West. They denounced the booster attitude as being myopic and contradictory because it encouraged the same behavior in Alaska that had annihilated the environmental conditions that nurtured the pioneering spirit in the West and shaped American virtues and attributes. As far as these conservationists were concerned, the booster mentality presented an even greater threat to the environmental raw material of the twentieth-century Alaskan frontier than it had to the West because it was wedded to more formidable technology. In common with their predecessors, certain pipeline critics in the conservationist community portrayed the history of the West as a warning. They sought (to adapt a phrase from Frederick Jackson Turner's frontier thesis) a "gate of escape from the bondage of the past;"[8] this past, in their view, forming a persistently tragic pattern of abuse of nonhuman nature. Their attitude toward the historic frontier process and its pioneering mentality was deeply ambivalent. Many pipeline critics venerated "the frontier" as an environmental condition in the nineteenth-century West and twentieth-century Alaska. They also celebrated the national qualities associated with pioneering on the frontier by Frederick Jackson Turner. Yet they denounced "the values of the frontier era,"[9] and contemporary manifestations of this ethos in Alaska insofar as these were destructive of this wildness.

Frederick Jackson Turner observed that the American nation was based on the idea of "perennial" individual and collective "rebirth."[10] As a literary historian has commented: "Whatever else the frontier contributed to American development, it gave people a great myth. The myth proclaimed that on the open frontier a person could be reborn: he could have a second chance. . . . In the American consciousness, the West symbolized hope. . . ."[11] Historically, would-be Americans had sought relief in the New World from Old World injustice, vice, and decay. In the West, Americans sought additional opportunity and a release from the confines of the East. For conservationists in the twentieth century, Alaska also symbolized hope and offered a "second chance." In the late 1960s, a time of acute national introspection and questioning of dominant values, the American past itself was put on trial by those estranged from the traditional attitude toward the environment. For preservationists in particular, Alaska offered a rare opportunity for national redemption by applying the lessons of the past. This meant the chance to protect substantial parts of the last frontier as wilderness.[12]

Though they rejected Turner's view that the frontier period of American history had ended in 1890, conservationists and boosters who embraced this "frontierism," whether or not they realized it, were the historian's lay disciples. This testifies to the enormous enduring popular appeal for Americans of the frontier thesis as a nationalistic way of looking at themselves and their past, regardless of the reservations of historians concerning its validity.[13] "The frontier," in this sense, was evidently a myth, a symbol, and an image. Accordingly, as literary historian Henry Nash Smith famously argued, these "products of the imagination" can be dealt with on two, though not entirely separable, levels. Firstly, they can be treated as representations (distortions) of reality existing "on a different plane" to "empirical fact," wherein resides the chief interest and importance of frontier symbols, myths, and images. Secondly, they can be approached, as Smith also approached them in *Virgin Land: The American West as Symbol and Myth,* despite an unconvincing disclaimer; in terms of the relationship between these myths and symbols and what actually happened, that is, their authenticity. Smith also commented that myths and symbols, "as I have tried to show, sometimes exert a decided influence on practical affairs."[14] Like Smith, Ray Allen Billington grappled with the slippery question of the impact of myths and images on events. Billington's final book, *Land of Savagery, Land of Promise: The European Image of the American Frontier in the Nineteenth Century* (1982), discussed the image-making role and influence over Europeans of travelers, novelists, and colonization agents. Writing on this subject elsewhere, he concluded that "the image makers . . . played a larger role in [the] history than they have been accorded."[15] American images of Alaska, like European images of the West, were of a cold hell and a wonderland.

Throughout this book, frontier myths, symbols, and images are studied as a rhetorical force and a technique of persuasion, which were designed to create public sympathy, influence decision-makers, and glamorize and embroider more prosaic arguments. Yet it is likely that these images were not just manipulated cynically but were genuinely important to many on both sides. However, it is hard to isolate this factor and assess its relative weight as a determinant of attitudes and behavior, let alone measure its impact on the decision-makers.

1.1 Images of *The Last Frontier*

The image of *the last frontier,* still the most popular American image of Alaska, was not the first—or even the second. The first image was the *"icebox."* The second image was the *storehouse.* The image of Alaska as the last (that is, unfinished) part of the national frontier came afterwards. These positive second and third images originated as booster protests against the negative appraisals of Alaska's value that the icebox image embodied. In 1867, the leading opponent of acquisition in the press had been Horace Greeley's New York *Daily Tribune* (11 April 1867). Its reaction to the Senate's ratification of William Henry Seward's treaty provided an exemplary exposition of the icebox image: "We may make a treaty with Russia," it declared, "but we cannot make a treaty with the North Wind or the Snow King. . . . We simply obtain . . . the nominal possession of impassable deserts of snow, vast tracts of dwarf timbers, frozen rivers, inaccessible mountain ranges. . . . Ninety-nine hundredths of Russian America are absolutely useless. No energy of the American people will be sufficient . . . to reclaim wildernesses which border on the Arctic Ocean."[16]

The exact origin of the icebox image is obscure, but it was doubtless derived from the ice trade with Alaska that flourished in the 1850s and

1860s. In 1850 ice was a luxury item in gold rush San Francisco. The boom town was supplied from Boston via Cape Horn until a group of San Franciscan businessmen organized the American-Russian Commercial Company in 1851 and signed a contract with the rulers of Russian America for a supply from the colony.[17] Some Americans considered the value of this ice trade alone sufficient justification for purchase.[18]

While many Americans appreciated the potential value of other Alaskan resources such as furs, fisheries, whales, coal, and lumber, most popular opinion and the bulk of the press agreed that Alaska was too far away and too inhospitable for traditional American colonization. Alaska's minimal agrarian potential had been one of the main arguments of those who had opposed appropriation. Cows in Alaska, wrote *Harper's Weekly* sarcastically, would give ice cream.[19] Other unflattering terms applied to Alaska were "Walrussia," "Icebergia," and "Frigidia."[20]

Seward himself, who included Alaska in his description of the Northwest Pacific as "in the near future the great fishery, forestry and mineral storehouse of the world," was naturally the first to emphasize Alaska's potential.[21] In the summer of 1869, after his retirement from office, the former secretary of state visited Alaska as part of a world tour and gave a speech at Sitka, the capital. He asserted that "no man can exaggerate—the marine treasures of the territory."[22] Seward, however, was not an archetypal booster. It is not necessary to dissect his personal and political motives for engineering the Alaska Purchase. Suffice to say that he paid greater heed to Alaska's strategic value and the opportunity to extend republicanism in the Americas than he did to the value of Alaska's natural resources.[23]

Senator Charles Sumner delivered the prototypical booster oration during the debate before the ratification of Alaska's purchase. This speech set the tone for a century of recurring themes, images, and rhetoric.[24] Dismissing the reservations of the sceptics, he emphasized the mildness of the coastal climate in southeast and southcentral Alaska, which he compared favorably with winter temperatures in New England, Washington, D.C., and many northern European capitals. He also dwelt on the summer heat in the interior (often above 90 degrees Fahrenheit). He admitted that most of Russian America was unsuitable for raising wheat and rye. However, he stressed the possibilities for cultivating barley and potatoes, even in subarctic regions, and for grazing livestock on Pacific islands like Kodiak.[25] He identified copper, coal and gold as the minerals of greatest promise. Among renewable resources, he highlighted the value of timber in the southeast and furs, especially in the Arctic. Above all, he drew attention to the abundant fisheries throughout Alaskan waters.

Such oratory aside, only congressmen and businessmen from California, the state with the closest commercial ties with Russian America, placed much emphasis on Alaskan resources as a justification for purchase.[26] One of the earliest boosters was probably Hubert Howe Bancroft, a historian and a wealthy businessman from San Francisco. In the preface to his pioneering American historical study of Alaska (1886), Bancroft commented that "my labor has been in vain if I have not made it appear that Alaska lacks not resources but development." Though conceding that "the greater portion of Alaska is practically worthless and uninhabitable," Bancroft stressed that the resources that were of value were virtually inexhaustible. He pointed out that much of Alaska was located in the same latitude as Scotland and southern Scandinavia, and looked forward to the day when the territory would support larger populations than these European counterparts.[27]

As the century progressed, others were to insist that much of Alaska was a frontier in the traditional mould. The ideas and rhetoric of Charles Hallock, the founder of *Forest and Stream,* and those of a Seattle newspaperman, John J. Underwood, were representative of this outlook. "When all the land is homesteaded," commented Hallock of the West in the 1880s, "men will look to Alaska."[28] Underwood had close ties with the Seattle Chamber of Commerce. In 1913 he described Alaska as "a land of plenitude—bounteousness. In the years yet to be . . . her verdant fields will be harvested; her cereals will be ground into flour. . . . The sturdy men and women who conquered the great Northwest, who pierced the back-bone of the continent with railway tunnels, who made productive millions of acres of desert land, were of the same hardy stock who, today, by their endurance, energy and industry are slowly converting the vast wilderness of Alaska into an Empire."[29]

The promotion of the storehouse image had reached a peak during the Alaska-Yukon-Pacific Exposition that was held in Seattle during 1908 and 1909. One objective of the exposition, as the editor of *Alaska-Yukon Magazine* explained, was "to tell the truth about Alaska," which inevitably involved the Scandinavian analogy.[30] These protests against the icebox image recall the efforts of railroad companies and western chambers of commerce in the nineteenth century to replace the myth of the "Great American Desert" with the image of the garden. In fact, Hallock and Underwood drew comparisons between the debilitating icebox and desert images.[31]

The exact origin of the epithet "the last frontier"—now a cliché that graces Alaskan automobile license plates—is difficult to trace. (The title of a book published in 1903 by Thomas Willing Balch, *The Alaska Frontier,* was misleading. Balch used *frontier* in the European sense and this was a study of the boundary dispute between Canada and the United States.)[32] But in 1909 *Webster's New International Dictionary of the English Language* listed as one of its definitions of *frontier* "the border or advance region of settlement and civilization; as, the Alaskan *frontier. Chiefly U.S.*"[33] Also that year, E. S. Harrison, a leading booster of Alaskan economic development and editor of *Alaska-Yukon Magazine,* referred to Alaska as "the last frontier of the United States" in a promotional guide to the territory.[34] The best-known poet of the Klondike gold rush, Robert Service, described the Yukon region as "the last lone frontier."[35] At the beginning of the century, the brochures of the Grindall Development Company of Seattle referred to Alaska as "Our Frontier Wonderland."[36] The Alaska Bureau of the Seattle Chamber of Commerce called Alaska America's last "Big Out of Doors."[37] By the late 1930s the precise use of the term was well-established and no longer exclusively boosterist.[38]

For the most part, however, enthusiasm for the term "last frontier" did not extend to historians of the American frontier. Alaska played no part in the genesis of Turner's thoughts and was rarely mentioned in his writings. Yet in 1914 he did recognize it as part of the final North American geographical frontier.[39] In spite of these later references, prominent "Turner school" historians ignored Alaska. For Frederic Logan Paxson the "last frontier" was the Far West.[40] In the first edition of *Westward Expansion: A History of the American Frontier* (1949), Ray Allen Billington referred to the homesteading "rush" to the Canadian "Prairie Provinces" of Alberta, Manitoba, and Saskatchewan between 1900 and 1920 as a movement to "the last best West."[41]

Boosters and conservationists (especially) tended to use the terms *frontier* and *wilderness* interchangeably with reference to Alaska. In the 1930s an advance guard of conservationists also began to draw attention to Alaska's importance, more specifically, as the last American *wilderness* frontier. For some of them, wilderness *per se* was Alaska's most valuable frontier resource.[42] At the height of the TAPS controversy, the Sierra Club—one of the oldest, largest, and most influential of the American conservation organizations—evoked a heady and seductive image of Alaska as a reincarnation of the New World encountered by the first European settlers; "One contemplates Alaska today with a numbing sense of historical perspective. *We have seen this pristine land before*—The United States at its birth two centuries ago."[43]

Wilderness, like *frontier,* is a cultural concept (and therefore ethnocentric) as much as an empirically identifiable physical entity.[44] Michael McCloskey, who became the executive director of the Sierra Club in the 1970s, explained in the 1960s that wilderness was valued "more as a mental image than a physical reality."[45] As the first European settlers perceived it, the entire New World was wilderness.[46] Wilderness for them meant the uncultivated abode of wildlife and the domain of savages where the hostile forces of nature were beyond their control. This pioneer perception contrasts sharply with the increasingly appreciative conservationist attitude toward wilderness since the turn of the century. The veneration of Alaskan wilderness is arguably the most dramatic manifestation of what Roderick Nash has called "a national intellectual revolution."[47] In 1974 the radical and recent nature of this shift in attitude toward American wilderness in general, and Alaskan wilderness in particular, struck one journalist while he was on a pipeline assignment in northern Alaska. Huddled in a tent in the Brooks Range during a blizzard, he reflected: "History is a record of people trying to get out of places like this . . . trying to come in from the cold, get out of caves and tents. . . . [wilderness] is what has been snapping at our heels like a wolf at a caribou."[48]

Ironically, the promotion of Alaska as the last American frontier spawned fresh *idées fixe,* sometimes as frozen and uncompromising as the icebox myth. The most zealous boosters marketed Alaska as a limitless storehouse of resource abundance and a cornucopia of wildness so overwhelming and inexhaustible no amount of human effort could alter its essence. At the other extreme, some conservationists (notably preservationists) ardently portrayed Alaska as a pristine and indivisible wilderness, a mental entity as much as geographical reality, which a single act of development—such as TAPS—threatened to violate irrevocably in both a material and, just as importantly, a symbolic sense.

Particularly germane in connection with this discussion of what Henry Nash Smith called collectively held "intellectual" and "imaginative" "constructions" is Smith's remark that "history cannot happen—that is, men cannot engage in purposive group behavior—without images which simultaneously express collective desires and impose coherence on the infinitely numerous and infinitely varied data of experience." These indispensible images and myths, as Smith observed, "are never, of course, exact reproductions of the physical and social environment. They cannot motivate and direct action unless they are drastic simplifications." Then he issued a warning: "If the impulse toward clarity of form is not controlled by some process of verification, symbols and myths can become dangerous by inciting behavior grossly inappropriate to the given historical situation."[49]

The application to Alaska of images and attitudes derived from earlier frontiers violated what James C. Malin, the ecological historian, once called "natural regional adequacy." In a striking passage, Malin asserted that no ecological region was intrinsically superior or inferior to any other. He stressed that they were all "biologically complete products of nature" and that each one was "never super or sub anything for its native fauna or flora." He condemned how "deciduous forest man" in North America had defined the trans-Mississippi West in terms of what was absent by the standards of the temperate, humid, thickly forested region with which the colonist was familiar. According to Malin, this attitude and prejudice bred certain behavior: "In its relations with other regions, the tendency has been for this deciduous forest man to attempt to impose his particular culture upon them irrespective of its adaptability and to expect a uniformity of results in all parts of the world. Anything that resisted that pattern was assumed to be inadequate, inferior or deficient, even nature had blundered."[50] Malin went on to argue that the recognition and acceptance of regional ecological completeness and distinctiveness ("adequacy"), and the limitations this imposed, were the criteria for the successful "transplantation" of any type of civilization evolved elsewhere.[51]

Malin was a complex, often contradictory thinker, and these particular views do not reflect his general attitude toward the natural environment. He firmly believed that man, with the assistance of technology, would "continue to reappraise the properties of the earth, bringing new ones continuously into the horizon of his utilization." Accordingly, he denounced Turner's ideas about the closing of the frontier as an insult to human ingenuity and creativity. Because he had such boundless faith in the potentialities of science and technology ("the properties of man as an inventive animal") to enlarge the sphere of human opportunity and to turn the earth to humankind's advantage, Malin insisted that practically no part of the globe, even Antarctica, "sometime," was uninhabitable by whites.[52]

Not surprisingly, Malin was an admirer of the twentieth-century Canadian explorer-anthropologist Vilhjalmur Stefansson, who dreamed up many lofty schemes for colonizing the Arctic and living off the fat of the land. Malin probably knew little about the Arctic aside from what he had learned from Stefansson's writings. Stefansson was probably the best-known crusader against the icebox image of the Arctic and the most vocal debunker of popular prejudices against the region in North American history. The kernel of his argument was summed up in the title of his book, *The Friendly Arctic* (1921).[53] Stefansson was also one of the most visionary and evangelical boosters of his day.[54]

Stefansson, who believed that cows and prairie-style ranching could replace caribou in the Arctic, had a questionable understanding of arctic ecology, and his plans for making over the region in the image of more southerly lands certainly violated the principle of natural regional adequacy. For, taken at face value, this lays down that the ecology of a particular region sets definitive boundaries for human activity. Malin's own call for a reappraisal of the so-called arctic wasteland (as quoted at the head of the chapter) did not indicate a sensitivity to the peculiar ecological value and intrinsic worth of the Arctic. Malin sought to rehabilitate this "maligned" region for the same boosterist ends that Stefansson pursued.[55]

No matter that Malin usually violated his own principle of natural regional adequacy.[56] The idea is useful to bear in mind as we turn to events themselves because it provides a tool for measuring the "adequacy" of images, attitudes, expectations, and behavior imported to Alaska.[57]

1.2 Environmental Issues in Alaska, 1867–1930

Alaska has never really been peripheral to the history of conservation in the United States. During the Progressive Era, Alaska, with the West, was the prime testing ground for the ideas and principles associated with President Theodore Roosevelt and the architect of his conservation program, the chief forester, Gifford Pinchot. Utilitarian conservation, typified by Pinchot and the policies of the Roosevelt administration, is one of two points of view identified in early twentieth-century conservation by historians.[58] Its aim was to curb the traditional practice of short-term exploitation of natural resources for private profit and to promote their efficient and scientific management by experts and their wise use on a sustained yield basis in the long-term public interest. This utilitarian, interventionist type of conservation dominated federal conservation policy. The other strand of conservation, exemplified by John Muir and the Sierra Club he founded in 1892, has become known as preservationist. Preservationism was an emotional force imbued with romanticism and transcendentalism that sought to protect nature for various nonmaterial reasons.[59] In contrast to utilitarian conservation, preservationism was noninterventionist in the sense that it believed that humans could not improve upon nature, which, therefore, was best left alone. While there was often considerable overlap, the difference between the two approaches was dramatically demonstrated by those national issues, such as the purposes of national forests,[60] and Hetch Hetchy, which ranged them on opposite sides. Both strands of conservation can be recognized in Alaska at this time. In fact, abusive practices precipitated early conservationist efforts in Alaska that predated the emergence of conservation as a national phenomenon.

1.2.1 RESOURCE EXPLOITATION

On the eve of purchase, many of Alaska's land-based, inanimate resources such as trees and minerals were relatively unscathed. However,
this was not true of nonterrestrial wildlife. Russian interest had been
attracted to Alaska in the 1740s for the same reason imperial control
had expanded into northern Siberia—the fur trade, which became the
mainstay of the Alaskan economy.[61] The image of the "sucked orange,"
which was projected even by some who supported purchase because
of Alaska's potential future value, reflected the Russian traders rapacious
treatment of furbearers. As the New York *World* (1 April 1867) pointed
out, supplies of the only Alaskan resource it considered to be of any
immediate value—seals and sea otters—were already imperiled.[62] The
sea otter and a terrestrial furbearer, the blue fox, were virtually extinct
by 1867. Fur fox farming was introduced on the Aleutian Islands in
the 1880s.

The reality of overexploitation was exposed by two notable individuals
long before 1907, when the term *conservation* became popularized by
Pinchot and his colleagues in Washington, D.C.[63] From 1872, Henry
W. Elliott served as special agent for the United States Treasury Department on the Pribilof Islands of St. Paul and St. George. These islands
were the location of the world's largest fur seal rookeries, which had
been ruthlessly exploited since 1786. Elliott's job was to monitor the
federal lease under which operated the San Francisco-based Alaska Commercial Company, the monopolist successor to the Russian-American
Company. According to various sources, seal numbers had declined from
5 million when the rookeries were discovered in 1786 to a low of 70,000
by 1830. Although stocks had recovered to near 3 million by 1867 they
had dwindled by 1872 to around 450,000, thanks to pelagic (open seas)
sealing.[64] William Healey Dall, the naturalist, known among the scientific
elite of the Cosmos Club in Washington, D.C., as "the dean of Alaska
experts," also displayed an early interest in the fate of Alaska's marine
mammals.[65] Dall was alarmed at their increasingly jeopardized status
throughout Alaskan waters. At high risk, he contended, in addition to
the Pribilof seals, were the elephant, ringed, harp, and harbor seals;
so were the sea lion, the walrus and, of course, the sea otter.[66]

Dall and Elliott were no foes of Alaskan economic development. Dall
subscribed as much as any empire-builder to the belief that Alaska constituted a storehouse of resources for the nation to draw on. He differed
in calling for patience. In a book (1870) that remained the standard
reference work for Alaska until the early twentieth century, Dall denied
that Alaska was ripe for large-scale development in the 1870s. However,

he did hope that in two and a half centuries, "there may be a new New England where there is now a trackless forest." He drew an extensive analogy between the climate and terrain of the Scottish Highlands and islands and certain parts of Alaska, expressing confidence in the agrarian potential of the Aleutians.[67] By contrast, Elliott was sceptical of Alaska's overall potential. His *A Report Upon The Condition Of Affairs In The Territory Of Alaska* (1875) was representative of the restraint and sobriety that characterized most federal government reports on Alaskan resource potential in the late nineteenth century. His retort to those who clung stubbornly to visions of Alaska's agrarian promise was sharp: "there are more acres of better land now lying as wilderness and jungle in sight on the mountain-tops of the Alleghenies from the car windows of a Pennsylvania [rail]road than can be found in all Alaska."[68]

A word of caution is in order here. There is a tendency for environmentalists and some environmental historians to be overzealous in the quest for harbingers, mentors, and antecedents of the modern era. As a result, Henry David Thoreau has been claimed as a pioneer environmentalist and John Muir has been appropriated as a latent "deep ecologist." The avant garde nature of Thoreau and Muir's thinking should certainly be granted. But it would be a gross distortion to see Elliott as an antihunter, like Muir, or a biocentric progenitor of a modern environmental organization like Greenpeace. In 1906 Elliott explained his rationale for protecting seals. Had they been "blocking the settlement of a new domain, or in the way of railroads or mines . . . then by the law of our civilization," he argued, there would have been no justification for their existence. He contrasted the situation in Alaska with the position of the buffalo in the West, which "did block the settlement of a new domain; it had to go; there was no alternative." He advocated the management of the seal herd "to the great annual gain and good of all mankind."[69] His campaign to save them from extinction continued unabated until Great Britain, the United States, Russia, and Japan reached an international agreement at the North Pacific Sealing Convention of 1911, by which time seal numbers had fallen below 100,000. This treaty was a classic application of the basic principle of utilitarian conservation.[70]

Wildlife exploitation also involved an assault on Native Alaskan cultures.[71] On the continental frontier, dispossession of the Native American had occurred indirectly, by undermining forces like trading and disease, or directly, by the United States Army, through a series of "Indian Wars." In Alaska the *promyshlenniki* (as the fur traders were known) did not come to settle and Russian influence barely extended beyond the coastal fringe. However, Russian enslavement and eradication of the Aleuts in the nineteenth century through guns and disease conformed to the pattern in the rest of the nation. In 1848 Yankee whalers passed through Bering Strait into western Arctic waters. In 1854 they penetrated the Beaufort Sea. By 1867 the bulk of the New England whaling fleet was operating in these Arctic waters.[72] Whalers coveted the baleen of the bowhead whale for corsets and prized the walrus for its ivory tusks. (John Muir compared this latter practice to killing buffalo for their tongues.)[73] The shrinking numbers of these creatures jeopardized the existence of the Eskimos, who subsisted on them. This was one of the reasons for the introduction of reindeer (the domesticated European and Asian cousin of the North American caribou) from Siberia in the late nineteenth century. Instigated by Sheldon Jackson, a Presbyterian missionary and general agent of education for Alaska, this scheme was designed to provide the indigenes with an alternative source of food and income.[74] Whatever the socioeconomic and cultural impact of this scheme, it was ill-fated from the standpoint of the environment and produced what one historian has recently referred to as "a little-publicized ecological disaster."[75] The imported reindeer competed with the native caribou for food, which led to overgrazing of the range.[76]

The first terrestrial resource (animate or inanimate) to attract outside attention was gold.[77] Discovery of the Klondike goldfields in 1896 precipitated a sudden invasion at a time when the existing white population of Alaska was little more than 10,000. One source estimates the number who set out through Alaska for the Canadian Klondike at 100,000.[78] The exact increase in the size of the white population of Alaska as a result is difficult to ascertain. At least the census figures for 1900 record that the total population had virtually doubled since the last census (1890), which had showed a net decline since the first Alaskan census was held in 1880.[79]

In the meantime, other Alaskan resources were attracting a different kind of attention. John Muir, who first visited Alaska in 1879, became a great publicist for Alaskan "nature tourism."[80] By the 1890s, the glaciers and fjords of the panhandle had become a fashionable cruise destination for well-heeled tourists on the grand tour that took them to Yellowstone and Yosemite.[81] The grandest and most famous of these cruises took place in 1899, at the height of the gold rush. The railroad magnate, Edward H. Harriman, organized a pleasure cruise-cum-bear hunt-cum-expedition. Harriman invited many of the nation's Alaska experts, such as Dall and Muir.

As the expedition's ship steamed north from Seattle to southeast Alaska and on to Kodiak Island and the Aleutians, the expedition's members witnessed considerable evidence of resource abuse. According to its historians the expedition saw "two Alaskas:" "One, the stunning, pristine land of forests and mountains and magnificent glaciers, the other, a last frontier, being invaded by greedy, rapacious, and sometimes pathetic men, often living out a false dream of success. The vision of the 'two Alaskas' was in many ways a reprise of the frontier experience in the lower forty-six states where the land was exploited as rapidly as possible and the Indian was dispossessed and trampled under the juggernaut of 'civilized' progress."[82]

It seems, however, that the members neglected the signs of exploitation when they recorded their experiences. Muir confined himself to the occasional grumble over the activities of the stampeders who were frantically spilling into Alaska en route to the Yukon goldfields. He described the miners as "a nest of ants, taken to a strange country and stirred up with a stick," and remarked "they'll be spoiling our grand Alaska."[83] Such anecdotal complaints aside, a more thorough examination of Muir's attitude to the Klondike-Alaska "scrambles" indicates that he did not regard these events as a serious threat to Alaska's pure and wild integrity. Unruffled, Muir concluded that "comparatively little harm will be done." He saw the rush as an isolated and temporary disturbance and did not fear that the pattern of permanent settlement and environmental despoliation that had characterized the history of the West would be repeated. Because of Alaska's harsh climate and unproductive terrain, Muir predicted that "the miner's pick will not be followed by the plough." Indeed, he regarded the stampeders' road-building as a boon that enhanced the value of Alaska's natural wonders. These arteries would lead "many a lover of wildness into the heart of the reserve, who without them would never see it."[84]

John Burroughs, the nature writer, first visited Alaska as a member of the Harriman expedition, of which he was the official historian. Burroughts felt intimidated by what he saw as the rawness and violence of the Alaskan wilderness. His home ground was the gentle, humanized landscape of the upper Hudson in New York State. The only part of Alaska where he felt at ease was Kodiak, whose lush pastures and bucolic verdure, the scene of some pioneer cattle ranching, reminded him of familiar, tamer places. Burroughs even referred to the island as a "pastoral paradise" whose "smooth rounded hills as green and tender to the eye as well kept lawns" suggested "natural sheep Ranges."[85] Kodiak was one of the few places in Alaska that could have evoked such a response. At the turn of the century, the interest of nature writers and preservationists was largely restricted to the "monumental" scenery of southeast Alaska (often compared to Switzerland, Norway, and Colorado), which conformed to prevailing canons of taste. This was the Alaska that Muir knew best and that inspired his best-known Alaskan writings.[86]

With other regions of Alaska Muir had little acquaintance. He ventured into Arctic waters twice but had little contact with the land. However, Muir appreciated its future significance. In the 1890s he remarked with regret that not even the desert regions of the Southwest—which conservationists generally considered to be the last substantial tract of wilderness and frontier left in the continental United States—had proven to be indestructible and untameable. So he pinned his final hopes on the "vast tundras of Alaska" as the "most extensive, least spoiled, and most unspoilable of the gardens of the continent."[87]

Well aware from his political engagements in the continental United States that "nothing dollarable is safe,"[88] Muir had exclaimed with relief that the Alaskan Arctic was "Nature's own reservation," guarded by climate and terrain from a tragic repetition of the devastation that his beloved Sierra Nevada had suffered from miners, loggers, and sheep and cattle grazers.[89] "Fortunately," he reflected elsewhere, "nature has a few big places beyond man's power to spoil," among which he included "the two icy ends of the globe."[90] The concern of conservationists like Henry Elliott for seals and other marine mammals did not extend to an appreciation of their arctic habitat (sea or land), which Elliott dismissed as "cheerless and repellent at any season." Elliott described the tundra "moorland" rimming the coast of northern Alaska with phrases such as "monotonous desolation" and "desolate sameness."[91] Yet a severe climate and a dearth of imposing topography and flora proved no barrier to Muir's poetic vision and nature mysticism, which encompassed the Arctic's subtle and neglected charms no less effusively than they did the granite domes, waterfalls, canyons, giant sequoias, and gentle climate of the Californian Sierra Nevada. Muir directly attacked the icebox image. "Nowhere in my travels," he enthused, "have I seen so much warm-blooded, rejoicing life as in this grand Arctic reservation, by so many regarded as desolate."[92]

Muir's appreciation of the Arctic "steppes" was unusually advanced. A decade later, another prescient thinker appeared. However, the special value Ernest Thompson Seton saw in the Arctic was cultural as well as ecological. The nature writer, artist, and pioneer of the American Boy Scout movement also rejected the image of the Arctic as barren and lifeless. He was one of the first North Americans to see the relevance of the Arctic to the American past. Writing over a decade after Turner's address announcing the closing of the frontier, Seton reveled in the "miracle" of "a backward look" to the old days of the frontier that was possible in northern latitudes. In the book with the revealing title, *The Arctic Prairies* (1911), he dwelt upon the delicious privilege of being able to see again "the great Missouri while the buffalo pastured on its banks, while big game teemed in sight and the red man roamed and hunted, unchecked by fence or hint of white man's rule." In 1907, when Seton visited the far northwest of Canada, white men's influence was restricted to "scattered trading posts, hundreds of miles apart, and at best the traders could exchange news by horse or canoe and months of lonely travel." Here he found "hoofed game by the million . . . where the Saxon is as seldom seen as on the Missouri in the times of Lewis and Clark." These caribou, Seton declared, "outnumber the buffalo in their palmiest epoch." He agreed with Muir that the Arctic, which he portrayed as an almost benign place, was safe: "there is no reason to fear in any degree a repetition of the Buffalo slaughter that disgraced the plains in the U.S."[93]

Nonetheless, some eastern sportsmen were concerned about the future of Alaska's terrestrial wildlife. Members of the prestigious Boone and Crockett Club, who did not share Muir and Seton's optimistic assessments of the chances for avoiding a repetition of frontier tragedies, were most active in opposing abusive practices.[94] The purpose of this New York City-based organization was to promote "manly sport with the rifle." A related objective was the exploration of "wild and unknown country." The quarter century since the club's foundation in 1887 had seen "the old, wild frontier of the limitless prairie . . . gone forever." So its leaders felt that the club's principal efforts should be devoted to measures for the preservation of stocks of large game in perpetuity.[95] One of the club's first preoccupations had been the threat of extinction facing the buffalo in the 1880s; hence its interest in Yellowstone National Park as a sanctuary for remnant herds. An authority on the history of Alaskan wildlife has commented that the awareness of the near-extinction of the buffalo provided "perhaps the most powerful emotional prop under the whole conservation movement."[96]

At the time of the Klondike-Alaska gold rushes, Boone and Crockett Club members turned their attention to Alaska. Edward W. Nelson, a specimen collector for the federal Bureau of Biological Survey, warned that the influx of gold seekers "threatened early extinction of bear, sheep, moose and caribou."[97] Determined to prevent a reprise of the Great Plains massacre, Madison Grant, the president and secretary of the New York Zoological Society (founded by Boone and Crockett in 1894), and Representative John F. Lacey (Iowa), also a club member, secured the passage of Alaska's first game law in 1902. Earlier laws (1900) had only protected birds.[98]

It was another prominent member of the Boone and Crockett Club, Charles Sheldon (after Theodore Roosevelt, the best-known big game hunter in American history—and a far better shot), who led the campaign to create Mount McKinley National Park (1917) in the Alaska Range.[99] The gold miners who flooded into the Kantishna region (now just outside the park) at the beginning of the century slaughtered local game, notably Dall sheep. This provided the impetus for the establishment of Alaska's first national park as a wildlife refuge with a backdrop of scenic splendor formed by North America's highest mountain.[100] The imminence of the Alaska Railroad (authorized in 1914) promoted visions of a repetition of the game slaughter that had accompanied westward expansion.[101] One of these concerned individuals was Belmore Browne, the mountaineer, artist, and hunter. Browne founded the Camp Fire Club of America, which had assisted Elliott's seal campaign. Haunted by the memory of the transcontinental railroads and their role in the "winning of the West," especially the demise of the buffalo, Browne emphasized the need to avoid a repetition of the mistakes of the past in Alaska.[102] Conservationists like Browne worried about the rise of a new generation of professional market hunters in Alaska. Hunters like Bill Cody, better-known as Buffalo Bill, had slaughtered bison to feed the transcontinental railroads' construction crews, to supply the dining cars, and for sheer amusement.[103]

The chairman of the Boone and Crockett Club's Game Preservation Committee was the ethnologist, George Bird Grinnell. Grinnell was the editor and publisher of the first American conservation journal, *Forest and Stream* (1873). He also provided the impetus that eventually led to the formation of the National Audubon Society in 1905 in response to the massive slaughter of wild birds by plume hunters who sought feathers for hats. Grinnell (like Edward W. Nelson) was a key supporter of the McKinley Park bill. He had also been the member of the Harriman expedition most outspoken about resource abuse. He focused his attention on fisheries and was most trenchant in condemning the excessive competition between San Francisco-based canneries, which had invaded the traditional fishing grounds of the Haida and Tlingit in the 1870s and, by the 1880s, had seriously depleted salmon stocks. He denounced the cannery policy of erecting dams and barricades across the mouths of rivers, which prevented fish swimming upstream to spawn. This practice had been technically outlawed by Congress in 1889 but had persisted. Grinnell referred to the situation on Afognak Island near Kodiak. Here the law was still violated, although the island's cannery had been shut down when it was set aside as a forest reserve.[104]

Grinnell found an ominous analogy for these wasteful practices in recent frontier history. In Alaska, only the choicest salmon meat, the belly, representing only 10 to 20 percent of the entire fish, was taken. The remainder was discarded. This reminded Grinnell of buffalo that had been slaughtered just for their tongues. He remarked sardonically that when an Alaskan is asked about the supply of a resource, "the same language will be used that was heard in past years with regard to the abundance of wild pigeons, or of the buffalo, or of the fur-seals of the Bering Sea."[105]

Similar sentiments were expressed in 1913 by an associate member of the Boone and Crockett Club who had been a prominent supporter of Elliott's seal crusade. William T. Hornaday was the leading American wildlife protectionist of his day, and the first director of the New York Zoological Society's Zoological Park (1896–1926). An expert on the buffalo and the history of its near-demise, he had been in the vanguard of efforts to save it. With reference to the situation in Alaska, he declared: "The preservation of the Alaskan fauna on the public domain should not be left unreservedly to the people of Alaska, because, as sure as shooting, they will not preserve it!"[106] He meant that Alaskans were too intimately locked in day-to-day confrontation with the land and its fauna to appreciate them in anything but a utilitarian sense.[107] To the exasperation of many Alaskans, this attitude would gain increasing currency among conservationists.

The next major federal conservation measure after the creation of McKinley Park was the establishment of Katmai National Monument at the base of the Alaska Peninsula in 1918. This area was withdrawn to provide a sanctuary for the giant coastal brown bear (the world's largest carnivorous mammal) and for scientific purposes; to study botanical and biological adaptation following the eruption of Mount Katmai in 1912. Scientific considerations also played an important part in the establishment of Glacier Bay National Monument in 1925. The entrance to Glacier Bay, one of the highlights of the cruise along the "Inside Passage" in the 1890s, was blocked by ice following an earthquake in 1899. The driving force behind the creation of this reserve for the study of plant succession and glaciology was William S. Cooper of the Ecological Society of America. Many residents of southeast Alaska protested loudly against President Coolidge's action because the area of the monument was not restricted to ice and rock. They claimed that it violated mining claims and infringed on potential agricultural lands.[108]

Goetzmann and Sloan consider the hallmark of Alaskan history to be the plunder of natural resources: "in human times Alaska had never been anything but the prostrate victim of man's pillage."[109] With the exception of the gold rushes, however, this saga of plunder had not significantly affected inanimate terrestrial resources by the turn of the century. Besides increased hunting pressure, the major ecological consequence of the Alaska-Yukon gold rushes was timber destruction through fire, both accidental and set. Miners used smouldering timber (and steam generated by wood fires) to burn off the thick vegetative mat and melt the frozen ground so they could work the ore. They also lit smudge fires to relieve the mosquito menace, built fire rings for hunting moose, and burnt down trees to clear ground for pasture and small-scale agriculture.[110]

Federal conservationists seized the opportunity for preventative action to forestall these wasteful traditional (individualistic) practices elsewhere in Alaska. More urgently, these officials wanted to harness the absentee corporations that, more than anyone, were alive to Alaska's resource potential. Two of the nation's largest national forests were created in Alaska. In 1902 the Tongass National Forest was carved out of the public domain in southeast Alaska. The Chugach National Forest was set aside five years later. Of the 148 million acres of American forest reserved by the end of Theodore Roosevelt's administration (1909), 26.26 million were in Alaska. The only state or territory that contained more acres of reserved forest was California.[111]

In 1906 coal-bearing lands in Alaska were withdrawn by executive order from all forms of entry under the public land laws. This was designed to foil the efforts of corporations like the Morgan-Guggenheim Alaska Syndicate to circumvent the acreage limitations on coal lands entry imposed by the 1873 Mineral Leasing Act.[112] The syndicate had a monopoly over transportation to Alaska from Seattle. It also owned twelve of the forty canneries in the territory. The Alaska Syndicate's interest in Alaskan coal sparked one of the best-known conservation conflicts of the Progressive era; the Ballinger-Pinchot affair (1909–11). Progressives viewed Richard Ballinger's appointment as interior secretary by President Taft as a retreat from the conservation policies of his predecessor, James Garfield. Pinchot, among others, accused Ballinger of collusion with the Alaska Syndicate in filing illegal claims while he was in charge of the federal Land Office in Seattle (1907–8). Ballinger resigned in 1911, and Taft dismissed Pinchot.[113]

Many Alaskans saw Ballinger as a friend of Alaskan development and believed he was unfairly hounded from office.[114] Alaska's congressional delegate, James Wickersham, was a notable exception. Wickersham, a personal friend of Pinchot, was a vociferous critic of Morgan-Guggenheim's Alaskan activities and joined the Bull Moose Republicans in 1912.[115] The delegate approved of federal conservation policies as a restraint on outside interests that creamed off Alaskan wealth. However, he opposed conservation when it also excluded the small-propertied individual from access to Alaskan resources.[116] While admitting that conservation policies might be appropriate to the depleted state of resources in older, more settled parts of the nation, some Alaskans considered them vindictive and absurd where resources seemed barely touched.

The coal lands controversy (like the Alaska-Yukon-Pacific Exposition of 1908–9, which 3.5 million visited during its four-month run in Seattle) focused some national attention on Alaska's natural resource potential. In 1910 a journalist for *Collier's* reported that the corporations viewed

Alaska as "A Land to Loot" with more gold than California and more coal than Pennsylvania. "The only natural commodity which Alaska may not possess in sufficent amount to make her a greater Pennsylvania is petroleum," he concluded. Like Wickersham, he claimed that what Alaska needed instead of corporations was the "gritty pioneer" who knew "how to get into a great new country, battle with it and make it surrender."[117] Not that they and others like them underestimated the nature of the task. "We have reached our last frontier," explained *Alaska-Yukon Magazine*, "and notwithstanding all the things we have received from science and invention during the last half-century, it will be our most difficult frontier to subdue."[118]

Alaska was not the only part of the nation affected by federal conservation policies. Coal lands—a total of 66 million acres—had been withdrawn throughout the West as well. Also part of a broader western pattern was the Alaskan reaction to conservation. The most famous Alaskan protest was the "Coal Party" that took place on 3 May 1911 in Cordova, the port closest to the rich deposits of the Bering River coalfields. Incensed citizens warmed up by burning the much-reviled Pinchot in effigy. The highlight of the day was a reenactment of the Boston Tea Party, which involved, according to sympathetic accounts, one quarter of the town's population of 1,200. The insurgents, whose leaders are said to have included the president of the local chamber of commerce, an ex-mayor, and several members of the city council, ceremonially dumped Canadian coal imported by the Copper River and Northwestern Railroad Company into Controller Bay.[119] Pinchot was burned in effigy again the following day at Katalla, the settlement nearest the Bering River coalfields. Demonstrators placarded the town with posters reading:

> Pinchot, my Policy:
> No Patents to Coal Lands;
> All timber in forest reserves;
> Bottle Up Alaska;
> Put Alaska coal in forest reserve;
> Save Alaska for all time to come.[120]

Pinchot visited Alaska for the first time in the fall to acquaint himself with the situation. Further clamor and some secessionist rumblings accompanied his brave appearances at Katalla and Cordova.[121]

These Alaskans failed to appreciate that Pinchot and his supporters were not opposed to the economic development of Alaska. A case in point was William B. Stephenson, a former U.S. commissioner (federal lower court official) at St. Michael on Norton Sound.[122] Stephenson, who stressed his Pinchotian credentials, was an exponent of appropriate development for Alaska at a proper and realistic pace. His book, *The Land of Tomorrow* (1919), was addressed to American soldiers returning after service abroad during the First World War. He invited them to accept the further challenge of settling in Alaska, which offered "untold opportunity for him who is willing to fight."[123] Yet he appended the reminder: "There is no excuse for a repetition of the blunders the mother-land may have made during the days of her youth. . . ." For Stephenson, conservation was not at odds with development. The national past still offered a blueprint for contemporary Alaska: "What will the future reveal? *Read the answer in the history of the American people!*"[124]

The aspect of the Alaskan past with which advocates of Alaskan development wanted to break was the established pattern of resource use. Ever since the arrival of the first Russian fur traders, the conquistador, "hit-and-run," extractive mentality had been the dominant ethos. In Alaskan history, *promyshlennik* had become a byword for one who overexploited a natural resource for maximum short-term gain. At the turn of the century, Alaskans frequently applied it to outside corporations.[125] Surveying Alaskan history in 1953, Ernest Gruening, who served as territorial governor of Alaska from 1939 to 1953, identified a basic difference between the Alaskan frontier and the national frontier; the absence of an agrarian form of colonization in Alaska.[126] Also struck by the dichotomy between agrarian and nonagrarian forms of frontier societies, economies, and cultures was Vianna Moog, the author of a comparative study of the American and Brazilian frontiers. *Bandierantes and Pioneers* (1964) distinguished between a "settler" variety in the United States and the "conquistadore" brand of pioneering that prevailed in the Portuguese colony. Moog, whose alternative title was *Conquistadors and Colonizers*, attributed the differences in national progress and success between the two countries to this dichotomy.

The accuracy of Gruening and Moog's characterization of the frontier in the continental United States could easily be disputed. Both clearly oversimplify; the conquistadore mentality was present in the West where Kit Carson and Jedediah Smith were surely the counterparts of the *bandierantes*, as were the French-Canadian *coureurs des bois* (runners of the woods). More important, however, is the possibility of a comparison between Brazil and Alaska and *promyshlennik* and *bandierante*. Blatant climatological and environmental differences notwithstanding, the Alaskan experience is certainly closer in some ways to that of Brazil than it is to the history of the United States east of the Mississippi. The impenetrability of the equatorial rain forest meant that Portuguese influence in Brazil, like Russian influence in Alaska, was restricted largely to the coast. Likewise, those who did travel inland in Brazil and Alaska usually went in search of gold, not agricultural opportunities.[127]

The behavior of all resource exploiters in Alaska during the period from 1867–1940—whether *promyshlenniki* in search of furs, stampeders in pursuit of gold, corporations attracted by coal and copper, or the "fish trusts"—conformed to a standard cycle of boom and bust. This cycle failed to provide the means for creating a stable economy and society (based on land ownership) comparable to those that arose in other, more temperate parts of the nation. Outside interests backed by big capital left little room to maneuver for the individual frontiersman of booster hope and rhetoric and who existed in the popular imagination.

In connection with this gap between popular expectations and images and historical reality in Alaska, the nineteenth-century American West—and even Australia—offer further illuminating economic and environmental comparisons. The arid heartlands of the American West and Australia were better suited to sheep and cattle ranching than to crops and homesteading. The typical western American and Australian frontiersmen were shepherds, cowhands, and wage-workers, not independent agriculturalists. The Alaskan situation was also more conducive to corporate rather than yeoman pioneering. The Alaska Syndicate and the fish trusts have more in common with the "collectivist" cattle and sheep stations of the Australian bush than they do with the pioneer archtype stemming from the American frontier experience east of the Mississippi.[128]

The explanation for what boosters saw as Alaska's deviation from national history, though they themselves resisted it, was to a large extent environmental. Turner himself observed in 1903 that any twentieth-century pioneering in the United States would be collective in view of the physical obstacles in the remaining frontier lands. "It is true that vast tracts of government land are still untaken," he conceded, but added that "they constitute the mountain and arid regions, only a small fraction of them capable of conquest, and then only by the application of capital and combined effort." He appreciated that successful pioneering beyond the ninety-eighth meridian required large-scale irrigation and technology beyond individual means: "the physiographic province itself decreed that the destiny of this new frontier should be social rather than individual," demanding "cooperative activity."[129] The same applied to Alaska, most of which Turner would probably have included among the large fraction of "untaken" government land that was incapable of conquest—regardless of the territory's undisputed frontier status in at least one significant respect; the number of people who lived there. Population density (under two per square mile) was Turner's classic measure of frontier status and the basis on which he announced its death in 1893. When the landmark eleventh national census was held in 1890, Alaska's total population was 32,000. This amounted to 18.3 square miles per person, 90 percent of whom were Native. Fifty percent of a sample of historians surveyed in 1941 accepted sparse population as part of the definition of *frontier*. Thirty-three percent believed this criterion was indispensible to its meaning.[130]

1.2.2 INNOVATIONS IN TECHNOLOGY AND COMMUNICATIONS

The environmental impact of resource exploitation was reinforced and extended by internal improvements and the penetration of evolving technology. Innovations in transportation and communications, in addition to their own direct environmental impact, contributed to further change by improving access to natural resources, both animate and inanimate. Technological advances enhanced human capacity to exploit these resources and intensify control over nature. A basic ingredient of the nineteenth-century American frontier—isolation—was reduced during westward expansion by the telegraph, roads, canals, and finally railroads. No freight-carrying waterway was dug in Alaska but all other avenues of intrusion played a role.

In the 1860s the only form of communication between Alaska and the rest of North America was by sea. The Western Union (Collins Overland) Telegraph Expedition of 1865–67 was an investigation of the possibilities for connecting the United States with Europe via British Columbia, Russian America, the Bering Strait (submarine cable), and Siberia.[131] Some historians have interpreted this enterprise as an expression of the same American continentalism that encouraged the purchase of Alaska.[132] Many miles of line were laid in British Columbia and some posts were erected on the Seward Peninsula in western Alaska. Hubert Howe Bancroft quotes the tribulations of one surveyor who related: "the country is a complete bog. If you dig down on the hills there two feet, you strike ice. . . . Our poles were on an average 15 feet long. . . . We dug them three feet into the ground, which consists of frozen dirt. In summer when the surface thawed, we found many of them, which we supposed to be very firmly erected, entirely loose."[133] In the rest of Alaska, however, this grand design did not proceed beyond clearing a right-of-way and cutting posts. The enterprise was abandoned when the Atlantic cable was finally successfully laid. Nevertheless, it marked the first systematic American exploration of the Alaskan interior, and Elliott and Dall first became acquainted with Alaska as members of the expedition.[134] This visionary undertaking was excelled by Edward H. Harriman's extravagant and equally fruitless scheme for an intercontinental railroad linking New York City with Paris via the same route.[135]

The first enterprise to materialize was the Washington-Alaska Military Cable and Telegraph System (WAMCATS), built by the U.S. Army Signal Corps. Before this rapid communications system was established (1900–1902) between the army's Alaskan outposts (set up in the aftermath of the Klondike gold rush) and Washington, D.C., communication was by horse, mule, canoe, steamer, or foot in summer and by dog team, sleigh, or snowshoe for the rest of the year. Using these methods, it could conceivably take up to one year to send a message to the capital and to receive a reply.[136] The most demanding section of the 1,500-mile line, which stretched as far as Nome and St. Michael on Norton Sound, proved to be the 420-mile stretch between Fort Egbert at Eagle, a village on the Yukon just inside the Canadian border, and Fort Liscum near the port of Valdez on Prince William Sound.[137]

Lieutenant William L. "Billy" Mitchell, who supervised construction from Eagle to Valdez, faced supply and construction problems peculiar to telegraph building in northern lands. During summer, mules and horses packing in equipment and cargos of 150-pound reels of wire faced swollen glacial streams that constantly shifted course, and the animals often sank to their knees in the swamp-like ground. To aggravate matters, soldiers and beasts of burden were preyed upon by hordes of mosquitoes. In springtime, they faced avalanches, and during winter they were vulnerable to extreme cold and snowdrifts. In winter, it was difficult to drive telegraph poles into the frozen ground. Mitchell realized that, despite the winter cold, mules and horses (able to haul much heavier loads than dogs) could adapt to pulling sleds and that equipment and supplies could be moved most efficiently at this time of year. He found it equally prudent to concentrate building activity in the summer months when the upper layer of ground thawed out.[138] Conventional telegraph pole construction methods were hard to follow because of the freeze and thaw cycle characterizing much of Alaska's terrain. When the upper layer of ground freezes, ice forms and pushes the surface of the ground upwards. In most cases, a structure driven into this type of terrain will not return to its original position once the surface layer thaws again. This net upward movement is called *jacking* or *heaving*. The U.S. Army Signal Corps and the infantrymen working on the line dealt with this problem by erecting tripods around the telegraph poles to anchor them and to prevent the related phenomenon of tilting, or by using tripods instead of poles. By July 1903 all Alaskan military stations had been connected by telegraph and the line was opened for commercial and private messages too. From Valdez, the line ran via Eagle and Dawson to Vancouver, terminating in Seattle.[139]

Participants were quick to claim the accomplishment for the American frontier tradition. Mitchell reminisced in the 1930s: "to a young lieutenant in the U.S. service, it seemed as great an undertaking as the Lewis and Clark Expedition. Men were back to conditions that existed west of the Alleghenies two hundred years ago."[140] But, thanks to WAMCATS, "America's last frontier had been roped and hog-tied."[141] Contemporaries portrayed the construction of WAMCATS as an epic struggle with the wilderness that brought forth Nietzschean superhuman qualities. And it might be argued that WAMCATS was the first technological breach in the natural defenses of Alaska's wilderness. However, the chief physical legacy of construction was abandoned wire, often miles of it. Between Valdez, Eagle, and Nome, Natives used it for snare fences to drive caribou into corrals. One prominent naturalist reported seeing wire fences up to six miles long.[142] Otherwise, only a few tripods and sections of wire survive.[143]

The U.S. Army Signal Corps appreciated the strategic value of Valdez—the most northerly year-round, ice-free port in North America—as the natural gateway to the interior of Alaska. This status was immediately confirmed by the gold seekers headed for the strikes in the vicinity of what became the town of Fairbanks (1900–1904). In 1905 Congress created a road-building agency for the territory, the Alaska Road Commission (ARC), within the War Department.[144] Alaska's climate and soils caused peculiar problems for road builders who soon discovered that the least effective way to build a wagon road that could be used in summer was to strip away the thick mat of vegetation (known as *muck*) that insulated the frozen ground. The best method was the corduroy technique, which involved laying logs on top of the vegetation to form a roadbed. These logs were then covered with gravel and earth.

By 1910 the work of the Alaska Road Commission had made travel possible on the Valdez Trail (Trans-Alaska Military Road), Alaska's principal wagon road, as far as Fairbanks—by horse and buggy in summer and by sleigh in winter. In 1913 the first automobile was driven from Valdez to Fairbanks over what was by then called the Richardson Highway.[145] During the next decade, Keystone Canyon was the scene of further intrusion and disturbances when rival railroad interests fought for the right-of-way to the Alaska Syndicate's copper mines at McCarthy, Chitina, and Kennecott in the Wrangell Mountains.[146] The 200-mile Copper River and Northwestern Railroad, which operated between 1911 and 1938, was built by the Alaska Syndicate to haul ore to the port of Cordova.

WAMCATS and the Valdez Trail illustrate how it makes little sense to treat Alaska as a single, integrated geographical entity. One-fifth the size of the rest of the United States, Alaska is best approached as a series of distinctive regions. It is more useful to talk of multiple geographical frontiers and to think in terms of local environmental impact. Southeast Alaska, "the panhandle," demonstrates this.

The panhandle, together with the Aleutians and Kodiak, was the scene of what Russian colonization did take place in Alaska. This is a most anomalous part of Alaska in a bioregional sense; a bioregion being an area defined according to its distinctive ecological features (topography, soils, climate, vegetation, and fauna) rather than political, administrative, or other man-made boundaries and definitions. Bioregionally speaking, southeast Alaska with its mild winters, high precipitation, and dense woods, belongs to the temperate rain forest zone of the Pacific Northwest. Southeast Alaska's isolation was broken by the Klondike-Alaska gold rushes and the White Pass and Yukon Route, built (1898–1900) to provide access to the goldfields. This first Alaskan railroad ran 112 miles from tidewater at Skagway over White Pass, where it entered Canada, to Whitehorse, Yukon Territory. Some contemporaries were convinced they were dealing with an unparalleled feat. In one opinion, the rugged mountainous terrain presented "more engineering difficulties than any other railroad in the world."[147] In vintage frontier-busting style, local boosters hailed it as "a marvel, the acme of engineering skill, the triumph of capital and labor in subduing and making subservient to man the heretofore impassable and barren vastness created by God, formerly preserved for Himself as it were, and visited only by the howling blasts of Boreas."[148]

The Alaskan booster conviction that railroads were precisely what they needed to push back the Alaskan frontier was based on an awareness of national history. In 1873, four years after the completion of the first transcontinental railroad, the superintendent of the ninth census (1870), had commented that "population has found its way into regions to which the rate of progress previously maintained would not in fifty years have carried it; into nooks and corners which five years ago were scarcely known to trappers and guides."[149] *Alaska-Yukon Magazine* had been calling for railroads with regularity since 1908, reminding readers that the West would have remained unsettled without them.[150] According to a headline by John J. Underwood, the Seattle newspaperman, "Population for Alaska Awaits Transportation Facilities (A Trunk-Line Railroad, Wagon Roads, and Trails Only Needed That Immigrants May Find in the North a Land of Promise the Equal of Any Section of the American Continent)."[151]

Private efforts during the first decade of the twentieth century to build a railroad connecting Fairbanks with the coast (the Alaska Central Railroad and its successor, the Alaska Northern Railroad) had failed.

Between them, these aborted ventures laid 72 miles of track from the port of Seward northward. Critics of the proposal for a federal railroad from the coast to the interior of Alaska (often from southeastern states) condemned the scheme as redundant, ill-advised, a waste of money, and, not least, socialistic. They continued to exploit the icebox image. During a congressional debate in 1913, Congressman Edward Watts Saunders (Democrat, Virginia) dismissed Alaska as "a frozen wilderness for the greater part of the year, and a fly-plagued sweat-bath for the balance of the time." "Why spend it at the North Pole?," asked Congressman Thomas W. Hardwick (Democrat, Georgia), about the money such a railroad would cost. "Why not spend it at home?"[152]

In 1914, however, Congress authorized what became the first railroad in American history that was financed, constructed, and operated by the federal government.[153] This decision was greeted euphorically in the communities affected.[154] At this time, access to and from Fairbanks in winter was by dog team or horse-drawn sleigh over the Valdez Trail. During summer, steamboats from Whitehorse plied along the Yukon, Tanana, and Chena rivers. Twice a year, between the departure of the last sleigh and the arrival of the first steamer after break-up, and the departure of the last steamer before freeze-up and the arrival of the first sleigh of winter, WAMCATS was the only link between Fairbanks and the outside.[155] The first train ran from Seward to Fairbanks in November 1921. On a regional level, the Alaska Railroad's completion in 1923 effectively eliminated the interior's isolation. President Harding traveled north to preside over the golden spike ceremony at Nenana, a new railroad townsite. This was the first time a president had visited Alaska.

In 1928, en route to his job as general manager of the Alaska Railroad, Colonel Otto F. Ohlson asserted in Seattle: "Colonization just as the trans-continental railroads colonized the western country is the next step in Alaska's development."[156] Aside from the incentive that Ohlson hoped the railroad would give to agrarian colonization in the Matanuska Valley and around Fairbanks, it was expected to release the mineral potential of the region it bisected. (Alaska's coal lands were restored to public entry in 1914.) Boosters had in mind specifically the coal seams around Healy, just north of McKinley Park, and in the Matanuska Valley. The railroad's coming also inspired dizzier visions of future glory. Scott C. Bone, a former chairman of the Seattle Chamber of Commerce, and territorial governor of Alaska from 1921 to 1925, saw "in Ketchikan a Seattle of Alaska to come, with its hills cut away for the making of a city; in Juneau a possible Vancouver or Portland; in Fairbanks, toward the top of the world, a future Winnipeg, or Minneapolis of the interior; in Anchorage, a Pittsburgh of fast-coming time. . . ."[157]

Not all Alaskans reacted exclusively in this giddy vein. John A. Clark, an attorney in Fairbanks, appreciated the railroad's benefits as much as anyone. After all, he had been one of the gold seekers who bicycled to Fairbanks over the Valdez Trail in 1902.[158] For Clark, the blessings, while overwhelming, were not unmixed. He feared that the easier access the railroad provided would seriously increase hunting pressure on the animals on which many Alaskans depended for food. He also compared pre- and postrailroad Fairbanks. He portrayed the town before its arrival as a stable, close-knit community happily devoid of crime and other social disturbances. "The advent of the railroad," he explained to novelist James Oliver Curwood, "has commenced to change all this. . . . it has introduced an alien element into our population that has not been for its betterment. For 16 years we never locked our doors at home. . . . During the past three or four months we have experienced practically our first housebreakings or burglaries committed by persons in their sober senses."[159] Among many Alaskans, what in modern parlance would be called social-impact concern preceded environmental-impact concern.

The introduction of new technology accompanied these advances in communications. During the Klondike era, gold mining had been mainly an individualistic, "pick and shovel" operation restricted to beaches, river bars, and creeks. Now the Alaska Railroad made it possible to haul in large pieces of equipment and machinery year-round. Once the application of the gold dredge became profitable and technologically feasible in the Alaskan interior, it was widely adopted there in the mid-1920s as the most efficient mining method. This transformed the search for gold into a large-scale, machine and capital intensive, highly organized, corporate undertaking.[160]

Since the turn of the century, it had been common practice to dig ditches to deliver large quantities of water under high pressure to mining sites where hydraulic methods were employed. Hydraulic mining used pressurized water to remove the thick layer of overburden, to disintegrate gravel banks, and to push the loosened gravel into the sluice boxes. Before a valley could be dredged, the frozen ground had to be thawed. One popular method was to steam-thaw the gravels using steam points driven into the ground. Another technique was to use water (hot or cold) brought by the ditches. The gold dredge (powered by wood or coal-fired steam) scooped up the thawed gravels in a series of buckets on a conveyor belt and then extracted gold with a series of screens and sluices. The gravel was washed through the dredge and discarded, via the conveyor belt, behind the operations on a tailings pile. To process gravel effectively, a dredge required a steady flow of water. Large and reliable quantities of water were not naturally available in the Alaskan interior, where precipitation is low. Moreover, slopes are often not steep enough to build up a decent pressure over a short distance.

In 1925 the Fairbanks Exploration Company, the local representative of a group of outside interests, began work on a 72-mile canal to bring water from the Chatanika River to its dredging operations (two of eight it controlled) at Goldstream and Cleary. Mechanical shovels (which had been used to cut the Panama Canal) and caterpillar tractors were employed to excavate the muck and the frozen ground. Some of the digging was also done by hand. The project included an additional 25 miles of feeder ditches, flumes, and siphons as well as a tunnel. The Davidson Ditch bears the distinction of including Alaska's first major pipeline for the project incorporated a total of 6 miles of steel pipe 46–56 inches in diameter, which carried the water across drainages. Water passed through this combination of sloped ditches (6 feet deep and 12 feet wide at the base) and pipes to the dredging sites where it roared out of a nozzle to wash away the overburden, excavate hillsides, and assist the operation of the dredges. At full capacity, the system, which took three years to complete, was designed to deliver 180,000 gallons of water each day.[161]

Discovery of the material value of Alaska's natural resources also continued apace. After the First World War, the American navy switched from coal to oil. In 1923 President Harding withdrew a block of land about the size of Indiana (23 million acres) in the northeast Arctic. Naval Petroleum Reserve Number Four (usually abbreviated "Pet 4") gave Alaska an equivalent to the petroleum reserves established in California in 1911 at Elk Hills and Buena Vista Hills, and in 1914 at Teapot Dome, Wyoming.[162] These were typical expressions of the attempt in the Progressive Era to build national strength and protect national security through the efficient use of natural resources.

The state of available technology confined the scope of the explorations that the U.S. Geological Survey began immediately (1923–26). In winter the only form of transit was the dog team. In 1924 exploration parties mushed up from Tanana into the Brooks Range, carrying all supplies, including canoes. After wintering in these mountains, they crossed the Arctic Divide and set up a headquarters in the petroleum reserve. When break-up came, the geologists switched to canoes, for the terrain became soft, boggy, and virtually impassable. This restricted their explorations largely to areas close to rivers. In 1925 and 1926 exploration parties mushed from Nenana to Kotzebue and then up the Noatak River. Penetration any distance inland by those who arrived by sea during the brief summers involved lengthy and arduous portages.[163] In their concluding report, the geologists noted the difficulty of shipping oil out by tanker since the earliest safe access by sea to Point Barrow was 1 August and the latest ice-free date for departure was early September. Moreover, the shallowness of the Beaufort Sea meant that vessels were unable to approach close to the shore to berth. Accordingly, an extension of the Alaska Railroad or a 1,000-mile pipeline to an ice-free harbor were seen as the key to transporting any oil that might be found there.[164]

1.3 THE 1930S: THE WILDERNESS ACQUIRES SOME DEFENDERS

Soon after the exploration of Pet 4 had begun, E. W. Nelson, now chief of the U.S. Bureau of Biological Survey (Department of Agriculture), was struck by its ominous implications, commenting that "the movement is already on to conquer our last American frontier."[165] The 1920s were certainly an important divide in the relationship between the agents of transformation and primitive conditions in Alaska. This was not only due to the Alaska Railroad, the gold dredge, and new roads, but also to the arrival of innovations such as the automobile, the tracked vehicle, and, not least, the airplane—not to mention a host of mechanical improvements like chain-saws and outboard internal combustion motors for boats.[166]

The airplane was the most sensational expression of the technological revolution and signaled the end of the dog team and steamboat eras. Airplanes flew into Alaska from the contiguous United States around 1920. The first commercial service was established in 1923 when Carl Ben Eielson began flying between mining camps around Fairbanks. In 1924 the U.S. Post Office awarded him the mail delivery contract between Fairbanks and McGrath on the Kuskokwim River in western Alaska. That year Noel Wien inaugurated the first service between Fairbanks and Anchorage. In 1925 Wien established a run from Fairbanks to Nome that reduced the journey to the western coast of Alaska from four to six weeks (by dog team) to three-and-a-half to four hours. By 1933 there were forty-two planes in the territory.[167]

Collectively, these advances produced a context far different from that which shaped the earlier responses of conservationists like Grinnell and Muir. The reactions of two particular conservationists in the 1930s to these accumulating changes are of seminal value because they span the divide between two eras of technology and the "old" and the "new" conservation. Frank Dufresne and Robert (Bob) Marshall were both natural resource managers in the federal government for a large part of their professional careers. Dufresne was a wildlife biologist and game warden who became chief executive of the Alaska Game Commission in 1935. Marshall was director of the Forestry Division of the U.S. Office of Indian Affairs from 1933 to 1937. Both were in the vanguard of Alaskan wilderness appreciation.

Dufresne was an agent in E. W. Nelson's Bureau of Biological Survey.[168] Like his superior, he knew what Alaska was like before the 1920s when the airplane had "annihilated distances."[169] But he was most concerned about the impact of roads on wildlife. In 1929 the Steese Highway was completed, connecting Fairbanks to Circle City on the Yukon.[170] Dufresne feared this road would provide car-borne hunters with easy access to the country northeast of Fairbanks (between Twelvemile Summit and Eagle Summit), which was heavily traversed by the Fortymile caribou herd during its spring and fall migrations.[171] The frontier analogy dominated his thinking. He believed that the game resources of Alaska were equal to the West's in the 1880s: "We still have our hundreds of thousands of caribou, as the West once had its buffalo, our willow ptarmigan of the northern tundras still darken the skies as their passenger pigeons once did."[172]

During the Second World War, Dufresne left Alaska to concentrate on writing. In 1946 he published a book based on his twenty years in the territory. In *Alaska's Animals and Fishes*, he saw impending tragedy if habits born of an overconfident belief in the inexhaustibility of the northland's resources were not broken. He was not optimistic about such a change, and the book's opening paragraph struck a solemn note: "It is our final frontier. When Alaska is tamed by man the subjugation of all our land possessions will be complete. We shall have conquered all. We shall have destroyed our last great wilderness."[173] At times, Dufresne's portrayal of Alaska's natural opulence almost sounded booster-like. However, he was sensitive to the changes that were taking place and realized that the comparison with the "old West" was not only facile but dangerous, because it encouraged unreasonable expectations. Like many, Dufresne was tempted by the buffalo-caribou analogy, but he stressed that Alaska's caribou herds were already dwindling in size. Much of their historic grazing range had been destroyed by the slash and burn practices of miners and Natives, in addition to natural fire. As a result, caribou were increasingly restricted to the Arctic.[174]

For Bob Marshall, the airplane facilitated the contact with the "true" wilderness that he had craved since boyhood. New York City–born and bred, Marshall was the son of a prominent constitutional lawyer, Louis Marshall, who had used his legal talents to defend New York's Adirondack State Park. Bob Marshall received a master's degree in forestry from Harvard University and a doctorate in plant physiology from Johns Hopkins.[175] In 1929 the airplane brought him to the mining community of Wiseman on the threshold of the Brooks Range, an extension of the Rocky Mountain Cordillera. He saw the Brooks Range as a recreation of the raw America explored by the heroes of his juvenile years, Lewis and Clark.[176] The airplane could return him within two hours to the nearest pavements, electricity, automobiles, and trains, 200 miles south in Fairbanks. He explored the central Brooks Range on four occasions. Between his first visit in 1929 and his last in 1939, he witnessed changes that he greeted with a mixture of indignation, anxiety, and resignation. The pace of interwar change made it difficult for Marshall to believe that the "good life" of the frontier, which he identified as Alaska's "outstanding quality," would remain inviolate.[177] When he made his second trip to the biracial hamlet of Wiseman, population circa one hundred, in the summer of 1930, his purpose was to study a "pre-industrial civilization"[178] "before it is too late."[179] The product of this thirteen-month stay was the acclaimed Literary Guild bestseller, *Arctic Village* (1933).

In this book, regarded as a pioneering piece of sociology, he described the villagers' reaction to the first airplane to arrive from Fairbanks in 1925. According to Marshall, Natives displayed excitement mixed with a little fear. As for whites, "they got not only the normal emotional reaction which any one received seeing his first airplane, but also they had the exceedingly practical sensation that civilization in an emergency was no longer three weeks to three months away, but only a matter of two or three hours. Notions set by nearly thirty years' experience were turned topsy-turvy in a moment."[180] Less sensational than the advent of the airplane was the introduction of the caterpillar tractor, universally known as the "cat," which made its debut in Wiseman in the winter of 1929–30. The tractor was used for hauling freight from Bettles on the Koyukuk River, which steamers could reach from the Yukon. The subsequent arrival of the first automobile, just prior to Marshall's departure in 1931, was something of an anticlimax, despite the fact that five whites and most of the Natives had never seen one.[181] During his long stay in Wiseman (1930–31), less than one plane a month landed at the airstrip. When he made his third visit in 1938, there was an average of two to three each week. He concluded gravely that year, "there had been more of a shift from the true frontier in the past seven years, than in the previous thirty years."[182]

A similar feeling that the last frontier was becoming less rugged and pure was expressed by some older white residents. In his autobiography, Frank Dufresne quoted an anecdote from a disgruntled "sourdough" living on the Yukon who complained: "the blasted things'll be buzzin' up an' down the river thicker'n mosquitos. Everybody'll start rushin' from one place to another. There won't be no time fer livin' an' I don't like it!"[183] In 1940 *Alaska Life* carried an article by "Skagway Bill." In "Alaska As I Knew It," this veteran of the Klondike complained that Alaska had "gone softy." The major reason for this change, he believed, was the airplane, thanks to which, "there just ain't any more real frontier left North of 54 degrees 40'."[184] According to figures quoted for 1938 there was a higher per capita rate of airplane use in Alaska than anywhere else in the world besides Arctic Canada and Russia.[185] In 1939 there were 175 planes operating on scheduled runs in Alaska. Fairbanks could be reached from New York City in twenty hours.

This "sourdough" attitude reflected the "Alaska for Alaskans" mentality. This outlook, *Alaska Life* explained, was interpreted in Washington, D.C., to mean "keep out population, keep out competition, keep out capital, and keep out development." In such ways, the magazine complained, the federal government received the impression that its passive Alaskan policies met with approval.[186] The federal attitude towards Alaskan development to which some Alaskans objected was represented in the National Resources Committee's report *Alaska; Its Resources and Development* (1938). This was the most comprehensive assessment of the territory ever undertaken.[187] A basic assumption of this cool appraisal of the immediate and short-term prospects for Alaskan economic development was the desirability of avoiding a repetition of the errors that had marred natural resource use throughout American history.[188] Introducing the report, Henry Wallace, the secretary of agriculture, agreed with its "emphasis of caution against the over-development of Alaska." He warned that the territory held no real potential as a new agrarian frontier. (In the Matanuska Valley, the Federal Emergency Relief Administration had recently sponsored the only significant attempt at agrarian colonization in Alaskan history. Critics denounced this experiment, whose degree of success is still a matter of debate, as vainglorious and a socialistic and paternalistic antithesis of true pioneering.) Wallace also denied that expansion of the major existing employment categories could provide the basis for a steady economy and a stable society. He explained that canning and mining, which were seasonal, male-dominated pursuits, "militate against the establishment of permanent communities," and discouraged "the building of normal family and communal life." Nor was he enamored of the "new Scandinavia" doctrine.[189]

The staff report "Wildlife Resources" was a collaboration of the Bureau of Biological Survey, the National Park Service, the Office of Indian Affairs, and the Forest Service. It remarked of Alaska's fauna that "no other feature of the Territory provides so much interest to the people of the United States, and it sends north a constantly increasing number of hunters, naturalists, photographers, painters, tourists, and other visitors." With this in mind, the authors asserted boldly that "a wholly satisfactory economy can be based in large part on the maintenance of large areas in a quite primitive state and the entertainment of visitors to these areas."[190] In view of its comparability with the buffalo of the nineteenth-century West, the authors singled out the caribou for special mention.[191]

One of the authors of the "Recreational Resources and Facilities" section was Bob Marshall, chief of the U.S. Forest Service's Division of Recreation and Lands—a position created for him in May 1937. This section extolled the "unusual opportunity" available "to avoid certain mistakes which were made in developing our own great west" and entered an avant-garde plea for the establishment of "wilderness areas." It countered the standard Alaskan objection that protection was superfluous due to the extent of wilderness in Alaska by referring to the western frontier, where history had shown that "wilderness conditions do not automatically preserve themselves."[192] Marshall's personal contribution to this section was the most uncompromisingly preservationist aspect of the entire report. It provides the best-known précis of his philosophy and his views on the future of northern Alaska, and these continue to provide the touchstone and inspiration for the policies of many national conservation organizations whose overriding Alaskan concern is the integrity of the Brooks Range. Marshall wanted to protect "pioneer conditions" and "the emotional values of the frontier." To redress the national imbalance between wilderness and civilization, he recommended that all Alaska north of the Yukon (with the exception of Nome) be kept free of roads and industrial development: "In the name of a balanced use of American resources, let's keep Alaska largely a wilderness!"[193] As we have seen, wildlife protection had been the goal of the first conservation efforts in Alaska. As Roderick Nash has commented, this was "the first direct and specific call for the preservation of *wilderness* in Alaska."[194]

The report of the National Resources Committee disappointed Alaskan boosters. It deflated hopes raised by the Matanuska project and reinforced abiding suspicions that the federal government was in league with conservationists to sabotage the territory's growth. Alaskans who had objected to the federal conservation policies applied to Alaska in the early twentieth century had not distinguished between utilitarianism and preservationism; they dismissed both as manifestations of an alien and undemocratic creed antithetical to local needs and interests, which they called "Pinchotism." In part, this reflected the absence of a rigid demarcation between these two aspects of conservation, which in Alaska were united in the belief that it was imperative to avoid past errors.

However, an examination of Alaskan protests against federal conservation policies during the 1920s and 1930s does reveal a discrimination between varieties of conservation. This was an important development. Henceforth, Alaskan proponents of projects and policies from which substantial material benefits were anticipated, of which some conservationists were critical, sought to portray them as examples of "true" conservation. Alfred Hulse Brooks—who was in charge of the United

States Geological Survey, Alaska Branch, from 1903 until 1912 (and for whom the Brooks Range was named)—made this point in 1925. In Alaska, he argued, conservation had been taken to mean "non-use rather than prudent utilization." He bemoaned the failure to cut mature timber in the national forests and cautioned that "the true conservationist must be distinguished from those who would permit perishable resources to go to waste in order to keep Alaska free from human beings."[195] The once-repugnant Pinchotism had soon become the pedigree from which Alaskan pro-development forces claimed their conservationist credentials.

The most vitriolic critic of federal conservation policies in the 1930s was John Hellenthal, a lawyer, who was the author of *The Alaskan Melodrama* (1936), a popular history of Alaska.[196] This account, which described Alaska as "prostrate" beneath the "feudal" rule of "the Conservationists,"[197] was a vintage exposition of what might be called the conservationist conspiracy theory of Alaskan retardation. In 1941, further incensed by the National Resources Committee's report (1938), Hellenthal updated and amplified these views in *Alaska Life*.[198] He interpreted the report as proof that "Czar" Harold Ickes's Department of the Interior, which had jurisdiction over Alaska, was a mere tool of the preservationist lobby. Predictably, Hellenthal was most outraged by Marshall's appendix, which he considered representative of the report's general thrust. For Hellenthal, its message was undisguised: "The idle rich, in the States, really need a playground and . . . Alaska can be made to serve this purpose admirably."[199] He argued that the desire to "freeze" the frontier in its original state embodied "orthodox" conservation at its most oppressive. This view, which persists in some Alaskan quarters today, held that preservationism amounted to a betrayal of the principles of genuine conservation. To reinforce his argument, Hellenthal pointed out that elsewhere in the nation the new style of conservation typified by the Tennessee Valley Administration (TVA) was being implemented. Only in Alaska, he maintained, did preservationism retain its grip. He described preservationism as a reactionary and crippling force, a gloomy philosophy that had "no confidence in future generations."[200] He considered this "fad," which vilified pioneers as "crooks," an insult to the American tradition. He also condemned it as antihumanistic because it glorified nature while denigrating man and his achievements.[201]

According to Frederick Jackson Turner, a fundamental ideal of the American pioneer era had been "individual freedom to compete unrestrictedly for the resources of a continent—the squatter ideal. To the pioneer government was an evil."[202] On the eve of the Second World War, many Alaskans saw themselves as frontier dwellers deprived of the rights traditionally enjoyed by American pioneers. In 1941 one writer attributed this historical anomaly to a radical political and ideological change, not climate or geography.[203] The major Alaskan conservation controversy during the 1930s concerned efforts to create a sanctuary on Admiralty Island to protect its huge brown bears, a favorite big game trophy, from commercial logging. The Boone and Crockett Club led the campaign to designate the island a national monument. The secretary of the interior in fact opposed such a reserve, but Louis R. Huber nevertheless denounced him for caring only about "the tufted puffins or the grizzly bears."[204] He thought Ickes's general attitude to Alaska was preposterous because "if you hark back to the days of the hell-roaring west, you find no record of a secretary of the interior trying to 'conserve' the frontier." In view of these man-made obstacles, Huber concluded that "being America's last frontier" was "a superhuman and thankless effort."[205]

The same issue of *Alaska Life* contained translated extracts from a book by a German Nazi, A. E. Johann, who had been dispatched to inspect the United States by Hitler's Geopolitical Institute.[206] Johann had been particularly enchanted by the *lebensraum* that Alaska offered. Whereas many Alaskan boosters were fond of describing Alaska as a nubile maiden yearning for the right suitor, Hitler's representative preferred the image of "a powerful, large and patient animal . . . still waiting for its master." With images reminiscent of Jack London, he depicted Alaska as "tremendous and virginal," a "Nordic land, created for Nordic people . . . the true home of the Nordic race . . . one of the great reserve territories." He speculated that the northland's resource "superabundance" could support a population larger than Scandinavia's. Paraphrasing Robert Service, he felt that Alaska was waiting for "the advent of a people who are a match for her." And, like Huber, he wondered why pioneers were not drawn to Alaska as they had been to the West.[207] For the editors of *Alaska Life,* the moral of the story was that a large undeveloped frontier region could not remain unused indefinitely without attracting the envious attention of more densely-settled, expansionist countries like Germany, "for these nations look on such unexploited riches as a prize which, seemingly, the owner does not care for enough to spend the necessary time, money and labor to develop."[208] Differences in political ideology notwithstanding, there was little to distinguish Johann's tract from the bombast of Alaska's keenest interwar boosters.

2

The Impact of War

The Alaskan historian, Clarence L. Andrews, dedicated his book *The Story of Alaska* (1931)

> To those who are endeavoring
> to develop the last of Uncle Sam's
> farms and who are conquering the
> last frontier, giving their lives to that service.[1]

In his preface, Andrews lamented that prospective settlers could not drive wagons to Alaska, as their ancestors had done to the Far West on the Oregon Trail.[2] Nostalgia notwithstanding, innovations in transportation, communications, and man's technological capacity to transform the environment set mid-twentieth-century Alaska clearly apart from the nineteenth-century western frontier. The differences between them can be seen as a reflection of global changes whose impact James Malin recognized in the 1940s:

> Water was the primary dependence of man for communications, on a world basis, until the mid-nineteenth century when the advent of mechanically-powered wheel communications superseded it for most inland services. In the second quarter of the twentieth century, both water and wheels are being superseded, in part, by air communications. The occupation of the grasslands of the world by modern civilization during the late nineteenth century was associated particularly with mechanically-powered wheel communications, and the arctic is being brought into the orbit of modern civilization by air communications. In North America the wheel pointed the advance westward, while air power points it northward.[3]

Before 1940, "mechanically-powered wheel communications" had played a minor role in the development of Alaska, whose status was practically insular. Water remained the "primary dependence" since the territory was accessible from the rest of the nation only by sea. In 1940 this physical isolation was reduced and the historic American sequence was broken when Pan-American Airways introduced the first commercial air link between Seattle and Alaska.[4] By 1943 it had also become possible to travel overland to Alaska, though not by railroad.

2.1 FROM ALASKA HIGHWAY TO DEWLINE

During the early decades of the century, Alaskan boosters had pinned their hopes on the Alaska Railroad as the decisive breakthrough in the liberation of Alaska's natural resource potential. However, they soon realized that the railroad's value was limited without surface communications extending to the lower forty-eight. By the late 1920s their largely unfulfilled yet dogged expectations for the "occupation" of Alaska had swung emphatically to the proposal for a road to the rest of the United States. In 1929, Donald MacDonald, a veteran of the Alaska Engineering Commission and locating engineer for the Alaska Road Commission in Fairbanks since 1922, formed the International Highway Association (IHA) in Fairbanks. This promotional organization consisted mainly of local businessmen. The Alaskan and the United States chambers of commerce endorsed the lobby, which established further branches in Dawson City, Yukon Territory; Vancouver, British Columbia; and Seattle.

In 1930 President Hoover established a three-member Alaska-Canadian Highway Commission (ACHC) to study the possibility of such a highway.[5] The Canadian government declined to set up a counterpart but a study group was established in British Columbia in 1931. In 1933 the ACHC submitted to Congress a favorable report on the feasibility and economic practicality of a highway through British Columbia. The report estimated that the project would cost $14 million. It also stressed the role the road would play in "the opening of new country" for settlement.[6] However, no action followed by either government.[7]

From 1930 onward, MacDonald and Anthony J. Dimond, Alaska's territorial delegate (Democrat) from 1933 to 1944,[8] lobbied indefatigably but fruitlessly for an international highway. In 1934 Dimond introduced the first highway bill, which called for $100,000 to fund a joint American-Canadian study and further funds to build a road. The Bureau of the Budget objected to these expenses. The bill, as passed in August 1935, empowered President Roosevelt to reach an agreement with Canada for the study of the location and construction of such a road—85 per cent of which would be in Canada—but did not include any financing provisions.

The Canadian government, fearful of American military domination, was unenthusiastic about such studies.[9] Despite lack of progress, the awareness that innovations in transportation and communications had always faced resistance from sceptical and unimaginative people consoled

Dimond: "We must remember that when the great railroads of the United States were first thought of, the more conservative members of the community dismissed the projects as a mere dream."[10] In 1935 he explained to the president that a road would "mean more for the development and permanent settlement of Alaska than all the other things which have been accomplished for Alaska in the last twenty years."[11] The following year, attempting to woo the secretary of state, Cordell Hull, Dimond argued that "the construction of such a highway ranks in importance with the building of one of the great railroads in a former generation."[12]

Dimond's frequent analogies to the transcontinental railroads did not of course involve faith in railroads themselves in what he called "the motor age." A few, notably James C. Rettie, counselor of the federal National Resources Planning Board (NRPB) office in Juneau, did want a railroad connection between Alaska and the lower forty-eight. Rettie proposed to link the Alaskan interior with the contiguous American network via Canada's Mackenzie Valley. He even speculated about the possibility of extending such a line to Barrow, the northernmost settlement in Alaska.[13] Edward H. Harriman's grandiose, turn-of-the-century vision of a railroad from New York City to Paris via the Bering Strait was often referred to as a precedent by both rail and road advocates in the 1930s and early 1940s.[14]

The analogy with the western frontier loomed large in the rhetoric of advocates of "the last great trail."[15] Borrowing terms from Frederick Jackson Turner, the International Highway Association demanded "free access to free land, free access to Nature, [free access] to the resources of a Northern Empire." The IHA emphasized the importance of trails and roads in the process of continental expansion. However, it explained that twentieth-century pioneering in Alaska must adapt to changes in transportation technology and peculiar geographical circumstances, which made it as hard to "[en]vision a Northern migration without the automobile as it would have been for our forefathers to plan the crossing of the Great American Desert without the prairie schooner." The IHA wanted the road to bring the backward territory ("still in the dog team stage") into conformity with the rest of the nation ("synchronize the North with the age"). With an international highway, the IHA hoped that Alaska would become "merely one of the United States."[16]

Like the supporters of the Matanuska Valley agrarian colonization project, highway proponents dwelt on the significance of the Alaskan frontier as a safety valve to alleviate the socioeconomic plight of the 1930s. In 1935 MacDonald tried to sell the president an image of Alaska as an untouched land of plenty; "a veritable paradise of game," off which settlers could live in time-honored style.[17] On another occasion,

MacDonald boasted of "game everywhere existing in primitive abundance: caribou, moose and sheep by the uncounted hundreds of thousands. Myriads of lakes that are full of fish that have never known the touch of an artificial lure . . ."[18]

The impetus behind the highway proposal was slowly gathering political momentum. In 1938 Congress created the Alaskan International Highway Commission (AIHC) to comunicate with the relevant Canadian authorities and study the problems of location, surveying, and funding.[19] The AIHC's report to the president in 1940 concentrated on the road's economic value to Alaska rather than its military value to the nation.[20] The AIHC chose a route close to the west coast of Canada, which the International Highway Association (IHA) preferred too. The AIHC echoed the IHA's views on the decisive influence of roads in pioneer America. It gave the example of northern Indiana and Illinois, which did not advance beyond a fur trapping economy until the coming of the first wagon roads in the 1830s. The AIHC also agreed with the

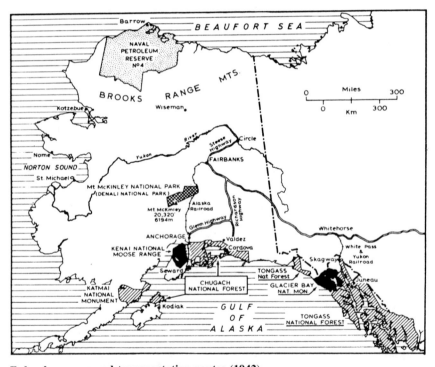

Federal reserves and transportation routes (1942)

Alaskan lobby's assessment of the potential for opening vast new areas for settlement in Alaska.[21] FDR submitted the report to Congress but the crucial support was not forthcoming. The State Department and the Interior Department, which both supported the construction of a road to Alaska, told AIHC member Ernest Gruening, territorial governor of Alaska since 1939, that plans could not proceed without the support of the U.S. Army, which saw no strategic importance in a road to Alaska.[22]

With the arrival of the Second World War, Alaskans began to promote the highway by drawing attention to Alaska's pivotal role in national defense. In August 1940, Canada and the United States established a Permanent Joint Board on Defense to consider mutual measures to protect North America. At the end of September, Germany, Italy, and Japan formed the tripartite alliance. In November the Permanent Joint Board on Defense authorized construction of a series of airbases east of the Canadian Rockies. These airbases at Fort St. John, Fort Nelson, and Watson Lake, known as the Northwest Staging Route, were ready for use by the summer of 1941.[23] However, the U.S. Army still denied the military necessity of a road to Alaska.[24]

In February 1941 Dimond unsuccessfully introduced another highway bill. That year, the situation in the Pacific grew more serious for the United States when Japan established bases in southern Indo-China and a Japanese occupation of eastern Siberia became a possibility. In the wake of Japan's attack on Pearl Harbor and its occupation of Guam and Manila, oceanic supply lines to Alaska became vulnerable to air and submarine attack. A surface link with the exposed and strategically important northern territory suddenly became expedient.

During peace-time, Alaskan proponents had discussed at length which route would best serve Alaskan economic development. The International Highway Association and the Alaskan International Highway Commission considered four routes to Fairbanks. Route A was the most westerly. Route B overlapped with A as far as Prince George, British Columbia, where it bent inland. Both Routes C and D, which ran east of the Rockies, terminated in the south at Edmonton, Alberta, where established routes led to the American Midwest. Both Routes C and D traversed more level terrain than A and B and provided a more direct link with the Midwest and the eastern seaboard. C and D (the so-called prairie routes) also furnished access to Alberta's oilfields. D possessed the additional advantage of passing close to the oilfield at Norman Wells in northern Yukon Territory. The U.S. Army believed that C and D were better protected from air attack and more practical since most of the military supplies for delivery to Alaska would originate east of the Mississippi.[25] Alaskan boosters, whose fundamental objective was to establish an effective link with Seattle, the historic "gateway to Alaska," wanted Route A. This was also the only route which, because of its proximity to the coast, provided the opportunity to create branch lines to settlements in southeast Alaska that were inaccessible by road. The Alaskan International Highway Commission also favored A.[26] After Pearl Harbor, the president appointed a committee consisting of the secretaries of war, the navy, and the interior to select a route.[27]

In February 1942, in view of the new political situation, Dimond resubmitted the bill he had introduced the previous February. His bill asked for $25 million to build a road "through the northwestern part of Canada." The exact route, that which best served national defense needs, would be chosen by the president.[28] At the hearings on the bill held by the House Committee on Roads, of which he was a member, Dimond's lobbying efforts again relied heavily on frontier images. He explained that the American pioneering impulse had been thwarted in its persistent search for wider scope by the peculiar handicaps, notably inaccessibility, from which Alaska continued to suffer: "The pioneer spirit, which plowed the plains and settled the west has no opportunity for expansion or development in Alaska because the present difficulties and delays involved in getting to Alaska are, to most of such prospective settlers, well-nigh insurmountable." But after a highway had been built, he looked forward "with confidence to the experience which this nation witnessed in the last century when our flag was carried from the Alleghenies to the Pacific Ocean."[29] Dimond, Magnuson, and Gruening stressed the superior economic and military value of Route A.[30] The hearings on the bill adjourned with no action taken because the military was making its own plans. Later that month, the Permanent Joint Board on Defense recommended Route C, a choice largely predetermined by the location of the Northwest Staging Route.[31]

Construction of the Alaska-Canada Military (Alcan) Highway had already begun on 7 March 1942. "Its the demand for speed," commented one reporter, "that makes the problem. Everything goes on at once. There's no time for desk planning. The surveyors are only thirty miles ahead of the clearing crews."[32] The armed forces dispensed with standard preparatory measures such as thorough surveying and proper staking. The U.S. Army Corps of Engineers originally planned to push through a rudimentary "pioneer-type" route, which the Public Roads Administration would later use to construct a permanent artery. The army engineers soon found this procedure too time-consuming and within three months all contractors were assigned to the U.S. Army road; the pioneer trail became the road itself.[33] Eleven thousand soldiers finished the job in under nine months.[34] By 20 November they had laid 1,422 miles of road from Dawson Creek, British Columbia, to Big Delta, Alaska, where the new road connected with the Richardson Highway for the final 95 miles to Fairbanks.

The breakneck pace of construction (an average of eight miles daily) captured the popular imagination and stimulated the first rediscovery of Alaska by the press since the Klondike. Journalists generally presented the pell-mell project as another round in the brutal and ennobling confrontation between man and nature on the frontier, which many of them felt lay at the heart of the national experience. One reporter described it as a "titanic struggle between human beings and a wild, treacherous, unknown tundra."[35] William Gilman, a freelance journalist covering the war from Alaska, rejoiced: "There was something as fresh as the evergreen forests and mighty rivers it crossed in the building of the road. This was the drama of combat engineers on the move, of logging and pioneering, of men riding upon the stout necks of snorting engines which charged through the impossible."[36]

At the highway's opening ceremony on 21 November 1942 at Soldier's Summit, Kluane Lake, Yukon Territory, Major-General James A. O'Connor of the Northwest Command, who supervised construction, remarked that the soldiers who built the road were "every bit as rugged and resourceful as the men who thrust the first railroads across the North American continent."[37] O'Connor officially named it the Alaska Military Highway on 17 June 1943. Its name was formally shortened to the Alaska Highway in July 1943. However, the military acronym *Alcan* continued (and continues) to receive considerable popular use. The security crisis in the Pacific had eased by 1943, when the Public Roads Administration built a final road, using private contractors, along a route approximating to the pioneer trail.

The language of frontier conquest used by participants and contemporary commentators and their celebration of technological potency conformed to well-established patterns.[38] A popular contemporary account of the construction of what soldiers simply called "The Road" was dedicated to "the men who smashed through a back door to Alaska."[39] The image that captivated journalists was that of a bulldozer toppling trees with impunity. Spruce and other trees growing on permafrost have shallow roots that branch out in fans. This made it easy for the "land battleships," as bulldozers were called in military jargon,[40] to knock them down when they hit the trunks with their blades raised high. In this way, the "cat skinners," as the operators of caterpillar tractors were known, flattened "thick timber as if it were no more than a field of cornstalks."[41]

As well as being portrayed as a helpless victim, the wilderness was depicted as "a battlefront, exacting its toll of [human] life the same as any other field of combat."[42] Nature was seen as a fierce adversary for man and machine. In the summer months, vehicles sank to their axles in the muskeg ("reputedly bottomless swamps").[43] To cope with this problem, soldiers used the time-honored corduroy technique. American troops had resorted to this method during the War of Independence and the Civil War. In fact, it had been adopted by military forces throughout history wherever swamps and marshes prevented the movement of heavy equipment. In the boreal forest of Alaska and Canada there was no shortage of trees to form a roadbed. In some spots, multiple layers of brush, trees, soil, and gravel had to be laid down before vehicles could pass over. Additional difficulties were presented by permafrost, which soldiers dubbed "the North's secret weapon."[44] At first, some troops had thought that frozen ground was an unexpected blessing. They believed that if they simply scraped off the vegetational cover, the solid frozen ground would provide a ready-made, smooth and firm roadbed when covered with gravel. This hope was quickly dispelled when the exposed permafrost began to melt. Thereafter, the road builders treated permafrost with respect, like muskeg. Leaving the thawed top layer of ground undisturbed, they overlaid the frozen ground with brush, corduroy, and gravel.

Without exception, those who covered the story of the construction of the Alaska Highway at that time, and those who wrote about it in the 1950s, saw it as one of the greatest engineering achievements in history. One writer believed that, in view of the "apalling and almost insurmountable obstructions" faced, such as permafrost, the highway was the most impressive accomplishment on the continent since the Panama Canal was completed.[45] Philip Godsell of the Hudson's Bay Company compared the new road to the Oregon Trail. Godsell, who claimed to have covered the entire route of the road at various times on snowshoe, by canoe, or dog team, offered this assessment: "As Lewis and Clark opened up the mighty trans-Mississippi empire so, out of the destructiveness of war, there may yet emerge a new industrial empire in this Last Northwest. . . . historians of the future will enter new names in the book of pioneers. Along with Lewis, Clark and Mackenzie; Fraser, Fremont and Carson, will be written O'Connor, Hoge and Sturdevant, the builders of the Alaska Highway."[46]

This inflated rhetoric was no consolation to Alaskans and others who wanted a highway primarily for economic reasons. As early as June 1942, Senator William Langer (Republican, North Dakota), a member of the Alaskan International Highway Commission, sought the appointment of a committee to investigate the U.S. Army's decision to build the road along Route C. If this committee decided that the military's choice was the wrong one, Langer believed it should recommend the relocation of the Alaska Highway along Route A. Langer denounced the road being built as "an engineering monstrosity" and "an economic absurdity."[47] The report (17 June) of the subcommittee of the Senate Committee on Foreign Relations, which was set up to deal with Langer's resolution, dismissed outright these claims. The subcommittee recommended against a full investigation as a waste of time and money and an interference with the existing effort to defend Alaska.[48]

The disappointment and dissatisfaction that Anthony Dimond expressed when the road was finished was typical of Alaskans who had lobbied for an international highway. While acknowledging that "any highway to Alaska is better than none," Dimond emphasized that "the location of the highway on that route as a supply line for Alaska was a mistake."[49] Donald MacDonald was equally outraged, for both economic and security reasons, by the choice of Route C.[50] These views were shared by Ernest Gruening, territorial governor of Alaska, and the other members of the Alaskan International Highway Commission. In 1942 and 1943 Dimond fatuously submitted bills for the construction of another highway in the west of Canada that would link up with the Alaska Highway at Whitehorse, Yukon Territory.[51]

After the war, when control over that portion of the road within Canada was transferred from the U.S. Army to the Canadian government (1 April 1946), Alaskan politicians chafed against the restrictions imposed on its public use. The Canadian authorities issued permits, contingent upon vehicle inspection and possession of adequate supplies, to residents, miners, and others who had business in Alaska. E. L. "Bob" Bartlett (Democrat), who succeeded Dimond as Alaska's congressional delegate in 1944, wanted the road available immediately to all without restriction. Alaskan officials were also keen for the Bureau of Land Management (BLM) to open up adjacent land for tourist camps, cabins, gas stations, taverns, and other roadside attractions. Canada lifted most restrictions in February 1948, though it continued to require evidence of a minimum amount of cash from those traveling to Alaska.

The Alaska Highway was not the road that many Alaskans had wanted, and it was certainly not what David Remley has grandly called "the last of the important overland trails" in the United States.[52] However, it was part of a dynamic growth process that included the construction of two other roads. The Glenn Highway connected the Alaska Highway with Anchorage and the Corps of Engineers constructed a link between the latter and the port of Haines. The total population of Alaska on 1 October 1939 had been 72,500. At this time, there were only 524 military personnel in the territory. By 1 July 1940 the military population had risen to 9,000. The military population reached an all-time high of 152,000 in July 1943, when the total population of the territory stood at 233,000.[53] The year before Pearl Harbor, 1,255 workers were registered in construction. By the end of 1941 the number had risen to 10,521.[54]

The center of growth was Anchorage, which had originated in 1915 as a tent city housing 2,000 construction workers for the Alaska Railroad. A year later, its population had grown to 6,000. By 1940 it had shrunk to 3,500. However, the armed forces selected Anchorage as headquarters for Alaska's military installations. An army base was constructed four miles away at Fort Richardson and Elmendorf Air Force Base was established at a distance of seven miles from Anchorage.[55] In the mid-1950s, one non-Alaskan advocate of Alaskan economic development interpreted this growth in a comparative framework as an indication of a stabilizing and broadening economy: "The pattern seems to parallel that observable in the historical development of the West, which centered about the establishment of military outposts and garrisons."[56]

Nor was the Alaska Highway the only major war-related construction project in Alaska. The Canadian Oil (Canol) Project, which included Alaska's first oil pipeline, was part of the same response to the Pacific security crisis that led to the Alaska Highway's construction. In January 1942 the U.S. War Department became attracted to the largely undeveloped oilfields of Canada's Northwest Territories as a fuel supply for American forces in Alaska. In June 1942, Japan occupied the Aleutian Islands of Attu and Kiska and military strategists began to think of launching a counter-offensive against Japan from Alaska. If planning for the Alaska Highway was minimal and its construction ad hoc, then the Canol project appears even less deliberated and more precipitous. According to testimony delivered to a senate investigating committee during hearings in 1943, the U.S. War Department's decision to go ahead with the initial phase of the project (29 April 1942) was taken without consulting any other government officials. Neither Harold Ickes, the secretary of the interior and petroleum administrator for war, or the secretary of the navy were consulted.[57]

The first phase involved construction of a 557-mile, 4-inch diameter pipeline from Imperial Oil's Norman Wells fields to Whitehorse, Yukon Territory (Canol One). Rudimentary surveying was carried out using dog teams and tractors. Construction equipment was delivered by boat and barge via the Slave River, Great Slave Lake, and finally, the Mackenzie. To supplement summer barge traffic, Canol's builders decided to construct a road. Just as the Alaska Road Commission had learned earlier in the century and like their compatriots on the Alaska Highway, Canol's engineers found out that removing the insulating ground cover of spongy vegetation/moss (muck) in subarctic, boreal regions produced a quagmire that no vehicle could cross in summer. The Canol service

The Alaska Highway and Canol Project

road could only be used in winter. Permafrost rendered conventional burial impossible so the pipeline was laid on the surface, without supports. This left it vulnerable to heaving of the ground as a result of the freeze-thaw cycle and exposed to flooding. Some federal geologists warned that this could lead to severe erosion when the muskeg turned to swamp during spring melt.[58] The Truman Committee (as it was popularly known) heard that the U.S. War Department never carried out a comprehensive route survey and did not study climate or terrain.[59]

Instead, the U.S. War Department relied on the advice (rather misleading and simplistic in most later opinions) of the Canadian explorer-anthropologist and leading promoter of arctic development, Vilhjalmur Stefansson. Stefansson, in accordance with his theories of the "friendly Arctic," downplayed the difficulties of the operation.[60] At these senate hearings, it was revealed that, even one year into construction, the final route was yet to be determined. In December 1943 the petroleum administrator for war (Harold Ickes), the U.S. Navy, and the War Production Board recommended that the project, which now included three distribution lines, be "junked" because it was not worth the cost of completion. The senate investigating committee calculated that the U.S. War Department had already spent $135 million on Canol.[61] Thomas Riggs, a former territorial governor of Alaska and a member of the Alaskan International Highway Commission, did not want to rely on foreign oil to supply the new military bases at Fort Wainright and Eielson near Fairbanks. The source he preferred was the Katalla oilfields of southcentral Alaska, the scene of Alaska's only previous well in the 1920s.[62] Harold Ickes believed it would have been cheaper to ship Californian oil to Skagway.[63] However, construction continued and the first oil reached Whitehorse in mid-April 1944. Japanese troops had been expelled from the Aleutians in the summer of 1943 and the shipping crisis in the Pacific was over. So the project was shut down in June 1945.

Canol now consisted of four parts. Products from the Texan refinery capable of handling 3,000 barrels a day (mostly aviation fuel), which had been reassembled in Whitehorse, were distributed by three lines. Canol Two, a 4-inch pipe, ran 110 miles to Skagway along the route of the White Pass and Yukon Railroad. The second products line, Canol Three, was a 2-inch line that connected with Canol Two at Carcross, Yukon Territory, 45 miles southwest of Whitehorse. It then extended 266 miles to Watson Lake, Yukon Territory, following the Alaska Highway. Canol Four, a 3-inch line, carried the oil 596 miles from Whitehorse to Fairbanks.[64] Canol One, sold for scrap, was removed during the winter of 1947–48.

Meanwhile, the U.S. Navy had turned its attention to Alaskan oil reserves. The modern era of oil development in Alaska began in 1944. That year, the U.S. Geological Survey (USGS) and the U.S. Navy's Office of Naval Petroleum and Oil Shale Reserves launched the first systematic exploration of Naval Petroleum Reserve Number Four (Pet 4). The navy wanted to find out if there was a useful supply of oil in the event of a prolonged war with Japan and Germany, which might interfere with regular supplies from the Middle East and South America.

The airplane and tracked vehicle (caterpillar tractor) had transformed the nature of petroleum exploration since the initial reconnaissance efforts there in the 1920s. The C-46, the C-47, and the C-54 were used to line-haul equipment from Fairbanks. The C-46, a Curtiss-designed twin-engine freight and passenger plane, was most efficient for flights of up to 1.5 hours. The C-47 was a Douglas twin-engine plane, equivalent to the U.S. Navy's R-4D and the civilian DC-3. The C-54, a Douglas four-engine aircraft (equivalent to the DC-4), was the plane used most often for heavy loads over long distances. Smaller bush planes with skis, such as the single-engine Norseman and Cessna, facilitated extensive local travel in winter. In summer, when overland movement was precluded, regular bush planes could land on gravel bars or beaches. Equipped with pontoons, they could set down on and take off from lakes, rivers, and lagoons.[65]

Like their predecessors in the 1920s, those in command of the explorations considered various methods of transportation in the event of significant discoveries. They came to the same conclusion about the impracticality of transit by tankers from Barrow. An early report in April 1944 indicated the navy's clear preference for a pipeline to Fairbanks, where access to highway and railroad began.[66] The 1,000-mile chosen route crossed the Brooks Range at Anaktuvuk Pass. It then followed the John River to its confluence with the Koyukuk at Bettles, cutting through the heart of the country tramped and mapped by Bob Marshall in the 1930s. The navy did not discover valuable quantities of oil so the proposal never left the drawing board.[67] When the operation was terminated in 1953, nine years of exploration had yielded more knowledge of the Arctic itself and the logistics and practical obstacles facing an undertaking of this kind there than they had provided information about significant oil reserves.

Permafrost had caused major headaches for the operators of tracked vehicles such as the so-called weasel, which served as a personnel carrier. The caterpillar tractor was a heavier machine. Pulling a sled train like a line of boxcars, it could move large tonnages (up to 80 tons per tractor) over long distances (up to 80 miles per day). The report winding up the operation explained how the government oil seekers had learned the hard way that it was impossible for these vehicles to traverse the soft, waterlogged tundra in summer without operational difficulties and without damaging the vegetational mat that insulated the permanently frozen ground. Huge quantities of gravel were required to build airstrips that could be used in spring and summer. Unless the ground under buildings was similarly insulated with gravel, their foundations sank. The report concluded with a comment on the vulnerability of the tundra. It warned: "the Arctic holds many camouflage problems. Vehicle tracks are likely to remain easily visible for years because of the slow recovery of tundra vegetation. Even winter tracks may long be visible because the compacted snow affects the following summer's growth."[68]

The Second World War had demonstrated Alaska's strategic significance. The Cold War made it even clearer. The Distant Early Warning Line (DEWline) was a joint American-Canadian defense enterprise whose goal was to string fifty radar and communications stations across 3,000 miles of arctic terrain from Point Barrow, Alaska, to Broughton Island, Canada. The DEWline was the farthest north of three strategic defense lines. The others were the mid-Canada line and the Pine Tree line, which ran along the border between the two countries. The Western Electric Company won the construction contract and work began on the first part of the project, a test segment across Alaska, in 1953. For construction of the rest of the line, Western Electric received a thirty-two month deadline.[69] Work on the Canadian section began in December 1954 and the line went into operation in July 1957.[70]

The recently terminated Pet 4 exploration had been evaluated by its principal participants as "unique" in arctic history "in size, complexity, and accomplishment."[71] However, these epic dimensions were, in the opinion of the Western Electric Company, immediately surpassed by this new venture. Western Electric hailed it as a feat "unparalleled in military construction history."[72] The construction of the DEWline spawned a plethora of superlatives. The largest shipping convoys ever to leave Seattle in peacetime and the biggest commercial airlifts in American history were organized to supply construction sites in Alaska and Canada. C-124 "Globemaster" aircraft of the U.S. Air Force carried caterpillar tractors to construction sites. The planes disgorged the twenty-ton cats on airstrips, which had to be 6,000 feet long and 200 feet wide to accommodate these large planes.[73] As the naval oil explorers recently had, the DEWline's builders encountered many construction problems peculiar to the Arctic. They were obliged to install gravel pads 5 to 6 feet thick under roads, buildings, and all other structures to prevent the underlying permafrost from melting.[74] Tremendous quantities of gravel were excavated to meet these construction requirements for air strips, roads, work pads, and building foundations. According to one popular account, the amount of gravel needed for air strips alone was sufficient to lay a two-lane highway with a 12-inch thick roadbed from San Francisco to New York City.[75]

2.2 CONSERVATIONIST REACTIONS

As indicated, the Alaska Highway and Canol project did not meet with universal approval. Many Alaskans were dismayed by the road's location and, like many federal officials, saw no value in the pipeline. Some Alaskans were unhappy with the Alaska Highway for other reasons. Without doubt there was some largely unexpressed and unquantifiable hostility to an overland link among the "sourdough" element. A cartoon in *Alaska Life* depicted two men looking over the edge of a cliff at a bridge running over the top of an old cabin directly on the route of the Alaska Highway. The caption read: "That old sourdough wouldn't sell."[76] The magazine's pages were full of requiems for a vanishing way of life in 1942 and 1943.[77] There was also debate among many Alaskans over the road's benefits and disadvantages. Some saw no advantages. Not all Alaskans wanted more settlers in the north, especially since those Alaska was likely to receive in the early 1940s were European war refugees, mostly Jewish, and destitute "Okies" and "Arkies" then in California, whom many Alaskans feared as a socioeconomic liability. These Alaskans were not fundamentally opposed to a road connection with the rest of the nation. But they did feel that it was premature.[78]

Alaskan boosters were accustomed to encountering scepticism and inertia at home as well as intransigence in Washington, D.C. One boosterist tract scorned the "new crop of congenital unbelievers" raised by the highway proposal.[79] However, this "new crop" did not include the conservationist community. Only a handful of preservationists opposed the building of roads, anywhere, including the grand nature parks of the West. So the Alaska Highway was hardly an issue among conservationists—before, during, or after construction. The sole documented criticism of the idea of a highway on conservationist grounds came from Bob Marshall. Marshall, who was the most enthusiastic, dedicated, and effective campaigner for the cause of wilderness preservation since John Muir, did not share the latter's tolerance for road penetration into remote areas. The years between Muir's death in 1914, soon after the Hetch Hetchy defeat, and Marshall's rise to prominence in the U.S. Forest Service in the 1930s, had seen the open country of the West increasingly laced with roads.

Marshall's opposition to any road construction in Alaska north of the Yukon had been publicized in the National Resources Committee report on Alaska (1938). He had equally strong convictions about the "International Highway." Marshall died at the age of thirty-eight—two years before the attack on Pearl Harbor. In an unpublished memorandum, he listed six "major inimical aspects." Two are of special relevance here. Firstly, underlining that it was the "frontier aspect" itself, as opposed to merely magnificent scenery, which constituted Alaska's paramount recreational asset, he declared that "automobile roads and the frontier are incompatible." Secondly, he believed that a highway with its northern terminus at Fairbanks would have a drastic effect on socioeconomic conditions. For it would facilitate an "invasion" by the restless spirits of the lower states who wanted to live off the country like their ancestors on previous frontiers. Marshall argued that increased competition for resources, especially hunting and trapping opportunities, would reduce the quality of life for the territory's existing residents.[80]

Marshall did not offer the frontier life he so admired—and in which his book, *Arctic Village* (1933), sparked so much interest—as a solution to the woes of the Depression. He received many letters from those whose appetites for pioneering had been whetted by *Arctic Village*. Loathe to incite another stampede north, he was at pains to discourage prospective settlers by stressing the rigors and practical difficulties of life in Alaska when replying to requests for information. The pioneering he approved of was mostly vicarious. Whereas Donald MacDonald and Anthony Dimond spoke of freeing the noble pioneering impulse, which had been suppressed in the congested lower states, Marshall abhorred the prospect of the Alaskan interior being overrun by "miscellaneous game slaughterers." In his view, the road was "the most ominously dangerous of all the suggested developments of Alaskan resources."[81]

His distaste for "drifters" was repeated elsewhere. If there was a road, he worried that "just such people . . . would find somehow or other a broken-down fourth-hand automobile to get to what they would imagine would be this land of plenty."[82] This was consistent with his views as chairman of a subcommittee appointed by Mordecai Ezekiel, the executive secretary of the Department of Agriculture's Committee on Refugee Problems. The task of Marshall's subcommittee was to investigate the possibilities for resettling Eastern European war victims in Alaska. Marshall insisted that Alaska "does not lend itself to the covered wagon method of attaining population."[83]

Marshall's stance was *sui generis*. In both an organizational and an intellectual sense, he represented no one. Alaska, generally speaking, was a marginal field of interest for most conservationists in the 1930s. Only two controversies engaged their attention during the interwar period, and both arose in southeast Alaska; the attempt to prevent Glacier Bay National Monument being opened to mining (which it was in 1936),[84] and the campaign to establish a brown bear sanctuary on Admiralty Island.[85] Neither cause attracted Marshall's attention.

In 1935 he was instrumental in founding the Wilderness Society in Washington, D.C., the first American conservation organization established specifically for the purpose of wilderness preservation. For Marshall and the Wilderness Society there were many campaigns outside Alaska of far greater urgency: opposition to a "skyline drive" proposal for the Great Smokies National Park; support for the establishment of an Olympic National Park in Washington State; and support for protection of the Quetico-Superior lake country of Minnesota from logging, airplanes, powerboats, and roads.[86] A groundbreaking study of the last remaining "roadless areas" in the United States, which Marshall co-authored, entitled "Last Vestiges of the Frontier," did not cover Alaska. It singled out the desert canyonlands of southwestern Utah as the nation's greatest surviving expanse.[87]

However, Marshall's prewar misgivings were echoed by a few others once the road was a reality. Like Marshall, Frank Dufresne was apprehensive about the influx of aspiring pioneers that boosters hoped the road would bring. In 1942 he observed with some trepidation that "lately there has been a marked tendency for people to move to Alaska 'for the purpose of making their living by hunting and trapping,' in the spirit of Daniel Boone." Dufresne recognized the appeal but agreed that such behavior was dangerously anachronistic: "This class of settler is reluctantly, but positively, warned against the undertaking."[88]

Dufresne's boss at the Fish and Wildlife Service, Ira N. Gabrielson, was the next to sound a warning. The occasion was the National Audubon Society's annual conference in New York City in 1943. Gabrielson, a member of the Boone and Crockett Club, addressed himself to the question of repercussions for wildlife resources and frontier qualities from the population expansion that the territory was undergoing as a result of the wartime growth process. Drawing attention to Alaska's special status as "the last reservoir of the great game herds" in the United States, he denounced its image as a land of superabundance.[89] He explained that the ability of northern ecosystems to support large numbers of fauna was limited. He emphasized the low physical carrying capacity of the land in the so-called sportsman's paradise and the slow reproductive and restorative rates of fauna and flora. He was perturbed by the mounting pressure on fish and game in the vicinity of military bases and elsewhere as a result of new air and road access. Gabrielson was particularly disturbed by reports alleging the indiscriminate destruction of animals by soldiers.[90] At this time, however, such warnings were isolated and exceptional.[91]

The DEWline was a classified project and a virtual news blackout had been imposed during its planning and construction. There were a few adverse conservationist reactions, but they were faint and uncoordinated. Probably the leading critic was Lois Crisler, a member of the governing council of the Wilderness Society. The experimental Alaskan section of the DEWline was being built while she was in the northeast corner of Alaska with her husband, Herb, who was a wildlife photographer working on assignment for Walt Disney. Crisler communicated her anger to George Marshall, one of the late Bob Marshall's brothers, who was a prominent official in both the Wilderness Society and the Sierra Club: "The Arctic wilderness is folding up and shrivelling, as a live thing, like paper in a bonfire."[92] In the book *Arctic Wild* (1958), a product of her 18-month sojourn (April 1953–November 1954) in Alaska, Crisler attacked the DEWline as "the most destructive incursion the Arctic ever sustained." She claimed that earlier "incursions," such as the exploits of nineteenth-century Yankee whalers, paled into insignificance beside it. She also accused the DEWline workforce of wantonly destroying wildlife. She emphasized that it was "the biggest, best-armed invasion of this fragile life zone . . . ever performed."[93] In 1960 Crisler reviewed the film *Arctic Wildlife Range,* shot by H. Robert Krear during an expedition to the northeast Arctic in 1956. She drew attention to the struggle between human greed and the emerging ecological conscience, which she called "the cold war we do not talk about," but which was "perhaps more important in the long run than the well-known cold war," both of which the DEWline symbolized.[94]

As will be seen in the next chapter, *Arctic Wild* became a source of inspiration for conservationists eager to protect some of the Arctic comparable to *Arctic Wilderness* (1956), a selection from the journals and letters of Bob Marshall. In his foreword to *Arctic Wilderness,* A. Starker Leopold, a leading conservationist, reflected on the recent intrusion in the easily bruised Arctic (made possible by the airplane) in the shape of oil exploration and the DEWline: "Defense units are scattered along the coast of the Arctic Ocean, maintained from the air with a shuttle service as dependable as many suburban railroads."[95] In contrast, the impressions of those who constructed the DEWline conformed to the orthodox wasteland and icebox genre. One participant, not surprisingly, described the Arctic as "the bleakest, coldest, and most ruthless land on earth." The language of righteousness and aggression was unrestrained. Phrases such as "full-scale attack on the Arctic" and "assault on the Arctic" were not taboo in the late 1950s.[96]

3
Boosters and Conservationists in the Postwar Era

This last frontier is not to be likened to the Old West nor
even the Alaska of the early days. It is a frontier primarily
of city dwellers who expect and have the comforts of modern
American living. It is a frontier of skilled workers and engineers, with little need for the type of pioneer who hacked
his living out of the wilderness.
–First National Bank of Seattle, *Alaska: Frontier for Industry*, 1954

Chapters three, four, and five consider the aspirations of boosters and
conservationists and the conflicts between them in Alaska during the
postwar, pre-oil era. Nationally, this was a transitional period between
traditional conservation and the "new" conservation (environmentalism)
of the late 1960s. This chapter begins with a discussion of the booster
community's support for Alaskan statehood and introduces two leading
figures whose involvement in Alaskan affairs spans the period from the
Second World War to the present. It continues with an overview of
the statehood movement, which traces its ascendancy in Alaska and
political fortunes in Congress. The final aspect of this section is the
theme of tendentious frontier imagery, upon which statehood proponents leaned heavily. The second section and, more importantly, the
third investigate the first organized postwar conservationist responses
to change in Alaska (as opposed to the scattered reactions to military
developments discussed in the previous chapter).

3.1 STATEHOOD

During the first decade of the century, federal conservation policies
had fuelled demands in Alaska for more home rule. There was also
a close relationship between the gathering strength of the statehood
movement in the postwar era and deepening frustration with federal
unwillingness to promote economic progress at a pace agreeable to

development-minded Alaskans, especially those newcomers attracted north by the war and defense-related economic boom.

At this point, the term *booster* deserves some amplification. Daniel J. Boorstin, while recognizing the similarities between them, has drawn a clear distinction between *boosters* and *boomers*. The term *booster* dates from the 1840s and was immortalized by Sinclair Lewis in *Babbitt* (1922). Boorstin characterizes the booster as a loyal "community builder" with a genuine vision, a firm commitment to a particular place (at least for a time) and a belief in its special destiny. According to Boorstin, the booster spirit involved certain "simple faiths" such as "a belief that the future could hold anything or everything." He portrays the boomer (the term comes from the southern California land boom of the 1880s) as a more cynical frontier type, mainly a speculator and spectator, who was interested only in short-term profit.[1] Throughout Alaskan history, plenty of people who came to Alaska had been eager for swift profits (if not based on real estate values). Russian fur traders, Yankee whalers, gold miners, and the Alaska Syndicate were all boomers in this sense.

Whatever subtle differences might exist between the outlook of booster and boomer, their attitude toward natural resources was the same. Whether in the nineteenth-century West or post–Second World War Alaska, these entrepreneurs wanted unrestricted access to them. It might be helpful to approach Alaskan boosters and boomers in the postwar era in the context of recent theories of ecological imperialism. Here, in fashionable parlance, was a mid-twentieth-century case study of a pioneer species "invading" new "eco-terrain" and colonizing an "unoccupied" or "underoccupied" "econiche."[2]

It is definitely worth looking at these postwar boosters in the context of their nineteenth-century predecessors. There has been a popular infatuation with the "Jeffersonian" yeoman figure, and frontier historians used to be preoccupied with the agrarian aspects of westward expansion. This meant that for a long time the influential roles of distinctive frontier types such as the logger and lumber baron, the miner, the soldier, the Chinese "coolie," the railroad builder, the cattle baron, the land speculator, and the urban entrepreneur and booster, were overlooked.[3] Richard C. Wade and Earl Pomeroy have demonstrated that the urban developer—particularly in the West—was frequently in the vanguard of frontier intrusion and that urban areas often grew faster and earlier than the countryside.[4] The most important and lucrative "pioneer" environment in mid-twentieth-century Alaska was the "upstart"[5] city of Anchorage, which provided far more opportunity for economic success than the land itself, 99.8 percent of which remained federally owned until statehood. During the 1950s Alaska's population increase was the highest in the nation; 77.4 percent in comparison with the national average of

14 percent. Anchorage absorbed the bulk of it.[6] In 1939 only 6 percent of Alaska's population lived there. After the war, it was the territory's fastest growing settlement. In 1950 the city already contained a quarter of Alaska's population. By 1960 this proportion had risen to 37 percent.

"Like the musket," comments Boorstin, "the newspaper became a weapon and a tool, to conquer the forest and to build new communities."[7] The newspaperman was a distinct type on the nineteenth-century frontier and Robert B. Atwood, the editor-publisher of the *Anchorage Daily Times,* is a continuation of the tradition in a northern context. Born in Chicago in 1907, Atwood moved to Anchorage in 1935 with his wife Evangeline, a native-born white Alaskan. He was chairman of the Anchorage Chamber of Commerce (1944, 1948) and president of the Anchorage Hotel Association in the 1950s and 1960s. In Alaska, newspapers proliferated as they had done in the West. In the years after 1915, Anchorage had been served by six weeklies in addition to the daily *Cook Inlet Pioneer.* The *Cook Inlet Pioneer* later became the *Anchorage Daily Times,* which had a circulation of about 600 when Atwood took control in 1935.[8] At this time, the *Juneau Daily Empire* was the territory's leading newspaper. During the Second World War, Atwood absorbed most of the other Anchorage papers and his *Times* remained the only city daily until 1946, when the *Anchorage Daily News* began operating. By 1953 the *Times* had replaced the *Juneau Daily Empire* as Alaska's top seller.[9]

The second key figure, Walter J. Hickel, was the son of a tenant farmer who left Dust Bowl Kansas in 1940 at the age of twenty. Originally bound for Australia, Hickel made his way to Alaska by train, boat, and thumb, arriving in Anchorage with a legendary thirty-seven cents in his pocket. After the war, Hickel bought land and started putting up housing developments. In 1948 he formed the Hickel Construction Company, which built Alaska's first motel, the Anchorage Traveler's Inn, in 1953.[10] Soon afterwards, Hickel constructed the Northern Lights Shopping Center in Anchorage—another first for Alaska—in which he installed Alaska's first escalator.[11] In 1955 he built a Traveler's Inn in Fairbanks. Meanwhile, thanks to the Cold War–related military construction boom, Anchorage's popultion had increased to 70,000.

Noting that hotels were often the most prominent structures in emerging American cities, Boorstin comments that "in the period of most rapid urban growth, it was not by churches or government buildings but by hotels that cities judged themselves and expected others to judge them." Moreover, as "laboratories and showcases of progress in the technology of everyday life," they pioneered in the development of plumbing, heating, ventilation, and air conditioning.[12] Following the devastating earthquake of Easter 1964, which leveled much of downtown

Anchorage, Hickel built the multistory, luxury Captain Cook Hotel. In 1950 his political career began when he ran unsuccessfully as a Republican candidate for the territorial legislature. In 1954 he was elected Republican national committeeman for Alaska, a position to which he was reelected in 1956.

Boosters like Atwood and Hickel did not attribute what they saw as Alaska's underdeveloped economic condition and slow progress towards a greater measure of political self-determination to geography, environment, or climate. They blamed them on federal mistreatment and neglect, conservation policies, and the outside interests that controlled the colonial economy. In 1957 a pro-statehood pamphlet declared that "if Alaska becomes a state, other people, like ourselves, will be induced to settle here without entangling federal red tape and enjoy an opportunity to develop this great land."[13]

Like resentment against federal conservation policies, the conviction that the national government had been particularly slow in extending home rule to Alaska conformed to the traditional pattern of territorial-federal relations. Following purchase, the U.S. Treasury Department administered Alaska as a customs district, though it was effectively run by the U.S. Navy. The Organic Act of 1884 had conferred the status of a civil and judicial district with a presidentially appointed judge and governor. In 1912, the year the last two contiguous American territories, Arizona and New Mexico, were admitted as states, Alaska became an incorporated territory with an eight-member Senate and a sixteen-member House (under the Second Organic Act). In accordance with established policy, the federal government retained control of fur, fish, and game. Throughout American history, territorial status has been a transitional stage eventually leading to acceptance as a full member of the Union. However, this temporary status varied considerably in duration—from four years in the case of California to sixty-two for Arizona and New Mexico, which, like Alaska, had comparatively small populations with a large percentage of non-whites, and were far less developed than many existing states. In 1912 Arizona's population was roughly 200,000; New Mexico's was 300,000. Alaska's poulation stood at around 100,000 in 1947.

In 1943, urged by Delegate Dimond, Senator Langer (Republican, North Dakota), a member of the International Highway Commission in the 1930s, and Senator McCarran (Democrat, Nevada) introduced the first statehood bill since 1916. In December 1943 Dimond produced his own unsuccessful bill that sought to transfer control over most of Alaska's public domain, as well as its federal parks and monuments, to the new state. In 1944 the territory's governor and most of its legislators endorsed the cause. There was also growing grassroots support in Alaska. In 1945, Evangeline Atwood, the wife of Robert Atwood, formed the Alaska Statehood Association in Anchorage. Branches of this voluntary, nonpartisan citizens organization were established in most Alaskan communities. Their task was to lobby Alaskans in preparation for a referendum on statehood that Gruening had arranged. In October 1946, 9,630 voted in favor and 6,822 in opposition.[14] With the exception of Atwood's *Anchorage Daily Times,* pro-statehood since the Langer-McCarran bill, most of the territory's newspapers, notably the *Fairbanks Daily News-Miner* (owned by Alaska's leading industrialist, Austin E. "Cap" Lathrop), and the *Juneau Daily Empire,* were against home rule. These critics believed that the white population was still too small and the territory too underdeveloped to bear the cost.

In 1947 Bartlett introduced the first postwar statehood bill. It contained sweeping provisions for land transfer that, like those Dimond proposed in 1943, were unprecedented in American history. Bartlett's bill granted the state title to practically all public lands, including federal reserves (with the exception of McKinley National Park, Glacier Bay National Monument and Pet 4) as well as all areas claimed by Natives.[15] While no conservation organization took an official stand in principle against statehood, most did object to the extent of state control over natural resources. Fred Mallery Packard, representing the 3,000-member National Parks Association (NPA), was apprehensive about the bill's implications for Katmai National Monument and other federal reserves (including the national forests) and wildlife refuges.[16]

Fearful of higher taxes and other unfavorable state legislation such as the abolition of the notorious fishtraps, the most powerful anti-statehood lobbies throughout the 1940s and 1950s were the Seattle-based salmon canneries and steamship companies.[17] The other focus of congressional hostility was a coalition of southern Democrats and conservative Republicans. They opposed statehood because of Alaska's small population, which they felt would be disproportionately represented in Congress, and the large nonwhite component of the population. They also objected on account of Alaska's underdeveloped condition, its unstable, seasonal economy, and the territory's noncontiguity. Finally, they did not want Alaska admitted because it was likely to return liberal Democrats to Congress. The redoubtable icebox image also played its part.[18] Acting for these reasons, these groups ensured that no subsequent action was taken on Bartlett's bill.

In 1949 the Alaska Statehood Committee succeeded the Alaska Statehood Assocation. The committee was a bipartisan group of eleven appointed by Territorial Governor Gruening to mobilize support in Alaska and the contiguous states. Robert Atwood was appointed chairman, a position he held until 1959. Gruening, Bartlett, and Dimond became ex officio members.[19] Gruening also formed a hundred-member "national committee of distinguished Americans" to lobby for the cause. One of its members was the author, Rex Beach. Beach was a literary figure and sometime Alaskan goldminer, whose best-known writings belong to the same genre as Jack London's Alaskan and Klondike material. Beach had campaigned to open Glacier Bay National Monument to mining in the 1930s and was a loud critic of federal conservation policies in Alaska.[20] Other members were the booster historian, Jeannette Paddock Nichols, and the Canadian explorer-anthropologist, Vilhjalmur Stefansson. Ira N. Gabrielson, the director of the U.S. Fish and Wildlife Service, also became a member.

In March 1950 the House of Representatives passed Bartlett's latest statehood bill, which contained considerably more modest land grants than his earlier bill (1947). It also guaranteed Native land rights and did not claim federal parks and reserves. In April the Senate Interior Committee conducted its first hearings on Alaska statehood in Washington, D.C. Representing the pro-statehood Congress of Industrial Organizations (CIO) was Anthony Wayne Smith. Smith, who was executive secretary of the CIO's committee on Regional Development and Conservation, felt uneasy about the possible alienation of public lands. He wanted the national forests, Pet 4, and all minerals, fish, and wildlife to remain in federal ownership.[21]

Although the Senate reported the bill, there was no motion to proceed and the Eighty-first Congress took no further action. In January 1952, Bartlett introduced another statehood bill, which the Senate narrowly voted to recommit. The statehood cause received another blow when Dwight Eisenhower became president. He endorsed statehood for Hawaii, which was likely to improve his party's slender Senate majority by electing Republicans, but did not mention Alaska.[22] Eisenhower argued that Alaska's strategic geopolitical importance demanded complete federal control. B. Frank Heintzleman, who succeeded Gruening as territorial governor, opposed immediate statehood. Gruening now devoted his journalistic talents to promoting statehood fulltime.[23]

Claus-M. Naske has dated the "populist" phase of the statehood movement from the hearings conducted by the House Interior Committee in the spring and summer of 1953 in Washington, D.C., and Alaska respectively.[24] In Alaska, only one in seven witnesses expressed opposition. After the Alaskan sessions, the Atwoods founded another Anchorage citizen organization, Operation Statehood. In April 1954 the Senate approved by a wide margin (57 to 28) statehood bills for Hawaii and Alaska, cojoined to prevent the latter's veto. But the House did not act. Frustrated by this slow progress, the new territorial legislature's Democratic majority passed a bill arranging for a constitutional convention. In November fifty-five delegates convened at the University of

Alaska where they deliberated for three months. Delegates included the future governor, William A. Egan, a grocer from Valdez, and the future congressman, Ralph Rivers, the mayor of Fairbanks.

The natural resources section of the constitution that they drew up was designed to protect Alaska's natural resources from the extremes of "robber baron" exploitation and preservationism from which they believed Alaska had suffered in the past. Article 8 declared that the new state would "encourage the settlement of its land and the development of its resources by making them available for maximum use consistent with the public interest . . . [and] provide for the utilization, development, and conservation of all natural resources . . . for the maximum benefit of its people."[25] This vague general statement expressed the hope that the development of Alaskan resources in future would benefit Alaskans, not outside corporations, and that it would not be suffocated by conservation policies. While providing safeguards against corporate and conservationist greed, the framers of this section of the state constitution wanted to encourage brisk economic development that did not despoil the land. There was bound to be conflict over the definition of an abstraction like "the public interest" and over how to decide which group or groups best represented it in Alaska. In addition, it was unclear how the concentration of wealth in a few hands was to be prevented within a national economic system dominated by big capital; and, not least, how to protect the environment without hindering economic activity.

At statehood hearings in Washington, D.C., in 1957, few fresh arguments were heard from either side. Conservationists remained sceptical. Clinton Raymond Gutermuth, the vice president of the Wildlife Management Institute (WMI),[26] criticized the natural resources section of the Alaska Constitution adopted in 1955. He considered wildlife protection inadequate and objected in particular to the continued bounty on the wolverine.[27] Gutermuth also attacked what he saw as the powerful position of commercial interests in Alaskan fisheries management. He complained that this ran roughshod over the noncommercial national interest values represented by sportsmen.[28] He concluded that the natural resource management philosophy embodied in the Alaska Constitution did not "inspire confidence."[29]

These objections were endorsed by certain Alaskans. A. W. "Bud" Boddy, an Alaskan resident for sixteen years, was director of the Alaska Sportsmen's Council (ASC), an umbrella organization of clubs affiliated to the National Wildlife Federation (NWF), the nation's largest sportsman-conservationist organization. Boddy conceded his preference for continued federal management of natural resources. He believed that Alaskans, with their pioneer mentality, were not mature enough to be entrusted with this responsibility.[30] Such statements reinforced the impression of statehood proponents that conservationists wanted to keep Alaska as their playground.

George Marshall of the Wilderness Society feared that statehood would involve a massive transfer of public lands comparable to the nineteenth-century "giveaway" of the West.[31] From the perspective of 1969, he expressed regret that Alaska north of the Yukon had not remained exclusively under federal control.[32] However, he has denied that the Sierra Club or the Wilderness Society assumed an official position for or against statehood.[33] Nevertheless, allegations have abounded. Gruening accused conservationists of being in league with the canning industry in the effort to thwart statehood,[34] In his memoirs (1969), Ernest Patty, the president of the University of Alaska from 1953 to 1959, and former Dean of the university's School of Mines, contended that Wilderness Society officials opposed statehood.[35] To Alaskans who had devoted many years to the cause, conservationist attitudes struck at the heart of their bid for self-determination.

Statehood's prospects had improved in 1956 when Fred A. Seaton, a supporter, replaced Douglas McKay at the helm of the Interior Department. Advocates had been further encouraged in early 1957, when Eisenhower announced his willingness to relax his opposition to statehood on condition that he retained the authority to withdraw land for defense purposes in northern Alaska. In January 1958 Eisenhower announced his support for Alaskan statehood. In May 1958, Alaska statehood passed in the House by 210 to 166. The bill contained a land grant of 102.5 million acres—by far the largest amount of land a state had ever received—to be selected over a period of twenty-five years.[36] The bill passed the Senate and the president signed it early in July. At the beginning of 1959 Alaska was officially admitted to the Union as the forty-ninth state.

The mood of braggadocio that then prevailed in booster quarters was epitomized, in its most baroque form, by the rhetoric of Walter J. Hickel. Hickel boasted that "we are going to build a Fifth Avenue on the tundra." He saw Alaska accomplishing "in less than fifty years what the United States did in a hundred."[37] Frontier ideas and images had characterized the style and content of statehood advocacy. Since its purchase, boosters had promoted Alaska as an extension of the continental frontier. Robert Atwood had set the tone for the postwar era in 1947,[38] and the analogy was extended by Major Marvin R. "Muktuk" Marston.[39] After five and a half years of military service in the territory (organizing Eskimo defense units as part of the Alaska Territorial Guard), Marston had come back to homestead. He told how, when his grandfather moved to California at the time of its gold rush, no more than a few thousand whites lived there. Roughly the same number, 10,000, had lived in Alaska before its own stampedes. Marston emphasized how California's population had grown by leaps and bounds ever since and was increasing at a monthly rate of 20,000. He suggested that Alaska could emulate the successful transformation of the Golden State—an appellation that California received from its wheatfields, not the ore that attracted the argonauts—from a mining frontier to a broader and more stable agrarian economy: "What California was to my grandfather, Alaska is to you and me—a great undeveloped territory with great wealth."[40]

Alaska's eligibility as an organic part of the American frontier was popularized after the war, notably by Ernest Gruening, the territorial governor of Alaska. In 1944 Gruening delivered a Greeleyesque exhortation to returning soldiers: "Go North, Young Man!" Gruening claimed that the new air and road access to Alaska would open up "free land" as the transcontinental railroads had done.[41] He assured Americans that "there still is a frontier. Alaska, too, is a vast, empty land, waiting for people. . . ."[42] Though few, with the exception of Gruening and Atwood, had probably read Turner, most statehood proponents were his unreconstructed devotees, who dwelt on the concept of national "rebirth" on the frontier in a way that complemented the "second chance" preoccupation of conservationists. In 1947 Gruening paid homage to the American virtues Turner had extolled, such as the "spirit of youth and vigor," which was then on the wane "in the earlier, settled, longer established parts" of the nation. However, Gruening did reject Turner's grave conclusion that "the first period" of American history had ended decisively in 1890. Gruening claimed that these historic qualities flourished in Alaska. Here the nation had been given a new lease of "robust" pioneer life and a "characteristically American" form of civilization would again be shaped "in a battle with nature."[43]

Virtually paraphrasing Turner, Gruening explained at the third Alaska Science Conference in 1952 that "the westward trek of peoples in search of greater freedom and greater economic opportunity" was "undoubtedly the oldest American tradition." He told them that this tradition had "brought the Jamestown colonists, the Pilgrims, the Dutch," and "led them across the Appalachians . . . across the Plains, over the Rockies and to the Pacific Coast." He believed there was a contemporary American urge to migrate to Alaska which, once realized, would truly constitute "a final chapter in a great episode."[44]

Gruening's views should be set against the prevailing opinions of historians of the frontier, such as Walter Prescott Webb. Webb became the first historian to release the study of the frontier from its regional American straitjacket in a book (1951) that analyzed westward expansion in the context of the global Anglo-Saxon expansionist "boom" that took place over five centuries (the "Great Frontier"). He identified three types of post-1890 "new frontiers"—geographical, socio-economic-political, and scientific.[45] With specific reference to Gruening's slogan "Go North, Young Man!" Webb poured scorn on Alaska's qualifications for inclusion in the first category. Webb contended that climatic severity, unfruitful terrain, and a general dearth of valuable resources drastically restricted its appeal to potential settlers and presented unprecedented obstacles to the advance of colonization. This made Alaska a "fringe" country that had been, and would always remain, peripheral to the process of expansion that had carried European power and influence around the globe between the fifteenth and late nineteenth centuries.[46] However, Webb was not entirely consistent in his attitude toward Alaska. He actually went there a year before his death in 1963 to teach for a summer at the University of Alaska. He wrote back to the *Dallas Morning News* in his native Texas, "I am in a country that is the nearest thing to a frontier that I have ever seen. . . . most of the country is a true wilderness. . . ."[47] Webb's assessment illustrates that for him "frontier" and "wilderness" were largely synonymous.

The tone of statehood advocacy illustrated three things: the rhetorical ubiquity of the agrarian frontier image; the tenacity of the settler experience east of the Mississippi as the pioneer prototype in the popular mind; and its use to counteract the icebox image, which some statehood opponents still turned to. At hearings in 1950, ex-delegate Dimond, whom gold fever had lured to Alaska in 1905, invoked his ancestors' experiences on a previous frontier. He told how his forebears, who immigrated to New York from England before the American Revolution, "went into the forest and chopped down the trees . . . and made pastures and gardens and fields."[48] Dimond was adamant that this archetypal pattern could be repeated in Alaska. Another witness at the hearings who spoke in similar terms was George Sundborg. Sundborg was the manager of the Alaska Development Board (1945) from 1946 to 1953.[49] He had been an active member of the Alaska Statehood Association and a delegate to the 1955 constitutional convention. He saw in Alaska a "new West," as important to the nation as the old had been, where the success, the excitement, and the glamorous cut-and-thrust of westward expansion would be emulated.[50]

Robert Atwood's testimony in 1953 drew extensively upon Turner's ideas. Without mentioning the historian or his thesis, Atwood summarized the significance of the frontier in American history. Using many of Turner's phrases, the chairman of the Statehood Committee described how the "procession of civilization" had advanced west in "successive" waves to occupy "free lands": "The explanation of American development is in the existence of an area of free land—and its continual recession—and the advance of American settlement westward." Atwood compared this national history with the Alaskan experience, emphasizing that only the first two frontiers, based on furs and mining, had been duplicated in the north.

Then he explained why a settler's frontier had failed to develop in Alaska after the Bureau of the Census declared the western frontier closed in 1890. Atwood did not hold Alaska's harsh climate, hostile environment, and isolation responsible. Pointing to the recent construction of roads and the growth of cities, he assured the congressional committee that "men have proven they can conquer the wilderness in the far north." According to Atwood, man-made barriers, notably conservation, were much harder to overcome. He argued that conservation, which he explained as a reaction to the disappearance of the frontier in the United States outside Alaska, was responsible for the repulsion of westward expansion from the north. Like Vilhjalmur Stefansson, statehood promoters often drew attention to the Soviet Union's admirable policies in its arctic regions (though, unlike Stefansson, they were never accused of communist sympathies). Predictably, Atwood concluded by contrasting the federal government's unpioneer-like attitude toward Alaska with the frontier spirit and gusto with which Russia was developing Siberia. "The Soviets," he announced, "are doing what we talk about."[51]

The ax and rifle, according to Frederick Jackson Turner, were the symbols of the backwoods pioneer in his fight with "the unmastered continent."[52] In 1957, C. W. Snedden, the new proprietor of the *Fairbanks Daily News-Miner,* combined traditional and modern frontier images. He issued an invitation to "people from every state in the Union, people who have the deep desire to pioneer," to come to Alaska ("a wasted asset"), "be they individuals with an ax or large corporations."[53] Probably the most perceptive comment on the anomalous relationship between contemporary Alaska and the nineteenth-century frontier was made by a British journalist covering statehood: "they like to call this the last frontier. It is not: it is the new sort of frontier. . . . A young man cannot come here with his hands and his courage and carve an estate out of the wilderness. Alaska is proper meat for the great corporations with capital the size of national debts and machines and helicopters and dedicated graduates from mining schools."[54]

3.2 THE KENAI NATIONAL MOOSE RANGE CONTROVERSY

Two controversies that overlapped with the final stages of the statehood struggle highlighted the divergent interests of Alaskan boosters and the conservationist community. In the late 1950s boosters saw the Kenai National Moose Range as the most galling obstacle to "pioneering" and Alaskan economic advance. This range had been established in 1941 by executive order to protect the gaint Kenai moose, the largest antlered deer on earth. The range's 1.730 million acres were managed by the U.S. Fish and Wildlife Service's Bureau of Sport Fisheries and Wildlife. The range contained an estimated 10 percent of the total Alaskan moose population.

The Kenai Peninsula was the area of Alaska that boosters considered best-suited for agricultural development. Major Marston, now an Anchorage realtor, was one of a number of returning veterans who had contributed to a minor postwar boom in homesteading. In 1946 entries were filed to 13,952 acres of Alaskan public domain, the majority on the Kenai. Three years later, the Bureau of Land Management (BLM) reported entries to 42,169 acres.[55] The agrarian qualities of this homesteading rush should not be exaggerated. Most entries were speculative, indicating a boomer mentality, and not of the settler variety.[56] By the end of 1955, 59 percent of the land transferred into private hands on the Kenai was either unoccupied or abandoned. Another 31 percent was solely residential.[57] This weak agrarian character is hardly surprising when federal statistics are taken into account. In the late 1940s and early 1950s, fewer than one million acres, less than a quarter of a percent of the total land area of the territory, was classified as tillable farmland. The Department of Agriculture considered another 2 million suitable for grazing.[58] In 1963 there were only 10,680 acres planted in crops in Alaska.[59]

Land speculators inveighed against the barrier raised by the moose range. A crisis was precipitated in July 1957, when the Richfield Oil Corporation of Los Angeles struck oil on the Kenai some forty miles southwest of the booming city of Anchorage. Allusions to the gold rush were inevitable as an ecstatic local press greeted the news of the first commercially significant oil find in Alaskan history.[60] "I have reports," mused Secretary Seaton, "that things are almost back to the gold rush days."[61] Many boosters hoped this strike would hoist Alaska out of the slump that followed the defense-related boom. Atwood editorialized jubilantly, "Hang on, we're going around a curve."[62]

"Make these moose move over and make room for people," Major Marston had commanded at the statehood hearings in 1947.[63] This pithy slogan now entered public debate in a literal sense because the discovery well was on the Kenai National Moose Range. Under regulations drawn up by McKay, Seaton's predecessor at Interior, shortly before he left office (2 December 1955), Richfield had been granted operational leases within the range. The National Wildlife Federation[64] had opposed them on four grounds: the potential damage to the moose population, the dubious legality of the leases, the availability of alternative areas for mineral exploitation outside wildlife refuges, and the danger of establishing a precedent inconsistent with the original purpose of the range. The federation denounced the leases as "unnecessary, unwise and not in the public interest."[65]

Almost four months after the original strike, not one barrel of oil had been produced from Swanson River Unit Number One. In fact, Richfield capped this discovery well while it waited for additional leases that would permit further exploration and development. Atwood (who, like Hickel, bought oil leases for sale to corporations) was furious at the delay and denounced restrictions on the industry's freedom of action: "Anything that slows down oil development is bad." Alaska's immediate future, he fumed, was at the mercy of moose. To underline the tragic proportions of the financial loss resulting from this enforced inactivity, he estimated that production from the Kenai alone "could equal that of the entire oil-rich nation of Iran."[66]

Addressing the Anchorage Chamber of Commerce, Gruening advocated removal of all existing restrictions on oil exploration and development within the Moose Range, and a crash program of land lease processing for the rest of the Kenai. He drew upon Alaskan history to bolster his arguments. To emphasize how economic development had always been thwarted, he compared the difficulties facing the oil industry on the moose range with the obstacles that had confronted those who tried to exploit Alaska's coal resources at the turn of the century. To indicate the extremes of action to which Alaskans were driven in their protests against injustice, and perhaps as a warning, he recalled the Cordova "Coal Party."[67]

In response to protests from the oil industry, hearings were held in December 1957 to consider Seaton's new regulations. These regulations were designed as a compromise between oil development and conservation. They prohibited oil leasing in critical parts of wildlife ranges and refuges unless oil was being drained from within by nearby wells. There was a clash between Alaskan sportsmen and the national organizations with which they were affiliated. The Alaska Sportsmen's Council (ASC) was represented at the hearings by Burton O. Ahlstrom, the president of the Anchorage Junior Chamber of Commerce. Ahlstrom charged conservationists from the lower states with ignorance and insensitivity to Alaskan opinion: "Come up here and look around and become familiar with our problems before [you] testify against the wishes of the people who live up here." To support his argument that moose and oil were compatible, he cited the absence of any scientific evidence that human activities were deterimental to the animal. That 90 percent of the Alaskan moose lived outside the range, many in unprotected areas, belied for him the case for strict protection.[68]

The Wildlife Management Institute, the National Wildlife Federation, the Izaak Walton League of America, the Sport Fishing Institute, the Wilderness Society, and the National Parks Association supported the latest regulations as a barrier to "uncontrolled invasion."[69] According to Gutermuth "frenzied voices, reminiscent of the Alaskan gold rush days, are clamouring for the federal government to step aside so that an unlimited and unimpeded follow-up can be made to the recent strike of oil. . . . those voices speak less for all Alaskans than they do for a few political and business leaders who, exhilarated by the possible reincarnation of a frontier bonanza, are willing to gamble for the highest stakes in quick dollars that their vivid imaginations will permit."[70]

Gruening, who led the oil industry's defense, responded with umbrage to Gutermuth's chidings by underscoring the insignificance of the extent of the proposed development in terms of Alaska's overall size. He pointed out that only 6 percent of the range was likely to be affected, of which less than 1 percent (0.48, to be exact) would be needed for wells, roads, and other installations. Gruening claimed to steer an even course between development and conservation. He announced: "I had been led to believe that we were living in a climate friendly to business. That's what Alaskans would like." Reviewing the sad history of wasted opportunity in Alaska and efforts "to open up the last frontier," which had floundered upon the misconstrued and over-zealous application of conservation principles, he pleaded: "It [Alaska] lost the coal age. It has, so far, with slight exception, lost the hydroelectric age. *We do not want it to lose the oil age.* We do not want history to repeat itself."[71] At the same time, he was careful to stress the sincerity of his commitment to conservation of a rational and balanced nature. As proof of his concern for wildlife, he cited his long struggle while governor to eliminate the bounty in Alaska on the national bird, the bald eagle (won in 1945). For him, the finest example of the "ideal of conservation" was the successful management on a sustained-yield basis of the once-endangered Pribilof fur seals.

Gruening explained that moose were hardly scarce in Alaska where, if anything was in short supply, it was human beings. The foothold of what he liked to call the "two-legged species" remained tenuous in a vast land dominated by wilderness. He explained that no part of Alaska was better suited to human colonization than the Kenai Peninsula, where the range had reserved 500 acres for each moose.[72] He considered this a blatant example of the extreme bias toward wild nature in Alaska. Indeed, he believed there was a strong case for abolishing the range completely.[73] Atwood agreed that, given civilization's fragility, there was no justification for "squandering . . . Alaska's land through withdrawals for special purposes."[74]

Alaskan proponents of oil development found a welcome ally in Burton H. Atwood of Illinois, the brother of Robert Atwood. In his capacity as national secretary of the Izaak Walton League of America (IWLA), Burton Atwood undertook a five-day inspection of the moose range as part of a three-week visit to Alaska in the fall of 1957. His report, which was not endorsed by the IWLA, contended that "moose seem to enjoy a little civilization."[75]

Though it contained much *de facto* wilderness, the range was not comparable to a wilderness area. Its specific purpose was to protect moose habitat, thereby guaranteeing a perpetual supply of the large deer. Moose prefer browse species (willow, aspen, cottonwood, and birch) over spruce. Accordingly, controlled burning and cutting (as Gruening emphasized) were an integral part of managing the range to maintain optimum conditions for them.[76] In fact, most of the area had been burnt-over in the late 1940s, which had assisted this management goal. Boosters argued that carefully controlled oil operations would benefit moose because regrowth around drill sites and along roadsides would provide better forage. On the other hand, management for the sake of wilderness values would allow the climax vegetation (spruce) to recolonize.

Prior to the 1880s, there were no moose on the Kenai Peninsula. Only after extensive naturally caused fire, which destroyed large areas of spruce and tundra, permitting browse vegetation to colonize, did moose begin to replace the dominant caribou.[77] Boosters cited the Matanuska Valley as further evidence that human intervention enhanced conditions for moose. They claimed that there were few moose before the advent of agriculture in the 1930s but argued that since the beginnings of substantial cultivation, moose numbers proliferated to the extent that they became a nuisance to farmers.[78]

The moose range row was a defeat for conservationists since Seaton opened 50 percent of the range (the northern half) to oil exploration in August 1958. The controversy had revolved around wildlife rather than wilderness and ecological values. Sportsmen took the lead in the campaign instead of preservationist organizations. By contrast, the second dispute in the late 1950s marked the start of a new era of conservationist involvement in Alaska and the first of three successes.

3.3 THE ARCTIC WILDLIFE RANGE

In 1949, while the U.S.Navy and the United States Geological Survey (USGS) explored for oil in the northwest Arctic, George L. Collins and Lowell Sumner of the National Park Service's regional office in

San Francisco were conducting a different kind of survey in the northeast Arctic. This was part of the Park Service's first formal Alaska recreation survey and was carried out in cooperation with the USGS, the Office of Naval Research, and the Naval Arctic Research Laboratory. Its purpose was to investigate the suitability of this region for protective status.[79]

The first member of a private conservation organization to echo this interest and support these efforts was Olaus J. Murie, a personal friend of Collins and Sumner. Born and raised in Minnesota, Murie had been one of Dufresne's colleagues in the Bureau of Biological Survey in Alaska before the Second World War. Murie's area of expertise was caribou. He had first come to Alaska in what his younger brother and fellow naturalist recalled wistfully as the "blessed days before the airplane."[80] Between 1921 and 1925 he carried out what became a pathbreaking study of these animals in the northern part of the territory.[81] In 1946 Murie left the U.S. Fish and Wildlife Service (which had superseded the Biological Survey in 1940), to become director of the Wilderness Society. In this capacity, he became a leading national spokesman for conservationist causes, especially wilderness preservation, in the post-war decade.[82]

Murie rejected the traditional belief of boosters and an earlier generation of conservationists (John Muir, for example) that a combination of remoteness, inhospitability, and worthlessness would furnish informal protection for the Arctic in perpetuity. In 1950 at the inaugural Alaska Science Conference, Murie reflected on the upheavals he had witnessed in Alaska since 1920.[83] He described the changes brought by the influx of people during and after the Second World War and the effect of new technology. He drew particular attention to the airplane, thanks to which there were "no longer remote areas in the old sense."[84] He also spoke of the extension of the road network, especially the Alaska and Steese Highways.[85] Like Dufresne, he noted their detrimental impact on caribou. These technological changes and this improved accessibility convinced him that long-term land use planning, in which he believed Alaska could and should lead the nation, was urgently required. Otherwise, he did not know how to avoid the depletion of wildlife and desecration of wilderness characteristic of the nation's earlier frontiers.[86]

In 1951 Murie visited the region that Sumner and Collins were studying and was soon enlisted in the bid to protect it. Encouraged by Murie, a group of national organizations adopted the cause. In October the Wilderness Society revived Bob Marshall's ambition by suggesting the creation of a "large arctic wilderness," though the resolution did not specify where.[87] In May 1952 the Sierra Club urged "active study" by the appropriate federal agencies of an "Arctic Wilderness Area" stretching from the Arctic Ocean to the southern slopes of the Brooks Range.[88]

That fall the Izaak Walton League endorsed this stand.[89] In 1953 the Western Federation of Outdoor Clubs (WFOC) passed a resolution in favor of an "Arctic Wilderness Preserve."[90] When formulating policy for the area in 1952 and 1953, conservationists made extensive use of the wilderness concept. However, despite the National Park Service initiative, there was no decision to press for one form of protective category to the exclusion of others. At no point did Sumner or Collins officially recommend national park status. They were at pains to reassure Alaskan officials that they did not seek this form of protection.[91]

In 1953, following a second, fuller summer survey of the region, a report by Collins and Sumner and authorized by the Park Service was published in the journal of the Sierra Club. This was the first extensive discussion of the ecological, scientific, recreational and spiritual values considered to be at stake.[92] These pioneer advocates of protecting the Arctic made comparisons between mid-twentieth-century Alaska and the nineteenth-century West as frequently as statehood proponents were doing. However, whereas the latter emphasized only the applicability of the analogy, Collins and Sumner drew attention to its shortcomings as well. Like Murie, they recognized the dramatic impact of the airplane, which annihilated the apparent remoteness of an area hundreds of miles away from the nearest road. Nowadays, they mused, it was easier "to get into the heart of the Arctic . . . than it was to get into the heart of Yellowstone forty years ago." In the far northeast was "America's last chance to preserve an adequate sample of the pioneer frontier, the Stateside counterpart of which has vanished."[93] In an appendage to the article, they described the spectacle of a caribou herd (*La Foule*) on the move: "Now we know what it must have been like to see the buffalo herds in the old days."[94] The Wilderness Society also sought to protect "a bit of original Alaska . . . as the pioneers of 1898 found it."[95]

In 1954 Collins and Sumner formally recommended federal protection for the northeast corner of Alaska wedged against the Canadian border. They also hoped that in cooperation with Canada, an international reserve could be established, where mineral exploration and extraction would still be allowed and subsistence hunting by Natives remain unaffected. Enthusiasm for some form of federal protective designation united the entire conservation community both in Alaska and the contiguous United States. In the summer of 1956, Murie and his wife Margaret led an expedition (delayed two years due to Murie's ill-health) to the region in question. The expedition was sponsored by, among others, the New York Zoological Society and the Conservation Foundation.[96] It was joined for a week by the Supreme Court Justice William O. Douglas, a keen conservationist, and his wife. The expedition's purpose was to collect the data needed to launch a public campaign for an international reserve.[97] Olaus Murie wrote of the need to dedicate areas to perpetuate what the poet, Robert Service, had called "the freshness, the freedom, the farness."[98]

In December 1956 the *Sierra Club Bulletin* published Sumner's article "Your Stake in Alaska's Wildlife and Wilderness." Once more, the nature and validity of the frontier analogy figured prominently in his thinking. He argued that reckless and deleterious traditional practices had already marred Alaska and that the northeastern Arctic was the last ecologically undamaged part of the entire territory. Since much of Alaska's natural environment had already been compromised, he saw this as a final opportunity to preserve at least part of its "frontier atmosphere intact."[99] As evidence he cited a routine progress report (1951) on Alaska by the house committee with responsibility for the territory. The Hackett report was a positive appraisal of recent advances represented by new forms of communication and enterprises such as a new sawmill at Juneau. The report also noted with approval proposals for a cement plant near Cantwell, a tin mine on the Seward Peninsula, and an aluminum plant near Skagway. It concluded buoyantly that "whatever frontier there is remaining in Alaska appears to be rapidly vanishing under the impact of progress."[100]

Sumner insisted that the popular analogy between the buffalo in the West prior to the railroads, and the status of Alaska's caribou, while valid a half-century ago, no longer applied to many parts of Alaska. He repeated what Dufresne had explained in the 1930s. Caribou herds had been significantly reduced in size by over-hunting, the destruction of lichens by natural and man-set fires, and factors such as the introduction of Siberian reindeer. Sumner compared the detrimental effects of this latter policy to the overgrazing of the Great Plains as a result of large-scale cattle ranching. Echoing Dufresne, he warned that only in some arctic regions did caribou still thrive in anything close to "primeval" numbers. Sumner also found an alarming parallel from the history of the lower states for the decline in the size of the Alaskan salmon pack since the turn of the century due to commercial overfishing without respect for biological rates of reproduction. The example that came to his mind was the tragedy of the California sardine fisheries in the 1930s and 1940s.

Sumner pointed out that ecologically speaking the contiguous states contained no real equivalent to any part of Alaska. In particular, the frontier analogy disregarded the low physical and biological carrying capacity of the Arctic where it could take more than a century for lichens to recover fully after fire or overgrazing, compared to five to fifteen years for vegetation in temperate zones. The Arctic was simply less biologically productive than other parts of Alaska, let alone the rest of the country. What seemed to outsiders an unrestricted abundance of game was not evidence of the Arctic's inherent fecundity but a reflection of its hitherto relatively unaffected condition. He believed that facile comparison with the West overlooked another crucial difference; the technological revolution. "Changes," he commented, "whether constructive or destructive, that extended over months or years in the old-fashioned days of small, slow fishing boats, river transportation, hand shovels, and axes, sometimes are accomplished in an equal number of hours or days by today's bulldozers, airplanes, amphibious tractors, motorized floating canneries and motor-driven chain-saws."[101]

From 1957 onwards, the campaign launched by the two Park Service officials became largely the responsibility of Olaus Murie and the Wilderness Society. On one level, it was a battle against misconceptions of Alaska comparable to the struggle waged by boosters. Ironically, for their ultimate aims were quite different, conservationists and boosters shared the same immediate goal; the abolition of the icebox image. This limited common purpose found a focus in their stance against land withdrawals for military purposes, to which many Alaskans objected as much as they disliked federal removal of land from public entry for conservation.

In 1943 the Department of the Interior had issued Public Land Order (PLO) 82, which withdrew all of Alaska north of the crest of the Brooks Range (some 48.8 million acres including Pet 4) for the exclusive use of the military "in connection with the prosecution of the war." This closed the area to all forms of appropriation under the public land laws to allow the navy unfettered opportunity to explore for oil and gas. In 1953, when the U.S. Navy announced its intention to cease activities there, Dimond requested (unsuccessfully) that the order be revoked.[102] Murie was another prominent critic of such withdrawals. In 1955 he protested against the withdrawal of 137,000 acres in the Alaska Range for chemical testing.[103] In 1956 the U.S. Army withdrew a further 85,200 acres northeast of Fairbanks and 2.473 million acres west of Anchorage. "Alaska is not a big wasteland," Murie complained, nor was it "a dumping ground for the debris of civilization developed farther south."[104] In 1957 the Wilderness Society passed a resolution condemning the "disparaging of the country north of the Yukon River, as if it is an expendable, worthless part of Alaska which we can freely use for military devastation."[105]

That spring, at the invitation of various Alaskan groups and sponsored by the Conservation Foundation, the Muries returned to Alaska for extensive lobbying. In May the Muries made a presentation in Fairbanks to the Tanana Valley Sportsmen's Association (TVSA), an organization with 500 members. Soon afterwards, the TVSA made the first concrete proposal for an Arctic Wildlife Range, to be administered by the Bureau of Sport Fisheries and Wildlife (USFWS).[106] TVSA's resolution (passed overwhelmingly) stressed that the area contained "comparatively small amounts of known mineral resources." It urged action to protect its "Arctic frontier flavor" and "unique and increasingly necessary opportunities for recreational use." The resolution indicated that the creation of this range would not prevent mineral exploration and extraction, hunting, fishing, or trappping. However, the use of aircraft and tracked vehicles for these purposes would be prohibited.[107] James M. Lake, the TVSA's president, claimed the full support of the residents of the interior for this policy.[108]

In July 1957 Ross L. Leffler, assistant secretary of the interior for fish and wildlife (and a member of the Izaak Walton League), announced during an Alaskan visit that his department would seek legislation to establish an 8,000-square-mile arctic wildlife range. In November, D. H. Janzen, the director of the Bureau of Sport Fisheries and Wildlife, requested action on the range from the Bureau of Land Management's branch in Fairbanks. Citing the Hackett report's assessment of the Alaskan frontier's disappearance, Janzen saw "an ideal opportunity, and the only one in Alaska, to preserve an undisturbed portion of the Arctic large enough to be biologically self-sufficient."[109] A few weeks later, Seaton announced that he had taken the initial steps to modify Public Land Order 82, whose revocation was a precondition for any form of action in the Arctic, whether in the interests of mineral development or conservation. In mid-January, Janzen formally applied for the withdrawal of nine million acres to create an Arctic Wildlife Range (AWR).

The support of many Alaskans was largely predicated on the assumption that the region harbored little of economic value.[110] The Fairbanks Chamber of Commerce backed the proposal, but on condition that nothing would interfere with the use of valuable minerals, should they be located. Since the area might contain "the largest oil basin on the face of the earth," the chamber wanted the range to be managed according to the multiple use principle. It hoped that this form of management would prevent the recurrence of a controversy like the recent one concerning the Kenai National Moose Range.[111] The *Fairbanks Daily News-Miner* (17 March 1958) took a similar position: "We favor it on the grounds that some few sections of Alaska should be preserved for posterity just as God made them, and this is an appropriate area, being remote and unpromising as far as commercial and industrial development is concerned."

On the other hand, the proposal was greeted with dismay by the Anchorage Chamber of Commerce, and one Alaskan interest group in particular; the miners. In February, under pressure from the Alaska

Miners Association and Phil Holdsworth, Commissioner of the Territorial Department of Mines, Alaska's senior U.S. Senator, E. L. "Bob" Bartlett, urged the Department of the Interior to arrange hearings in Alaska. He wanted to provide critics with an opportunity to comment on the legislation to create the range, which would be introduced the following spring.[112] In March, the Alaska State Legislature, meeting for the first time, registered opposition to the proposal because it would "discourage industrial and mineral development."[113]

On 1 May 1959 legislation prepared by the Department of the Interior was introduced in Congress. An accompanying press release from the fish and wildlife service repeated the need to protect some of Alaska's "primitive wilderness frontier." The Fish and Wildlife Service emphasized that hunting, trapping, and fishing would be permitted under the laws and regulations of the State of Alaska. The bill also allowed the secretary of the interior to authorize mining in the range but precluded appropriation of surface title to land. Likewise, the range would not affect the conduct of defense activities in the area.[114]

Bartlett was pessimistic about the outcome of the hearings that Interior scheduled, and over which he, as chairman of the Merchant Marine and Fisheries Subcommittee of the Senate Committee on Interstate and Foreign Commerce, would preside. He suspected that conservationists would succeed in "jamming" them, effectively excluding any dissent.[115] In Washington, D.C., in June 1959, Governor Egan protested that 92 million acres in Alaska were already segregated in some form of federal reserve. Alaska's junior U.S. senator, Gruening, complained that over 14 million acres of Alaska were parks and wildlife refuges, many of them unnecessary, such as the Kenai National Moose Range. He believed that AWR was being offered as a sop to conservationists unhappy because the moose range had been opened to oil development.[116]

Opinion at the Alaskan venues in the fall was broadly split. The Alaska Conservation Society (ACS) had its origins in a group of advocates from Fairbanks who collaborated to coordinate their testimonies. This society grew out of an awareness that in the delicate atmosphere shortly after statehood, any outside efforts from conservationists to influence events in the new state were likely to receive a hostile reception.[117] Relations between Alaska's political and business interests and national conservation organizations were already less than cordial because of the moose range controversy. The Alaska Conservation Society's founders hoped that a homegrown organization, while maintaining close ties with national groups, would enjoy greater credibility among fellow Alaskans. The society wanted to provide an outlet for those "interested in speaking as Alaskans in Alaskan affairs, rather than yielding our right to speak for our beliefs to outside conservation groups."[118]

In 1959 their arguments for swift action on the range proposal were reinforced by growing publicity concerning the Atomic Energy Commission's preparations for a nuclear blast at Cape Thompson in the northwest Arctic. For Richard Cooley, a founder member of the Alaska Conservation Society who was director of the Alaska Natural Resource Center in Juneau, these AEC activities proved that natural factors no longer ensured *de facto* protection. "Today," he pointed out, "even remoteness has competing demands."[119] William O. Pruitt and Leslie Viereck, also founding members of the Alaska Conservation Society, were part of the AEC's bioenvironmental studies team at Cape Thompson. As such, these scientists from the University of Alaska were in an authoritative position to comment on the damage to the tundra perpetrated there by tracked vehicles.[120]

Alaskan sportsmen were much in evidence at the hearings. Many of them were members of the Alaska Sportsmen's Council (ASC), which represented 2,500 Alaskan members of thirteen individual organizations, including the Tanana Valley Sportsmen's Association. (However, *Alaska Sportsman* was doubtful of the range's merits.)[121] Sportsmen from all over the world were coming to Alaska in increasing numbers. Accordingly, Alaskans who were economically dependent on such visitors—like big game guides, lodge owners, outfitters, and taxidermists—were sensitive to Alaska's status as "the last of the wilderness and frontier of the U.S." These interests identified Alaskan wildlife as a valuable international resource.[122]

Most Alaskans whose livelihoods were to some extent dependent on wildlife and wild conditions—whether economically, scientifically, recreationally or spiritually—perceived a direct personal interest in the range. This position was based on a belief in the supreme value of wildlife and wilderness, potentially infinitely renewable resources, in the life, culture, and economy of Alaska.[123] Pruitt recommended Crisler's recently published book, *Arctic Wild,* as an exposition of the wilderness qualities now endangered in the northeast Arctic. He placed a higher premium on these wilderness values than he did on the riches of any gold mine that might be found there.[124] John P. Thompson, a local reporter, mountaineer, and future Alaska Conservation Society member, acknowledged the inspiration that Bob Marshall and his writings, notably *Arctic Wilderness* (1956), had given to range supporters.[125]

Opponents rejected the *cri de coeur* that here was the last opportunity to protect a large chunk of pristine Alaskan frontier wilderness. They considered this conservationist argument melodramatic to the point of hysteria and insisted that formal protection was superfluous. Clarence Anderson, commissioner of the Alaska Department of Fish and Game, spiced his reaction with a touch of sarcasm. In May 1959 he had claimed that the "only real threat to the wildlife and wilderness of the Alaskan arctic stems from the activities of a handful of wilderness extremists and Federal officials." The Arctic, he concluded, was "probably in little more peril of being trampled in future years than is the moon."[126] Gruening also dismissed advocates as scaremongers. The region, he vowed, would "remain a wilderness no matter what we call it, by virtue of its inaccessibility and bleak character."[127] Another critic exclaimed that "it is beyond my comprehension . . . that great areas of Alaska can become anything more than miles and miles of miles and miles."[128]

A basic question that absorbed the adversaries was how to ensure the survival of the American character and those distinctive cultural, social, and political values that the frontier had bred. Frederick Jackson Turner and his disciples had agonized over the future of a frontierless United States. In 1959 Stewart Brandborg, the director of the National Wildlife Federation, worried about the same situation: "We in this country have grown to love the out of doors partly because our forebears gained such an intimate understanding of it as they pushed our frontiers westward. Their experience, which gave them close contact with wilderness and unspoiled nature, is a source of much of the vigor, self-reliance, initiative and physical stamina of our people and our nation. When in our quest for a higher standard of living . . . we permit all of the wilderness to be destroyed, we rob ourselves of the experiences and conditioning that have contributed so much to the inner strength of our people and the achievements of our nation."[129] In language reminiscent of Theodore Roosevelt, Brandborg deplored the growing materialism, flabbiness, and softness of his fellow countrymen (particularly young Americans) and advocated the virtues of the strenuous life.[130]

Alice Stuart, a supporter of the range who had lived in Alaska since 1942, warned: "The American wilderness, to a great extent, made Americans what they are, with an outlook different from that in the older countries, not cramped and crowded. Destroy all wilderness, and you will have changed Americans. Let us Alaskans pass on the secret of Americanism to future generations."[131] Another woman who echoed this theme was Ginny Wood, a founding member of the Alaska Conservation Society. Wood was one of the first women pilots in Alaska. Since the early 1950s she had operated a wilderness camp close to Mount McKinley National Park with Celia Hunter, another former member of the Women Air Force Service Pilots (WASPs). Wood, who emphasized the influence that Lowell Sumner's articles had had upon her, gave another exposition of the frontier thesis with a conservationist flavor: "The wilderness that we have conquered and squandered in our conquest of new lands has produced the traditions of the pioneer that we want to think still prevail: freedom, opportunity, adventure, and resourceful, rugged individuals."[132] Supporters of the range were intent on preserving a cultural setting for vicarious pioneering as well as protecting an ecosystem. Their attitude toward the role of the frontier in American life was equivocal. They wanted to derive the historical benefits—emotional, spiritual, and physical—of confronting wild nature, without inflicting damage on the land itself.

In contrast, Alaskan opponents rejected this emasculated, surrogate form of "pioneering" in favor of what they perceived to be its genuine practice. They were convinced that the proposal endangered the intimate relationship between "free land" and the American character. Resentment of this kind was a further expression of the "land hunger" that surfaced during the recent debates over oil leasing in the Kenai National Moose Range. In the spring of 1958, Ernest Wolff, a future president of the Alaska Miners Association, complained that the range would be an anti-democratic "hunting reserve of the old European type."[133] At the hearings (1959), some range opponents explained that they had come to Alaska in search of the *lebensraum* the lower forty-eight could no longer provide. In the opinion of Wenzel Raith, an electrician who had come to Fairbanks from Duluth in 1952, the most admirable latter-day pioneer, and modern symbol of toughness and adventure, was the Alaskan bulldozer operator ("cat skinner") who spurned the beaten (rutted) track to blaze his own trail. Raith believed that these men had inherited the mantle of "rugged individualist" from such nineteenth-century pioneers as Daniel Boone, James Bowie, Kit Carson, Davy Crockett, and Jim Bridger. Without the services of these particular men, he wanted to know, "Where would we be today?" By "curtailing the opportunity for individual initiative" through conservation measures, he was afraid that "we dry up our vitality . . . as a people, as a nation . . . at its well-head." Frederick Jackson Turner had believed that the frontier served as a safety valve for the release of Eastern discontents and the defusion of threats to democracy born of overcrowding and competition for scarce resources. No matter that his thesis had been largely discredited by economic historians. Wenzel Raith still believed firmly in the safety valve theory and implored, "Leave us our wilderness to which to flee."[134]

Joe Vogler was a gold miner and property developer from Fairbanks who had been born and raised on a farm in Kansas. He explained that "I suffer from claustrophobia" and related how he had moved to Alaska in 1944 from Texas, where there had been "too many fences" for his liking. When he first came to Alaska, it had seemed a haven for the eleutherian spirit. Now, he remonstrated, with threats such as the Arctic Wildlife Range, not even Alaska was safe from fences. Vogler had resigned from the Tanana Valley Sportsmen's Association when it sponsored the range.[135] Charles F. Herbert, the president of the Alaska Miners Association, felt that the consequences of another "land grab" by the federal government in league with "professional conservationists" would prove inimical not only to the economic and political future of Alaska but to a national culture based on unrestricted access to the frontier and its resources.[136] On a fundamental ideological level, *Alaska Sportsman* magazine also believed that the wildlife range proposal damaged the basic historic American right to use "free land": "In Alaska," the editor explained, "the frontier American has come to the end of his trail. He no longer can leave the regulated area and end the attrition of his freedoms by moving to a new frontier."[137] For this sector of Alaskan opinion, the range proposal epitomized how the Alaskan frontier was being closed prematurely and artificially.

Nevertheless, both the *Anchorage Daily Times* and the *Fairbanks Daily News-Miner* continued to back the proposal. The latter was untroubled by its size, pointing out that there were already larger wildlife ranges in North America, namely, Canada's Thelon Game Sanctuary (9.7 million acres) and Wood Buffalo Park (10.8 million acres).[138] Yet in spite of his editorial support, Atwood commented sardonically that the range proposal gave impetus "to the report that the passengers on the first rocket to the moon will be Alaskans looking for a place to live without man-made barriers to land-use."[139]

The bill to create the Arctic Wildlife Range (AWR) won approval in the House of Representatives in February 1960. Now Olaus Murie's task was to persuade Gruening and Bartlett to remove their objections to the bill in the Senate. The Alaskan senators took particular exception to the size of the range and often quoted the statistic that it was larger than the combined areas of Massachusetts, Connecticut, Vermont, and Rhode Island. They reminded proponents that the two biggest American national parks, Yellowstone and Mount McKinley, were only two million acres apiece. Murie tried to explain that the low carrying capacity of the Arctic dictated the size of the range. He pointed out in simple terms that the further north an animal lives, the more space it needs. This applied especially to the "barren-ground" caribou, which roam extensively to conserve supplies of lichen, their major food source. This constant movement acts as a natural rotation system, prevents overgrazing in any part of the range and allows for recovery and regrowth. So, as Murie stressed, "it does not do to compare Alaska with the size of some small New England state, as is often done."[140]

Proponents denied that the absence of any immediate threat to the region, and the low present rate of use for scientific, recreational, and other purposes obviated the need for protection. AWR's full value, they argued, lay in the future. A. W. "Bud" Boddy, the director of the Alaska Sportsmen's Council, accused the senators of being shortsighted and equated their opposition to the resistance that had faced the imaginative proposal to create Yellowstone National Park. In the 1870s the Yellowstone region had been visited by few people except for scientists and a handful of explorers. In the meantime, he emphasized, it had become the nation's most popular park.[141]

Gruening and Bartlett were not impressed by these arguments and the evidence of widespread Alaskan support for the proposal. Although there were plenty of individual objections, no nongovernmental organization in Alaska, with the exception of the Alaska Miners Association and the Anchorage Chamber of Commerce, found fault with the proposal. Bartlett admitted to Murie that no newspaper in the state had taken a stance against it.[142] Bartlett took advantage of his strategic position as chairman of the Merchant Marine and Fisheries Subcommittee to kill the Senate bill in committee.[143] The statement he issued explaining his action, which was crucial to the bill's demise, attacked the proposal primarily as a betrayal of statehood that would perpetuate the historic federal mismanagement of Alaskan resources (especially fisheries). He also explained that protection was superfluous since the state already controlled hunting and because the area would always remain wild. Moreover, he felt that the low quality of the region's scenery and its inaccessibility to the vast majority of Americans made it unworthy of protection.[144] Not least, he felt that the purpose of the range was too ambiguous. On another occasion, he claimed that nobody at the department of the interior had anything "but the vaguest idea of what exactly it was seeking to protect," though he suspected that what most conservationists really wanted was a wilderness area.[145] At the end of September Governor Egan suggested a state-managed "wildlife area" within the boundaries proposed for the range.

Seaton had already given notice of his willingness, in the event of opposition, to create the range by executive order. This incensed Alaskan critics even further. Gruening denounced it as "the kind of bureaucratic, dictatorial attitude from which Alaska has suffered in the last forty years in the variety of absentee controls."[146] With only a few weeks left in office, Seaton issued two executive orders on 6 December 1960. The first (Public Land Order 2214) created an Arctic National Wildlife Range (ANWR) and closed it to public entry under the mining laws. A second order, revoking Public Land Order 82, restored to public entry twenty million acres of the North slope in the vicinity of Prudhoe Bay. The *Anchorage Daily Times* and the *Fairbanks Daily News-Miner* supported Public Land Order 2214 as being in the national interest.[147] However, the *Fairbanks Daily News-Miner* regretted that it had been necessary to resort to executive order; a bill would have permitted mining. Gruening condemned the order as a violation of states rights, regardless of its merits (dubious at best) as an act of conservation.[148]

Organizations like the Wilderness Society and the Alaska Conservation Society praised Seaton for taking advantage of a priceless opportunity to protect an entire, self-sustaining, self-regulating ecosystem within which, with the exception of the musk ox, the complete range of "precontact" fauna still thrived.[149] This shows how wilderness was being increasingly defined as an ecologically intact area that provided the conditions deemed essential for the success of certain animals, especially caribou.[150] Murie was particularly pleased that a wildlife range had been established instead of a national park. A national park, he feared, would involve its own form of development in the sense that national park status usually led to the promotion of recreation and the provision of visitor facilities. This would compromise the wild values he believed could best be sustained by the management principles of a wildlife refuge.[151]

In a second bid to gain Senate approval for the range, Seaton submitted fresh legislation in January 1961. In deference to miners, the new bill guaranteed access (nonmechanical) to minerals. But it was also unsuccessful. Soon afterwards, the Alaskan congressional delegation asked John Kennedy's interior secretary-designate, a fellow-Democrat, Stewart L. Udall, to give priority to the revocation of his predecessor's order. In late February, the Alaska Legislature passed resolutions authored by the Alaska House Speaker Warren Taylor, which called on the president to review Seaton's action, which Bartlett called "a dirty Republican trick."[152] Gruening claims that in June 1961 Udall agreed to consider a transfer of power over the range to the state.[153] However, Udall took no action on further proposals submitted by Egan. Bartlett explained to Egan that Udall was not cooperating because "he not only fears what the conservationists would do to him . . . he is one of them himself."[154]

The creation of ANWR was the opening round in a twenty-year struggle for the protection of wilderness, "frontier," and ecological values in Alaska culminating in 1980 with the Alaska National Interest Lands Conservation Act (ANILCA). The new value assigned to the Arctic was a seminal expression of a maturing ecological awareness and a cultural revolution in American attitudes toward wilderness. The rugged scenery of the Brooks Range, which had captivated Bob Marshall in the 1930s, invited comparison with the Rocky Mountain West a century earlier. However, what enchanted Marshall, a traditional "monumentalist" in outlook, were the geophysical elements of mountain scenery he often compared favorably with Yellowstone and Yosemite.[155] Marshall never visited or applauded the Arctic "prairie," which, ecologically, was as integral a part of the Arctic as were the 9,000-foot peaks in the eastern Brooks Range, the highest in the North American Arctic.

The Arctic consists of three physiographic divisions; the Brooks Range, the low, rolling hills and plateaus of the foothills, and the Arctic coastal plain or Arctic Slope. The Arctic (North) Slope is relatively flat. The foothill region and the southern coastal plain are dominated by the moist tundra ecosystem. The main type of vegetation is the cotton grass tussock; these clumps are 15-31 centimeters high and wide and are separated from each other by troughs of 30 centimeters wide. Other types of vegetation found there are mosses, lichens, and, in some sheltered spots in the foothills, dwarf shrubs. The wet tundra ecosystem characterizes the northern coastal regions. Here the landscape is a combination of wet-sedge meadow, standing water and (in some parts) shallow lakes. In some areas, 50 percent of the coastal tundra consists of surface water. In other words, the coastal tundra and the arctic foothills are devoid of dramatic topography and all monumental scenic ingredients. As two federal petroleum geologists remarked in a classic description of the Arctic Slope (1930): "Prominent landmarks are entirely absent . . . over these plains the winds sweep with unbroken severity. . . . [There is] no natural shelter. In summer the poorly-drained tracts of upland afford only spongy footing, which makes travel laborious and slow."[156] The geologists were describing Pet 4, and the coastal plain in the wildlife range is much narrower and not so flat. However, at least to the non-Native, nonresident, untrained eye, the appearance of the Arctic Slope does not vary much between Cape Lisburne in the west and Demarcation Point at the Canadian border.

Responsible in large part for the difficulties of movement to which the geologists referred were tussocks, still known as "niggerheads" in the 1930s. The "niggerhead," according to Bob Marshall, "among the gifts of nature, ranks as the most cursed. . . . Occurring once this merely provokes profanity. Occurring fifty or a hundred times to the mile it is likewise exhausting."[157] In 1938, Henry Wallace, the Secretary of Agriculture, noted in the National Resources Committee report that "this remote section of the territory appears to offer very little to attract white men."[158] While Wallace was thinking of its economic lack of appeal, his assessment was just as valid with respect to its recreational, aesthetic, and ecological attractions.

The campaign for ANWR provides some of the first evidence of substantial change, indicating that this unlikely region had begun to be prized for scenic and ecological reasons. The emerging ecological, more holistic perspective should be contrasted with the desire to enshrine scenic wonders that strongly influenced the creation of the great nature parks of the West in the late nineteenth and early twentieth centuries.[159]

A comment concerning the disappearance of California's grasslands casts a good deal of light on the nature of the change that was taking place in attitudes toward the Arctic. Raymond Dasmann, an ecologist, has observed: "It is relatively easy to promote public interest in many phases of conservation. Rugged wilderness country is spectacular and has its devoted protectors. The larger species of wild anaimals, or the more glamorous ones, have lots of defenders. Forests have their friends, and even individual species of trees have organizations to protect them. But it is difficult to find friends for grass."[160] Finding friends for tundra was not an easy proposition either.

The campaign also showed that the Arctic had begun to acquire a distinct cultural value. Culturally, the new concern for the Arctic can be seen as part of a need to reaffirm the nation's frontier identity. Conservationists, who needed to find a new place, were increasingly disposed to perceive and portray the Arctic as a re-creation of the horizonless expanses across which settlers had voyaged westward in prairie schooners. Conservationists were now describing as "fragile" and "delicate" a region, parts of which matched the historically sanctioned wilderness-as-wasteland image to perfection.[161] This tender regard was an indirect reflection of the almost total disappearance of the 700 million acres of prairie grassland that once grew ten to twelve feet tall in states like Illinois and Iowa. The proponents of ANWR displayed a keen sense of the significance of wild nature as a historical document. Emphasis on the importance of landscape as history was nothing new. However, this cultural value of the environment was usually seen in terms of the richness of the human record it displayed, not the paucity of this legacy. Conservationists considered northeast Alaska rare and valuable (priceless in fact) precisely because of the absence of a palimpsest—the accumulated layers of human history.[162]

The campaign for ANWR showed national conservation forces on the offensive in Alaska with significant local support for the first time in Alaskan history. All previous parks and monuments in Alaska were the result of pressure exerted by influential, non-Alaskan individuals; Charles Sheldon (McKinley Park, 1917); Robert F. Griggs (Katmai National Monument, 1918); and William S. Cooper (Glacier Bay National Monument, 1925). This latest bid for protection possessed its guiding genius in Olaus Murie. Nevertheless, his efforts were reinforced by a broad range of groups, most of whose members had never been to Alaska. In this respect, the ANWR campaign was a critical transitional episode spanning the divide between the old and new varieties of conservation.

The campaign also illustrated how conservationists used the terms *frontier* and *wilderness* in connection with Alaska. On the one hand, they justified the range bcause the northeast Arctic was the final material bastion of ecologically pure, frontier conditions in Alaska. In this way, they presented themselves as realists, in contrast to the range's opponents, who, they argued, clung to the myth of the inexhaustible, perennial wilderness. Certain remarks of Lois Crisler in 1960 epitomized the conservationist belief that an apocalyptical contest was taking place in northern Alaska. According to Crisler: "Here in the Brooks Range the biggest of all historical movements, man against nature, meets actual living wilderness making its last stand. Beside this contest all others—man-woman, fascist-communist, totalitarian-democratic—are only pressure seams on top of the tide. So far, man has always won; living wilderness has always perished into desert or mere scenery."[163]

At the same time, a simplified general image of Alaska prevailed within the national conservationist community. Alaska as a whole was percieved as a *tabula rasa* providing the nation with a rare opportunity to write a more gentle and thoughtful chapter in its history, though, if one accepted the claims of ANWR advocates at face value, this opportunity had already largely disappeared. *Alaska Sportsman* (subtitled *Life on the Last Frontier* since statehood), referred to the state as "truly America's (and perhaps the world's) last real wilderness frontier."[164] Over the next two decades, the image of Alaska as an untouched land, calculated to appeal and gratify at a deep emotional and pyschological level, would provide a variety of conservationist causes with a particular rhetorical panache, and would be exploited by them with increasing *éclat* to mobilize popular support.

4
Project Chariot, 1958–1963

Since 1941 the federal government had dominated the Alaskan economy, and Alaskans had grown accustomed to federal financial injections, often in the form of large construction projects. In 1957 the federal government provided 60 percent of all employment in Alaska; more than two-thirds of that was in the military sector.[1] However, from 1957 onward, defense spending fell sharply as Alaska's military population shrank from 48,000 at the end of fiscal year 1957 to 34,000 by 1959. Within the private sector, construction provided 6 percent of employment in Alaska. After reaching a peak of 10,475 in mid-1951, the number employed in construction fell to 5,539 by the end of 1959.[2] At the same time, the long-term decline of the traditional extractive economy based on fur, salmon, and gold continued. A prominent feature of Alaskan politics and economics after statehood was the search of local politicians and the business sector for a broader, more secure economic base. This produced some support for one spectacular federal project, Project Chariot, and a good deal more for another grand plan known as Rampart Dam.

Frederick Jackson Turner had used the glaciological concept of *terminal moraines* to describe the marks left after each phase of frontier recession in the nineteenth-century West.[3] Since the controversies engendered by Project Chariot and Rampart Dam ended in defeats for the propositions, and setbacks for the booster cause, the concept of terminal moraines can be applied to the advance of conservationist strength in Alaska. Since the advocates of both projects exploited the icebox image of the Alaskan environment with little success and their opponents continued the progress that the proponents of ANWR had made eroding it, the concept of moraines may also be applied to the retreat of this stubborn image. Arguments for environmental protection and ecological and cultural values that attained widespread acceptance among conservationists and environmentalists during the pipeline controversy can be observed in their infancy during these two landmark debates. These controversies also illustrated how Alaska firmly maintained its central position in the history of American conservation. Although their national significance has been largely unrecognized by historians, these controver-

111

sies were important illustrations of the shift from conservation to environmentalism. They also highlight the increasingly widespread belief that science and the technology it produced were "getting out of hand."[4]

In 1957 the Atomic Energy Commission's Division of Military Applications established Project Plowshare to investigate, develop, and promote peaceful uses for nuclear explosives. The Atomic Energy Commission (AEC) assigned technical direction to the University of California's Lawrence Radiation Laboratory (LRL).[5] For those working on Plowshare, the most attractive possibilities were for large-scale "excavation applications" neither technologically nor economically feasible with conventional engineering methods. The AEC envisaged the use of nuclear bombs to dig canals, dam and redirect rivers, cut tunnels through mountains, and release mineral wealth. Perhaps the most ambitious proposal was for a larger and deeper canal across the isthmus of Panama.[6] Under contract to the LRL, the U.S. Geological Survey (USGS), the United States Army Corps of Engineers, and the E. J. Longyear Company, an oil drilling concern, conducted site selection studies for what the AEC called an "excavation application."

The AEC established four criteria to govern selection: an American location; the need to protect humans and wildlife (1957 was marked by increasing public unease over nuclear testing in Nevada, especially among "downwinders" in Nevada, Utah, Arizona, Colorado, and southern California); certain geological and engineering features necessary to provide essential data; and, finally, the demonstration of "possible long-term utilitarian value."[7] On these grounds, the consultants ruled out the traditional Nevada Test Site, only 70 miles north of Las Vegas, which had been in use since 1951. They also eliminated all other locations in the contiguous United States. Instead, they identified eleven sites on the Arctic coast between Point Barrow and Nome that met the AEC's definition of remote—at least 20 miles from human settlement—and chose Cape Thompson on the Chukchi Sea as the most promising.[8] There, at the mouth of Ogotoruk Creek, 110 miles north of the Arctic Circle, the AEC proposed to excavate a deep-water harbor by detonating simultaneously two one-megaton and two 200-kiloton devices just below the ground.[9] The larger explosives were to be detonated 150 feet below the surface while the smaller ones would be set off 90 feet underground. They were intended to create a channel 400 yards wide and 1,200 yards long with a turning basin 1,000 by 1,700 yards.[10] Each of these smaller bombs was ten times larger than the bomb dropped on Hiroshima. The AEC named the plan Project Chariot.

The AEC promoted Chariot as a public works–style internal improvement. The first announcement in June 1958, a week after the House of Representatives passed the Alaska statehood bill, claimed that "the absence of harbors on the northwest coast of Alaska close to important mineral deposits has in the past hampered development of such deposits. Fishing in this area has also been impeded by lack of a safe haven." The AEC explained that it would conduct field studies that summer to verify that the explosions would not endanger humans or wildlife. The blast was scheduled for 1959.[11]

The pioneering nuclear physicist, Edward Teller, who was director of the LRL, was responsible for marketing Chariot. A week after Eisenhower signed the Alaska Statehood Act, Teller, popularly known as "the father of the hydrogen bomb," came to Juneau to introduce the peaceful atom to a hastily convened assembly of officials and businessmen. According to Teller, the AEC had slated $5 million for a nuclear engineering project in Alaska. He envisaged a partnership with industry, wanting this $5 million to be matched by a $5 million investment from private enterprise. He reassured his audience that an excavation application would not be performed until it could be economically justified.

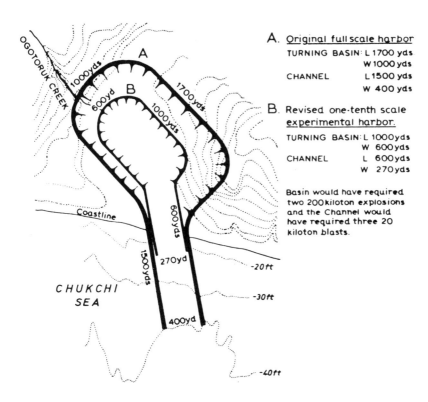

A. Original full scale harbor
TURNING BASIN: L 1700 yds
W 1000 yds
CHANNEL   L 1500 yds
W 400 yds

B. Revised one-tenth scale experimental harbor.
TURNING BASIN: L 1000 yds
W 600 yds
CHANNEL   L 600 yds
W 270 yds

Basin would have required two 200 kiloton explosions and the Channel would have required three 20 kiloton blasts.

Project Chariot: harbor configurations

As his colleague, Gerald W. Johnson, the associate director of the LRL, explained, "We don't just want a hole in the ground."[12] Teller then moved north to Fairbanks where he addressed a group at the university consisting of science faculty, university administrators, and local businessmen. His speech was replete with references to Alaska's physical size and frontier ebullience. In Juneau, he had professed that "We looked at the whole world, almost the whole world."[13] Now he explained that the AEC had chosen Alaska to host this daring act of nuclear pioneering because of its remoteness and sparse population ("you have the fewest people") and its inhabitants' receptivity to innovation ("you have the most reasonable people").[14] He stressed that here, at last, was a venture

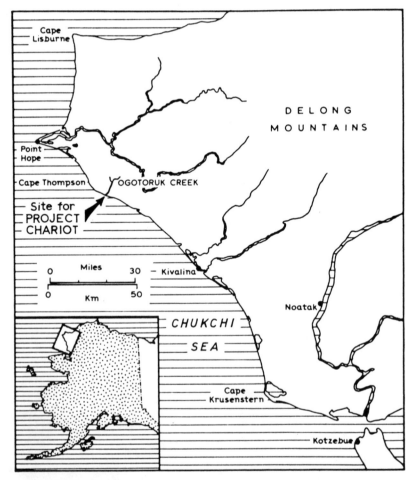

Project Chariot: Cape Thompson region

commensurate to the spirit and proportions of the land itself and its residents' pioneering aspirations.

The Alaskan response to Teller's hyperbolic offerings was mixed. Politicians in Juneau were not impressed by Chariot's economic value, though some expressed interest in the possibilities for nuclear engineering elsewhere in the state.[15] In contrast, most newspapers and chambers of commerce were delighted. Extending a warm welcome to the AEC, the *News-Miner* declared that "the holding of a huge nuclear blast in Alaska would be a fitting overture to the new era which is opening for our state." Its editor, George Sundborg, enthused that a nuclear engineering project, possibly resulting in a commercial harbor, would "center world scientific and economic attention on Alaska just at the time when we are moving into statehood and inviting development."[16]

Beyond these circles Chariot gained little tangible support. George W. Rogers, a prominent economist working for the State of Alaska, whom Teller had attempted to woo during his trip to Alaska, expressed early, private reservations over Chariot's feasibility and value. Rogers told Teller that the ocean at Cape Thompson was navigable for a maximum of two months per annum and explained that the coal the AEC wanted to mine was located beyond the Brooks Range—a considerable distance from Ogotoruk Creek. Undaunted, Teller retorted that the AEC would build a railroad and overcome the short shipping season by providing tank farms and warehouses to store coal, oil, and other minerals.[17] Group protest in Alaska originated in the fall of 1958, when a handful of biologists at the University of Alaska recommended studies to ascertain the project's social and environmental impact.[18]

In view of the LRL's limited success so far in selling Chariot as an economic boost, other officials from the LRL came to Alaska seeking a firmer committment. In January 1959 Harry B. Keller, LRL's test group director, and A. V. Shelton warned that the project would not materialize unless it received widespread Alaskan backing. They concentrated their attention on the business community as the most influential representative of Alaskan interests and opinion. The *Fairbanks Daily News-Miner* (10 January 1959) explained that "Alaskans are being given an opportunity to 'vote' on the matter through their Chambers of Commerce." The state chamber promptly announced its approval. Soon afterwards the chamber at Fairbanks, the closest major settlement to the blast site and a likely supply center, gave the project a unanimous vote of support.

Keller and Shelton's visit also served to solidify and extend opposition to Chariot on economic, social, and environmental grounds. A series of critical letters appeared in the *Fairbanks Daily News-Miner,* whose proprietor, C. W. Snedden, was a staunch Chariot supporter. One protest was from Irving Reed, a former member of the Alaska Game Commission (1927–40), who had been a prominent highway engineer in territorial Alaska. Reed, now a member of the Fairbanks City Council, had been raised near Nome (300 miles south of the blast site) in the early 1900s. He was convinced that there were better ways to spend $5 million of federal funds for the worthy cause of Alaskan development, such as paving the Alaska Highway and building a road to Nome.[19]

These critics believed that the project was hazardous as well as economic folly. Virginia (Ginny) Wood alleged that the AEC had "dumped" Chariot on Alaska because of protests against further testing in Nevada from the inhabitants of the Southwest.[20] Albert W. Johnson, a botanist at the University of Alaska, deplored the absence of a full discussion of the biological hazards of radiation.[21] In January, a group led by Carl Hamlin, the assistant manager of the Matanuska Valley Farmers' Cooperating Association Creamery, met in Fairbanks and formed the Committee for the Study of Atomic Testing in Alaska. This fifty-member group was affiliated with the Greater St. Louis Citizens' Committee for Nuclear Information (CNI), founded in 1958. CNI originated among scientists and medical faculty at the University of Washington, St.Louis, Missouri. Within a year, the CNI attracted a membership of 500, many of whom were scientists, doctors, churchpeople (especially Quakers), and civic leaders.[22] Through its bulletin, *Nuclear Information,* and a speakers' bureau, the CNI aimed to provide an antidote to AEC information about radioactivity in the form of scientific data on nuclear matters that were comprehensible to laymen. This was one of the first manifestations of what became known as the scientists' information movement, whose goal was to create an "informed citizenry" capable of making up its own mind about the benefits and risks from scientific discoveries and new technology. Groups like CNI and the Committee for the Study of Atomic Testing in Alaska believed that decisions about science and its relationship to the public welfare were too important to be left to scientific experts.

Meanwhile, international factors were contributing to Chariot's difficulties. Nuclear test ban talks were under way in Geneva, and the United States, the Soviet Union, and the United Kingdom had observed a voluntary moratorium on nuclear testing since November 1958.[23] The Soviet Union, whose territory lay only 180 miles from the site of a blast that was expected to hurl debris 30,000 feet into the air, opposed Chariot. America's western allies were sceptical about Project Plowshare in general. The Second United Nations International Conference on Peaceful Uses of Atomic Energy was held in Geneva in September 1958. American delegates, who included Edward Teller, argued that Plowshare should not be part of the test ban debate. Soviet representatives, on the other hand, claimed that Plowshare, especially Chariot, was a facade for military testing and could provide a way for the United States to continue weapons testing despite a ban.[24] These pressures undermined the project and during the fall and winter of 1958–59 the AEC made statements suggesting that Chariot might be delayed or even canceled.

In these statements, the AEC referred only to the problems it had encountered trying to promote the harbor at Cape Thompson as an economic project. One of the five AEC commissioners, Willard F. Libby, noted at the Geneva conference that "the only trouble with the plan is that we haven't been able to find anyone who really wants a harbor there."[25] In February 1959, John A. McCone, the AEC's chairman, reported to Congress's Joint Committee on Atomic Energy that the agency had set aside the harbor-blasting proposal. McCone, a Californian industrialist, gave the same reason Libby had, explaining that "we couldn't find a customer for the harbor." He told the committee that until the AEC found one, it would look for an alternative to Chariot that "would demonstrate exactly the same thing, and might be more useful."[26] Neither McCone, Libby, or any other AEC official offered an explanation as to why there had been no commercial interest. The U.S. Geological Survey provided a clue in a report recently released; minerals in the vicinity of Cape Thompson were of doubtful commercial value.[27]

But the AEC did not give up on Chariot because of the moratorium on testing, the project's unmarketability, and Alaskan criticism. The agency tried to revive the proposal by reducing its size from blasts totaling 2.4 megatons to 460 kilotons and reclassifying it as an "experiment" instead of a harbor.[28] The AEC's new stated purpose was observation of cratering effects of simultaneous detonation between 700 and 400 feet underground involving two 200 kiloton and three 20 kiloton bombs in a row. Building on theories and prediction models suggested by recent underground tests in Nevada, the AEC also wanted to obtain basic data on radioactive distribution and to measure the effects of fallout and those of air blast and seismic shock on local biota. The last tests carried out before Eisenhower's moratorium ("Neptune," Nevada, October 1958) were designed to yield data on the cratering effects of underground detonations. A row of simultaneous detonations at depths greater than 63 feet—and in permanently frozen mudstone—were unprecedented features of nuclear experimentation. The AEC expected bombs detonated deeper than 63 feet to trap more radioactivity.

In late February officials from the Lawrence Radiation Laboratory visited Juneau to introduce and promote the scaled-down project. They hoped that the new Chariot would be more acceptable internationally and would not arouse as much Alaskan opposition.[29] The Alaska House Legislature passed a supportive resolution, but the state's leading politicans were still less than enamored. Governor Egan and Senator Gruening were alert to potential abuses of federal power. Throughout the summer of 1959, they pressed the AEC to fund a state representative with the agency to ensure some degree of state oversight. The AEC refused, explaining that it had no authority to hire nonfederal employees.[30] Representatives of the State of Alaska requested "the full information necessary for them to perform their primary responsibilities to the citizens of Alaska."[31] The AEC replied that it had already gone beyond any of its previous efforts to inform officials in the states where its blasts were scheduled.[32] Robert Atwood criticized the State of Alaska's attitude, arguing that "it will take more than a liaison for state officials who are untrained in nuclear matters to understand the whole thing."[33]

Alaska's other U.S. Senator, E. L. Bartlett, had endorsed the original Chariot proposal on 9 February. Less than two weeks later, however, he disavowed the modified scheme. He urged the AEC, which he believed had no clear purpose, to conduct its experiments elsewhere.[34] While welcoming federal spending in Alaska, he was wary of any possible violation of states rights. His sensitivity was particularly acute immediately after statehood and applied especially (witness the Arctic National Wildlife Range debate) to land withdrawals. In April Bartlett called for hearings on the AEC's land withdrawal at Cape Thompson. His disquiet focused on its size, originally publicized as being 40 square miles but actually 1,600 square miles. (According to the AEC, this was a mistake.)[35] The federal Bureau of Land Management (BLM) declared that public hearings would be held "if circumstances warrant it." However, it received only two letters on invitation of comments. One simply requested further information. The other correspondent wanted to know more about the AEC's provisions for restoring the land to the public domain after the project. He was also concerned about the dangers of fallout and the threat to the Native community.

Nevertheless, Bartlett's attitude to Chariot (at least in his dealings with conservationists) remained ambivalent. He was unpersuaded by the economic benefits that were supposed to accrue ("industry will not be attracted to the distant Arctic coast merely because a harbor is created where one did not exist before"). But he remained enthusiastic about the opportunities for further "adventures" that might prove more beneficial, such as a waterway across the Alaska Peninsula, should Chariot prove to be a successful experiment.[36]

In addition to economic and political objections, Chariot still faced criticism on social and environmental grounds. The AEC and its Alaskan allies had always asserted that "the project is located in the wilderness, far away from any human habitation."[37] While this claim was possibly valid with respect to whites, it overlooked the Native villages of Point

Hope and Kivalina, 32 miles northwest and 40 miles southeast of "ground zero" respectively. According to AEC figures, Point Hope had 324 inhabitants and Kivalina had a population of 142. Point Hope was one of Alaska's oldest settlements, having been inhabited continuously for over 5,000 years. Some considered it one of the most successful Native communities in the Alaskan Arctic, in the sense that a harmonious blend of traditional and Anglo-American cultures had been achieved there.[38]

These Eskimos subsisted to a large extent on caribou meat. The major component of the caribou's diet is lichen, an organism consisting of algae and fungus living together. Lichens have no roots, often grow on rocks, and, unlike most other types of vegetation, receive nutrition directly from the air in the form of precipitation, dust, and other air-borne material. After the first atomic blasts in 1945, these nutrients became contaminated with radioactive fallout from Soviet and American tests. Lichens, in turn, became saturated with radioactive isotopes such as strontium-90 (Sr-90) and cesium-137. Their reception of nutrients straight from the air left them unprotected by the process of dilution, which occurs when radioactive fallout is absorbed from the soil through plant roots. The ability of lichens to concentrate radioactivity is increased by their slow rate of growth and longevity. As they passed up the short and relatively simple arctic food chain, these radionuclides attained even higher concentrations.

Concern over the project's impact on Natives featured prominently in a detailed list of questions the Committee for the Study of Atomic Testing in Alaska sent to the AEC in mid-February 1959.[39] The committee wanted to know why Alaska had been chosen for the experiment and why there was no provision for comprehensive social and environmental studies. A. R. Luedecke, the AEC's general manager, explained that the agency had chosen Alaska for the blast instead of Texas or California, where the committee had suggested it might prove more useful, primarily "for technical suitability and sufficient isolation." Chariot, he reiterated, was designed to "test theories and obtain data which are needed before nuclear explosives can be utilized in excavation work near population centers." To allay public fears, Luedecke declared: "We believe the the experiment could be conducted without hazard to the public and without serious disturbance to the marine and onshore life or food chains. . . . We are now proceeding to verify this opinion."[40]

The AEC had commissioned a one-year bioenvironmental research program in late February 1959.[41] The purposes of the Project Chariot Environmental Program were to discover the best time of the year for holding the blast and the optimum orientation of the fallout zone (i.e., when the least damage would be caused). A related purpose was to measure the biological effects of a nuclear explosion by contrasting the pre- and post-shot ecology of the region.[42] To carry out this program, the AEC's San Francisco Operations Office established the Committee on Environmental Studies for Project Chariot under the leadership of John N. Wolfe, chief of the Environmental Sciences Branch of the AEC's Division of Biology and Medicine.[43] The AEC negotiated contracts with the University of Alaska for some of the studies.[44] A group of biologists from the University of Alaska who had been in the forefront of criticism became part of the team. Albert W. Johnson was appointed senior scientist, Botanical Investigations, with Leslie A. Viereck as his assistant. An ornithologist, L. Gerard Swartz, was placed in charge of Sea Bird Cliff Investigations. William O. Pruitt, a mammologist, was assigned to lead Terrestrial Mammal Investigations. The AEC hired Don C. Foote, a twenty-eight year-old graduate student from McGill University in Montreal, Canada, to lead the investigations of human geography. The AEC allocated $100,000 for this program, which consisted of some forty individual studies.

Afraid that Alaskan support for Chariot would dwindle further, one AEC commissioner, Willard F. Libby, assured Senator Gruening that the experiment "would still provide an excavation suitable for a harbor, though a smaller one than originally considered."[45] Some boosters of Alaskan economic development remained captivated by this prospect and the project continued to enjoy the support of the state's leading newspapers—the *Anchorage Daily Times* and the *Fairbanks Daily News-Miner*. Six months after Chariot had been reclassified, journalist Albro Gregory, in a contribution to the *News-Miner* (24 August 1959), insisted that the harbor would be able to handle the largest ocean-going vessels. (As late as January 1962, the *News-Miner* persisted in referring to the harbor as the main purpose of the blast. The AEC's studied ambiguity served to keep these tenuous expectations alive. Vague allusions to the possibility of a useful harbor were frequent long after the official shift to experimental classification.)

In the spring of 1959, Harry B. Keller of the LRL had visited Kotzebue, the settlement with a significant white population which was closest to the blast site (125 miles southeast). In response to this first trip by an AEC official to the region, the Reverend William T. McIntyre, who was the president of the Arctic Circle Chamber of Commerce, sent a plea to John McCone. At its heart lay the storehouse image of Alaska: "Alaskans, who are all pioneers, if not in fact then in heart, should feel proud that our State has been singled out to play yet another leading role in the Drama of the Century. We who live in this part of the state . . . feel this project will be the means of paving the way for the future development of this area . . . possessing vast potentialities in the mineral, coal, oil and hydroelectric fields, indeed, want the original harbor plan rather than scaled down experiment."[46]

In contrast and for a quite different purpose, some conservationists also rejected the popular image of Alaska, and the Arctic in particular, as a worthless land of ice and snow. On 12 February 1959 the *Fairbanks Daily News-Miner* contained a letter from Olaus Murie, now resident in Wyoming. In a prophetic statement, Murie progressed from specific criticism of the planned detonation at Cape Thompson to a full-scale assault on what might be called the myth of the barren Arctic. Murie objected to Alaska being thought of as a "dump-ground" for projects unwanted elsewhere, "to do with as they please." Instead, he emphasized the rich and subtle beauty that he saw behind the Arctic's hostile and harsh facade.[47] The Wilderness Society passed an anti-Chariot resolution later in February—the first conservation organization to do so. The Sierra Club did likewise in March.

Teller's efforts to promote the new version of Chariot resembled his first lobbying campaign in many respects. Delivering the Commencement Day Address at the University of Alaska in May 1959, where he received an honorary degree, he flattered Alaskans, inviting them to rise to a momentous and revolutionary challenge: "Anything new . . . is repulsive and frightening; anything new that is big needs big people in order to get going. . . . And big people are found in big states." Teller told the audience that the AEC had the power not only to subdue nature but to "reshape the land to your pleasure." It could, he boasted, "dig a harbor in the shape of a polar bear if required."[48] Here was the tantalizing offer of a technology to extend human mastery over nature beyond the wildest imaginings of nineteenth-century pioneers who pushed the frontier westwards.

Seeking to demonstrate the wide benefits of Project Chariot, Teller introduced a fresh element. The AEC was finding it increasingly difficult to ignore the Eskimo people of Cape Thompson, whom the agency had never directly informed of its plans. Teller's strategy (returning to the original economic justification for Chariot) was to include them as beneficiaries. He argued that these indigenes, straitjacketed in poverty for centuries, stood to gain the most because Chariot would transform them into coal miners by making it possible to mine and ship coal.[49] When Teller addressed the Arctic Circle Chamber of Commerce in Nome later that summer, he added Japan to the list of beneficiaries. He maintained that the coal the former hunter-gatherers mined would help supply Japan's energy needs. He saw a measure of justice and atonement in this: "The Japanese were the first to suffer by atomic blast, how wonderful it would be if they might be the first to benefit."[50] In fact, Teller saw no disadvantages. In Anchorage, a few days earlier, in response to the Soviet charge that Chariot was a weapons test in disguise, he had denied that the Soviet Union would be affected; the blast would occur at a time when the wind was blowing inland.[51]

Even before the AEC announced its bioenvironmental studies some Alaskans were convinced that Chariot was safe as well as economically valuable. Robert Atwood believed that anyone who was not a nuclear scientist working for the AEC had no legitimate voice in the debate over Project Chariot. He illustrated his conviction with the following analogy: "Asking Alaskans for a decision on this proposed atom experiment is like a doctor asking his patient whether he wants an operation. The easiest answer is 'no.' The doctor usually tells the patient he must have one for his own good, and the patient does as the doctor says. The atom scientists are the doctor in this case. If they say that adequate safeguards for life and property hav been provided, how can laymen say otherwise?"[52]

William R. Wood, who had recently succeeded Ernest Patty as president of the University of Alaska, was another of the AEC's loyal local allies. Three months into the studies, Wood (whose academic field is English literature) echoed Atwood's faith in the AEC and his deference to its experts: "If the U.S. government decides that the project is a safe one, there is no reason for concern."[53] Albro Gregory, in a series of articles on Project Chariot in the *News-Miner*, indicated that the bioenvironmental studies were superfluous because "most scientists . . . already believe . . . [the project] to be feasible," and concluded that danger to humans was minimal since "only a few Natives pass by."[54]

On 7 January 1960, though field studies were still in the early stages, the Committee on Environmental Studies for Project Chariot supplied the AEC with its first report. Confirming the AEC's pre-study determination, the committee recommended that the blast take place in spring (March or April). At that time, it explained, few birds were in the area, little hunting took place, plants and many animals were under a protective blanket of snow, and the spring run-off would flush any radioactivity out to sea, where it would be diluted and decay harmlessly on the ice.[55]

However, at a meeting of the Committee on Environmental Studies in Anchorage in March 1960, Foote, the independent contractor, and Pruitt, Swartz, Viereck, and Johnson of the University of Alaska disagreed with these conclusions. Their research indicated that high winds of unpredictable direction and low precipitation meant that much of the exposed area around Ogotoruk Creek was bare in spring, when hunting by Natives on land and sea ice reached its annual peak.[56] They thought it unlikely that radiation escape ("venting") could be contained to less than 5 percent as the AEC predicted.

AEC releases and the Alaskan press did not address these arguments and the latter's coverage consisted predominantly of verbatim reprints of the former. Albro Gregory did mention that there had been some criticism from "misinformed" sources but ridiculed them as people who worried that "a caribou might be hit in the head with a rock."[57] In August 1960 the AEC invited reporters to the site for the first time. Though the commission stressed that the bioenvironmental studies had been an integral part of the project since 1958, Wolfe's comments to the press ("We are sort of riding the coat-tails of an engineering project of great magnitude")[58] tended to reinforce the impression of critics that the program was scrambled together in response to disapproval. Subsequent press reports did not mention any dissent from the official position.[59]

Chariot also faced mounting opposition from Alaskan Natives. On 30 November 1959 the Point Hope Village Council sent the AEC a petition condemning the project. In December the village council voted unanimously against Chariot. Governor Egan, Senator Bartlett, Commissioner Holdsworth, and James W. Brooks, the director of the Alaska Department of Fish and Game, urged the AEC to meet with the people of Cape Thompson. In March 1960 three officials made a one-day tour of the region.[60] At Point Hope, Russell Ball, the assistant manager for technical operations at the AEC's San Francisco office, who was in charge of general planning for Chariot, told an audience of one hundred Natives that "we no longer have any expectation that there will be any commercial value to the hole that will be produced." Daniel Lisbourne of the Point Hope Village Council suggested that, on the basis of remoteness, the coast between Point Barrow and Barter Island was more satisfactory. Ball, without explaining the problem, retorted that the climate there was too severe.[61]

As part of their presentation, the officials showed an eleven-minute film accompanied by a technical narration that was beyond the understanding of most of the audience in question—many of whom were not highly proficient in English—but that would also have been incomprehensible to anyone who was not a nuclear physicist. In response, a villager told the officials: "We really don't want to see the Cape Thompson blasted because it our home, homeland. I'm pretty much sure you don't like to see your home blasted by some other people who don't live in your place." Rod Southwick, who was publicity chief at the AEC's San Francisco office, tried to reassure the villagers by promising that "if we cannot find a time to do it safely we won't do it."[62] He added that cattle grazing in the vicinity of the Nevada Test Site had been unaffected by experiments there. He also pledged that, in the event of any property damage—the type he had in mind was broken windows and cracked plaster—the AEC would provide financial compensation, as it had done in Nevada.[63] Finally, Southwick informed the villagers that they would be invited to watch, as the people of Nevada had been in the past.[64] In June 1960, replying to the petition sent to them by the villagers the previous November, the AEC emphasized that no decision had been made to conduct the experiment and that Chariot would ultimately require presidential approval.[65]

In August the Wilderness Society registered the major conservationist protest of 1960. At its annual meeting, the organization's governing council passed a resolution that expressed anxiety over the threat to Alaska Natives and the blast's impact on the wilderness qualities of the De Long Mountains, which formed the most westerly extension of the Brooks Range. The society described this area, of which it considered Cape Thompson an integral part, as "our last great wilderness area in the world not in the tropical region."[66]

The AEC's Committee on Environmental Studies for Project Chariot issued its first summary report in December 1960. It conceded the existence of "certain bioenvironmental problems," notably the importance of springtime hunting by Natives on land and sea ice in the blast area, and admitted that "in some areas the data were incomplete." Nevertheless, it repeated the assurance that radiation effects would be "negligible, undetectable, or possibly nonexistent in the environment beyond the throw-out boundary."[67] Even before the report appeared, the *Fairbanks Daily News-Miner* hailed it as the best scientific analysis of any area ever undertaken. The newspaper ran an Associated Press story claiming that every conceivable aspect of the region and the behavior of its inhabitants had been painstakingly researched and faithfully recorded: "If an Eskimo takes a notion to go hunting, the investigators note and record where he went, what he bagged and when he returned."[68]

Others were not so approving. The report infuriated the project's critics, not least of whom were certain of its contracted university personnel. These scientists from the University of Alaska had taken the AEC contracts, as Leslie Viereck explained to Lois Crisler, "with the idea that we could do more to oppose the project in this way than if we sat on the sidelines and criticized from that position." Since then, however, it had become apparent to them "that the AEC is using us to their advantage. . . . The time has now arrived when we . . . must do something very drastic in nature and the question now is what would be most effective."[69] Nine days after the Committee on Environmental Studies issued its report, Viereck announced that he would resign from the project at the end of his current contract (July 1961). In his resignation letter to the University of Alaska's president, William Wood, he claimed that the AEC's manipulation of science for political ends was intolerable. He felt he could no longer maintain his "personal and scientific integrity and continue to work for the AEC project."[70] He cited the agency's conclusion that spring was the best time for the blast before any studies had begun, its suppression of contrary evidence, and public denial of the existence of any dissent among its researchers.[71]

Don Foote, the human geographer, had also arrived at a solid critical position. Foote explained that he was politically naive and had "absolutely no idea what Chariot was" when he joined the bioenvironmental studies team in May 1959.[72] In November 1960 he defiantly told a member of the Committee on Environmental Studies that he had launched a personal crusade to "lift the lid off Chariot," and expose a project "rotten to its very bottom."[73] Four months later, Foote explained to John Wolfe, the chief of the environmental studies committee, that his dissent emerged gradually as his "politically immature mind awakened to the modern interplay of politics and science."[74]

State and national conservation organizations intensified their opposition to Chariot in 1961. The recently founded Alaska Conservation Society, of which Viereck, Pruitt, Swartz, and Johnson were founding members, devoted its spring newsletter entirely to the controversy and printed far more than its regular mailing. Slowly, the Alaskan opposition's case was spreading. *Outdoor Life, National Wildlands News, Defenders of Wildlife News,* and the *Sierra Club Bulletin* published more or less critical articles or featured news items on the proposed experiment.[75] In June the Committee for Nuclear Information (CNI) released a major report, based in part on studies by dissenting Alaskan members of the AEC's Chariot research team. The contributions emphasized the dangers of radioactive contamination in the arctic food chain.[76] Barry Commoner, professor of plant physiology at Washington University, St. Louis, and one of CNI's founders, explained in detail the nature of the dust-lichen-caribou-man link. He cited evidence that as a result of this flow of energy, levels of strontium-90 in the bones of Native Alaskans in the Arctic were much higher than in those of other Americans, despite ground levels of Sr-90 that were actually lower in northern Alaska than they were in other parts of the United States.[77] In another contribution, Michael W. Friedlander, a physicist, argued that food chain contamination by Sr-90 as a result of Chariot would probably be ten times higher than the AEC's estimates.[78] Commoner and Friedlander subscribed to the view that Sr-90 causes leukemia, bone cancer, and, eventually, mutation of the gene structure. The spirit of inquiry also extended to the *Fairbanks Daily News-Miner,* which featured a series by Tom Snapp in August that was more sceptical than anything that had previously appeared on Chariot in its pages.[79]

Opposition to Chariot grew stronger in 1961 and during 1962. In March 1961 the villagers sent a protest to President Kennedy. They denounced Chariot as being "too close to our homes at Point Hope and to our hunting and fishing areas." Many villagers had received CNI information through tape recordings. "We know about strontium 90," they explained, "how it might harm people if too much of it get in our body."[80] In November 1961 Native Alaskan arctic peoples gathered in Barrow, in large part as a direct response to the Chariot proposal. This *Inupiat Paitot* was the first such political conference for these groups. The delegates denounced the permit the BLM had issued to the AEC for the use of Cape Thompson as a violation of legally recognized aboriginal land rights and demanded its revocation.[81] A year later, the conference recommendation that Natives establish their own newspaper bore fruit. The first statewide Native newspaper in Alaska, the bimonthly *Tundra Times,* was founded (with assistance from Tom Snapp) by Howard Rock, an Eskimo artist who had been born and raised at Point Hope.[82] Chariot's critics achieved a major breakthrough in April 1962 when *Harper's* published a comprehensive summary of their case by Paul Brooks, a Sierra Club director, and Joseph Foote, Don Foote's brother.[83]

In August 1962, after spending $3 million on Chariot, the AEC announced its decision to defer a recommendation to the president that the project proceed. Though it gave no reasons, the AEC stressed that this was not because the bioenvironmental studies, resumed for a third and final year, had suggested the likelihood of any danger to Natives or the flora and fauna on which they depended for food, shelter, and clothing.[84] The AEC explained that the anticipated data on crater formation, radioactive distribution, seismic shock, and the effects of simultaneous multiple detonations were already available or soon would be from other sources.[85] (The United States had resumed underground nuclear testing in May 1960.) Nevertheless, the *Anchorage Daily Times* (24 August 1962) felt that the AEC had capitulated to pressure and interpreted cancellation as "a victory over atomic science by Alaskan Eskimos."[86] According to an aide to the secretary of the interior, the Department of the Interior's disapproval had been the crucial factor in Chariot's defeat. This aide, who described herself as Stewart Udall's "confidential assistant," explained that Interior had an influential role in assessing the AEC's bioenvironmental studies. She told Don Foote: "Those involved with the project at the AEC and at Livermore knew that Interior was prepared to exercize severe judgment in the matter of Project Chariot. Rather than have us issue a counter-report to the President which we would have publicized, and which would have been most embarrassing to the

AEC, they withdrew early in the fray, and in the most face-saving man- ner."[87]

However, in the fall of 1962 there was no euphoria among the project's Alaskan critics, who remained suspicious of the AEC's plans as long as the land withdrawal remained in force.[88] The assistant secretary of the interior, John A. Carver, who had a particular interest in Native American affairs, also wanted the status of the withdrawal amended to specify that the AEC's sole purpose was biological research.[89] In January 1963 the AEC informed the Department of the Interior that its land needs at Cape Thompson could be reduced from one million acres to 96,000 acres. This placed the site on "caretaker status," deemed suitable for continuing scientific studies.[90] In April control over the AEC's research station at Ogotoruk Creek transferred to the Navy's Arctic Research Laboratory at Point Barrow. On 29 May 1963, under pressure from an increasingly impatient Department of the Interior, the AEC recalled in entirety its application for the remaining 96,000 acres. The land was restored to the public domain on 2 September 1963.

An event that received more attention that fall was the Senate ratifica- tion of the Partial Test Ban Treaty. Concluded in the summer between the Soviet Union, the United States, and Great Britain, it prohibited nuclear testing in the atmosphere, the oceans, and outer space. France and China, the other nuclear powers, refused to abide by the treaty and underground testing by its signatories continued at an even faster pace. The increasing likelihood of a test ban treaty—whether comprehen- sive or partial—that would constrain nonmilitary tests whose effects could extend beyond the United States, probably played a part in the AEC's decision not to press ahead with Chariot.

Many aspects of the Chariot controversy, notably its role in the awak- ening of Native political power and the contribution of Native protest to Chariot's demise, lie beyond the scope of this study. Within the frame- work of technological challenges to the natural environment and the evolution of an environmentalist consciousness locally and nationally, the main concern here is the relationship to earlier controversies and later developments. The Project Chariot story is a landmark in three major respects; in connection with the origins of environmentalism; in the context of the emergence of the now ubiquitous environmental impact statement; and as part of the evolution of a more appreciative American attitude towards the Arctic as a natural environment.

The late 1950s and early 1960s were a critical transitional period in American conservation history and the Chariot debate highlights the relationship between elements of continuity and change. The controversy over Chariot should be contrasted with the debate ten years later over the Trans-Alaska Pipeline (TAPS) proposal, when a project with powerful momentum and solid private, state, and federal backing was confronted by a wide-ranging coalition of conservationists and environmentalists. There were two major differences between the two debates. Firstly, no amount of promotion could disguise the original Chariot proposal's inherent lack of economic appeal to Alaskans. Moreover, the Lawrence Radiation Laboratory and the AEC had achieved no greater success convincing other federal agencies of the project's value in either its initial or modified forms. Senator Bartlett's comment (1959) on the demise of the harbor project was just as valid for the cancellation of the experimental plan. Chariot, he declared in his senatorial newsletter, "had failed for want of horses to pull it."[91] Secondly, it would be inaccurate to characterize the protest against Chariot as part of a "movement" in the sense that opposition to TAPS was an integral part of the environmental movement—arguably its *cause célèbre*. The leading historian of conservation in the Progressive Era has defined a movement as a popular force deriving from a broad social base.[92] This was clearly not the case with the opposition to Chariot. The anti-Chariot campaign was of modest size, fought principally by a cadre of committed Alaskans united by professional and personal ties within the Alaska Conservation Society and supported by the Committee for Nuclear Information and a scattered assortment of aroused individuals.

Many members of national conservation organizations were slow to respond to the call-to-arms from their Alaskan compatriots. Members of the Alaska Conservation Society on publicity missions to the lower forty-eight often despaired at the lowly state of awareness they found. "Chariot is an unknown," lamented Celia Hunter, one of the organization's founding members, toward the end of 1962.[93] Sportsmen tended to leave the issue alone and for those conservationists who did become involved, Chariot was an atypical cause. Some individuals felt strongly because they were directly acquainted with Cape Thompson or other parts of the far north. In the case of James and George Marshall, the two brothers of the late Bob Marshall, the Brooks Range adventurer of the 1930s, indentification with the Alaskan Arctic and concern for its fate was vicarious. These officials of the Wilderness Society, which their brother had founded, had never been to Alaska.[94] These alarmed conservationists encountered considerable apathy and resistance in their efforts to recruit their colleagues.[95] CNI and the Association on American Indian Affairs (AAIA) gave these conservationist critics the most support. Neither organization became involved in the struggle against Chariot for traditional preservationist reasons; namely, protection of scenic beauty, wilderness, and wildlife.

This relative lack of conservationist support is not entirely surprising. Even to its first Alaskan critics, the blast did not immediately commend itself as a conservation issue.[96] Nuclear matters, whether military or energy related, were not a major item on the agenda of conservation during the early Cold War years. If anything, some conservationists in the early 1960s were impressed by the potential of nuclear power as a clean, cheap, and renewable energy source that offered an attractive alternative to damning wild rivers and flooding wilderness to provide hydropower.

The Chariot controversy was not another familiar debate about protecting wilderness and attractive scenery. In 1960, Leslie Viereck had observed that wilderness qualities in the vicinity of Ogotoruk Creek had already been ruined by the construction of over forty buildings, including power and radio stations and two 2,200-foot airstrips, not to mention the extensive damage to the "delicate" tundra by tracked personnel carriers ("weasels").[97] However, there was little discussion of the nuclear experiment as a threat to a prized scenic and recreational resource; spectacular landscape features and dramatic topography were conspicuously absent at Cape Thompson, which conservationists had never proposed for protected status of any sort. Many conservationists questioned Project Chariot's merits and legitimacy as a conservationist issue. Allied to this question of suitability was the matter of competence. "We must be careful not to get into genetic and other fields we are not expert in," Richard Leonard, the secretary of the Sierra Club, warned its executive director, David Brower.[98] Leonard was evidently more comfortable with the customary pursuit of wilderness and scenic protection policies, for which be believed the club was better equipped.

Those conservationists who did become critics of Chariot were as concerned with how the proposal jeopardized the Native way of life as they were with the implications for the land and its fauna. Rarely did critics refer to one threat without discussing the other. This concern with the quality of the human environment in addition to the ecological health of the natural environment, and the emphasis on the inseparability of humankind from the rest of nature, gives the Chariot controversy much of its significance. The Chariot debate provides one of the earliest examples of the combination of the conventional nature protection concerns of conservation and preservationism with the larger and more complex issues engendered by mid-twentieth-century science, technology, and industrial culture. As the 1960s advanced, these latter concerns—especially the impact of pollution on people and the rest of nature, whether by pesticides, chemical fertilizers, crude oil, or radioactivity—became identifiable collectively as environmentalism. Only a few historians have recognized the importance of public concern over radioactive fallout as a factor in the onset of the "age of ecology."[99] Thomas R. Dunlap and Ralph H. Lutts argue that the debate in the 1950s over fallout played a special role in preparing Americans for Rachel Carson's environmental message and her revelations about pesticides in *Silent Spring* (1962) and helps to explain the book's enthusiastic reception.[100] Some historians find it convenient to date the beginning of the environmentalist era at the publication of *Silent Spring*.

By illustrating how traditional ways of thinking and established organizations first came to cope with some of these new issues, the Chariot debate sheds light on the roots of environmentalism and was a harbinger of what the media dubbed the "age of ecology" in late 1969 and early 1970. Barry Commoner of CNI was, for many, the prophet and leader of the environmental movement. Commoner, a pioneering advocate of "open," "restrained," and "socially accountable" science and technology, remembers Project Chariot as his personal introduction to ecology and the beginning of his career as an environmentalist. Referring to his newfound concern with ecological matters, he recalled recently, "It is absolutely certain that it began when I went to the library to look up the behavior of lichen in connection with the Chariot program."[101] Lichens, with their unique botanical and biological properties and striking connection with human life through the distinctive arctic food chain, provided tangible evidence of the interrelatedness of all forms of planetary life within a single community of air, water, soil, and other organisms.

The bioenvironmental studies for Chariot were a further indication of the controversy's seminal role. At the end of 1960, despite his criticisms and resignation decision, Viereck had already conceded the ground-breaking character of the study as a whole, which he praised as "one of the best ever conducted in an Arctic region."[102] This was the first time the AEC had conducted an ecological investigation of any significance prior to a detonation, and it had carried out over 150 tests between 1948 and 1958. (Limited studies of the impact on fisheries had been carried out prior to the U.S. Navy's Bikini atoll tests in the South Pacific in 1946, the first atomic explosions in peacetime.) Part of the explanation lies in Chariot's status as the first unclassified American nuclear experiment. After all, the bioenvironmental studies chief, John Wolfe, had described it as "the first opportunity to do a good biological study prior to a nuclear explosion and it may be our last."[103] The environmental research became a *de facto*, integrated environmental, biological, and social study of the region. *Outdoor Life* praised what it called the "census" as "the best overall fact-finding job that any of our government agencies has ever done before starting work on a large-scale project involving our natural resources."[104] John C. Reed, a scientist with the Arctic Institute of North America, felt that the Chariot studies could become a model for "coordinated investigations of the environments of other areas."[105] It was certainly the first environmental study to precede a proposed engineering project in Alaskan history.[106] In one sense, the final report (1966) was essentially a catalog of life forms designed to serve as baseline data for measuring the biological costs of a nuclear explosion. However, it might also be seen as a forerunner of the formal environmental impact statement that became mandatory with the passage of the National Environmental Policy Act (NEPA) in 1969 for any project involving the federal government and the use of public lands. Henceforth, all Plowshare proposals became subject to NEPA's environmental impact statement requirement.

In its official comments, the Committee on Environmental Studies for Project Chariot criticized the image and perception of the Arctic as a place that no human activity or technology could destroy—and where there was really nothing to destroy anyway—which the AEC and its local allies had exploited in their efforts to advance the project. The committee noted: "Not infrequently it is described as remote, desolate, barren, and climatically rigorous. Probably none of these adjectives is accurate, and very possibly they are misleading. . . . It is not remote to the Eskimo, the arctic fox, or the ptarmigan; the flowering plants . . . belie its barren-ness and desolation." The committee described Cape Thompson as a "pristine" area in a state of "dynamic equilibrium."[107] Conservationist critics in the private sector also saw Cape Thompson as part of a thriving ecosystem where "the caribou range at will, as did once the buffalo on the Great Plains."[108] These were early expressions of the belief that the Arctic is a complex, fragile, and still comparatively intact ecological whole, reminiscent in this respect to much of the West before the coming of the railroads.

Don Foote, Chariot's most persistent critic, was particularly sensitive to the global significance of what he saw as an increasingly embattled far north and the region's meaning for future generations: "This area [Cape Thompson]," he wrote to his brother Joe, who lived on the East Coast, at the end of 1961, "as the entire Arctic, is on the verge of a tremendous transition. The Arctic is important not only to the people who now live in it but to the people who are your next door neighbors." He did not assign value to the Arctic as a storehouse of mineral resources or as an outdoor laboratory for nuclear physicists. Foote wanted the Arctic left alone because it exemplified "a sane and peaceful world where people love and respect beauty, cherish life and the natural world." Moreover, he saw a further threat in the Arctic—petroleum development: "The new explorers up there are after oil. . . . we must prevent the destruction of the Arctic as wilderness."[109] James Marshall once mused wryly that Edward Teller was "gambling on creating more" wilderness in Alaska than already existed there, in the sense that wilderness was historically synonymous with wasteland. In Marshall's view Cape

Thompson was not aesthetically, biologically, and ecologically redundant. The wasteland would be created *by* the nuclear experiment.[110]

The AEC continued to present what some conservationists perceived as fresh threats in Alaska. Reflecting on Chariot's cancellation, Senator Gruening remarked, "If they wanted to blow a hole in the ground they should have picked an uninhabited island where there would be no possible danger to anyone."[111] Its cancellation of Chariot did not mean that the AEC had lost interest in Alaska as fertile ground for nuclear experimentation. As if taking Gruening's advice, it shifted its attention to Amchitka Island.[112] Amchitka, 1,340 miles southwest of Anchorage and 270 miles west of the nearest Native settlement (Atka, a village of eighty-eight Aleuts), is part of the Aleutian Islands National Wildlife Refuge. The executive order creating the wildlife refuge in 1913 contained a proviso that this designation would not interfere with the use of these islands for military purposes. On Amchitka, despite rising local, national and international indignation, the AEC conducted underground tests in 1965 (Long Shot), 1969 (Milrow), and 1971 (Cannikin).

These tests, conducted in conjunction with the Department of Defense, were not part of Project Plowshare, which was terminated in 1973. According to the AEC, Long Shot (80 kilotons) was designed to provide data on the differences between a nuclear blast and an earthquake or other natural shock. Improving the American capability to detect underground tests by the Soviet Union and China, the AEC claimed, would help to enforce the test ban treaty of 1963. The AEC described Milrow (one megaton, under the direction of the Los Alamos Scientific Laboratory) as a 'calibration' test designed to see if larger tests were possible. Critics claimed that its purpose was to test Spartan missile warheads. At 5.2 megatons, Cannikin was the largest underground blast in American history. Partially classified, it was justified on national security grounds (testing of Spartan antiballistic missiles). Cannikin illustrated how nuclear testing in Alaska continued to play a catalytic role in the history of conservation. Concern that the blast would trigger a huge tidal wave (tsunami) and seismic activity, and result in large scale radioactive contamination of the marine ecology led to the formation of Greenpeace, the direct action international environmental organization, in Vancouver, British Columbia. A clique of activists, which included some of the American founders of the recently established, Vancouver-based Sierra Club of British Columbia, attempted to enter the test zone in a converted fishing boat. After detonating Cannikin, the AEC announced that it had no plans for further tests in the Aleutians. In 1973, the agency "evacuated" Amchitka.

The Lawrence Livermore National Laboratory, the former Lawrence Radiation Laboratory, directed Cannikin, for which the Batelle Memorial

Institute's Columbus (Ohio) Laboratories organized bioenvironmental studies in 1967 that were modeled on the Chariot program. These studies, published in 1977, concluded that the ecological impact of the testing on Amchitka was slight and that any adverse effects had already been reversed.[113]

Recently, the attention of an influential member of the American nuclear establishment has shifted back to the Alaskan Arctic. In 1987 Edward Teller returned to Alaska to promote the Strategic Defense Initiative (SDA) of which he is a chief architect and leading proponent. Addressing Commonwealth North, an Alaskan economic development lobby in Anchorage, Teller argued that Alaska would make an excellent site for launching laser defenses against Soviet missiles. He gave three reasons: Alaska's location near the likeliest route of any missile attack, over the North Pole; its abundant energy resources, which could fuel laser technology; and the low temperatures of the Arctic, which would aid superconductivity.[114] Alaskans, he proclaimed in tones reminiscent of the late 1950s, are "reasonable" people open to new ideas and sympathetic to the scientific duty and national security imperative "to find out what kinds of weapons are possible."

A question from a member of the audience who had heard Teller speak about digging a harbor "in the shape of a polar bear" at the University of Alaska in 1959 led Teller to reflect on his involvement with Project Chariot. He restated his faith in the potential for peaceful uses of nuclear explosives and recalled Chariot as a fine engineering project, "It should have been done." He offered an explanation for its demise that was entirely at odds with the official explanation the AEC gave in 1962 and in its final report on the project (1966). In a concession to the project's conservationist critics, he explained that "the environmentalists" stopped the project "with no good reason."[115]

5
Rampart Dam, 1959–1967

Despite the advent of commercial oil exploration in the Arctic circa 1960, many Alaskans still identified hydroelectric power as their outstanding unexploited natural resource. The Flood Control Acts of 1948 and 1950 had directed the Alaska Branch of the United States Army Corps of Engineers to investigate the potential uses of Alaskan rivers and harbors, with a view to "improvements," including navigation and hydroelectricity as well as flood control. The Department of the Interior in 1948 had begun to fund its own dam-building agency, the Bureau of Reclamation, to survey and report on Alaskan water resources.[1] The Bureau of Reclamation, the Corps of Engineers' direct competitor, identified two hundred potential hydroelectric sites in a report that amounted to a distinctly boosterist general discussion of the state of Alaskan development and why economic progress was so slow.[2] In 1950 the bureau built a dam at Eklutna to serve the booming city of Anchorage, Palmer in the Matanuska Valley, the army base at Fort Richardson, and Elmendorf Air Force Base.

In 1954 the Corps of Engineers started work on plans for hydroelectric developments in the Yukon and Kuskokwim basins. It selected nine potential sites along the Yukon, the last great American river that flowed free from source to mouth.[3] One site appealed to the army engineers in particular. From its headwaters near tidewater in British Columbia, the Yukon flows northwest. Four hundred miles beyond the international border, it reaches Rampart Canyon, a 100-mile gorge just downstream from the massive Yukon Flats of east-central Alaska. The head of this gorge forms what an engineer or a reclamationist would see as a natural damsite. The corps proposed to build a dam there with twice the capacity of Grand Coulee on the Columbia, the nation's biggest.[4] The project would, over twenty years, flood the Yukon Flats, an area 200 miles long and between 40 and 90 miles wide, creating a reservoir with a greater surface area than Lake Erie.

The Corps of Engineers also had plans for Bradley Lake on the Kenai Peninsula, while the Bureau of Reclamation was promoting projects at Snettisham, near Juneau, and Devil Canyon on the upper Susitna. Sena-

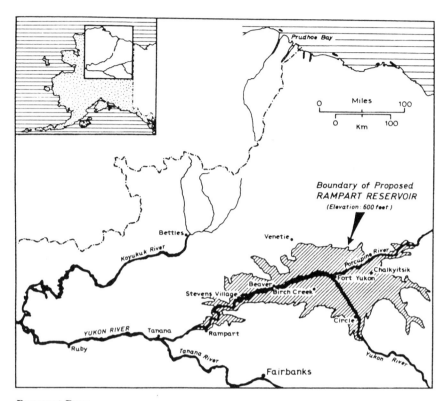

Rampart Dam

tors Bartlett and Gruening also wanted to see the Eklutna dam enlarged.[5] Each proposal received some support from a section of the community of Alaskan hydro advocates. Rampart was the particular favorite of Gruening, who adopted it as a major personal political priority. His early involvement was evident in his efforts to secure funding for the corps to undertake the first feasibility studies. In 1960 the Senate Public Works Appropriations Subcommittee, which Gruening chaired, allocated an initial $49,000 to this end.[6] That same year he presided over Senate subcommittee hearings held in Alaska to discuss the state's hydro power needs.[7]

In Anchorage and Juneau respectively, Bradley Lake and Snettisham were the most popular projects. The Bureau of Reclamation backed Devil Canyon as being best-suited to Alaskan needs. In Fairbanks (population 13,000), the city closest to the Yukon, Rampart received more attention than the other proposals. The representative of the local chamber of commerce claimed that the power produced—seventeen times the amount then generated in Alaska from all sources—would facilitate large-scale industrialization.[8] Ivan Bloch of North Pacific Consultants, a leading hydroelectric power proponent in the Northwest for over twenty years, argued that the "cardinal principle" of Alaskan hydro development should be "power availability in advance of actual need," and projected massive growth in power demand both in-state and nation-wide.[9]

In his résumé of the hearings, Gruening acknowledged the value of dams at Bradley Lake and Snettisham and recommended their construction as interim measures. However, in accordance with the principle Bloch had enunciated, he dismissed their importance as substantial long-term contributions. He rejected Devil Canyon outright for being too modest for the interior. He asserted that Rampart alone could do justice to the region's true potential and fulfill the future needs, not only of the state, but also of the nation.[10] In addition, he claimed (without citing evidence in support) that the enormous impoundment would enhance the agrarian potential of this part of the sub-Arctic by raising the local temperature.[11] Gruening's main official recommendation was the intensification of efforts to win approval for Rampart, which became the only project to which Alaskan hydro advocates gave serious attention.[12]

The 1960 Rivers and Harbors Act included a $2 million appropriation for a four-year feasibility study by the Corps of Engineers of all aspects of the project, including its impact on fish and wildlife.[13] In February 1961 the corps created the Rampart Economic Advisory Board (REAB), whose members included Irene Ryan, a mining engineer and state senator, William Wood, the president of the University of Alaska, and Stanley McCutcheon, a former territorial legislator. In April the corps hired the Development and Resources Corporation (DRC) to undertake the economic and market aspects of the study. This corporation was led by David Lilienthal and other veterans of the Tennessee Valley Authority. It became the advisory board's main task to review the progress of the DRC study. The reservations of the U.S. Fish and Wildlife Service (USFWS) emerged at the first of three conferences held by REAB.[14] The DRC report, issued in April 1962, endorsed the view that hydroelectric power in general, and Rampart in particular, offered Alaska the best basis for a secure economy.[15]

Stressing the Scandinavian experience, the consultants claimed that the power would attract light metals and alloy industries to the region and help process local iron, copper, lead, and zinc.[16] They also anticipated the development of a wood pulp industry at the damsite, a cement industry, and the stimulation of local agriculture.[17] Japan was pinpointed as a prime export outlet for these products.[18] The authors estimated that these various enterprises would directly create 19,746 jobs, exclusive of the number of workers who would be employed during construction.[19] The Alaska Railroad would be extended to the damsite and a road would also be built. The report estimated that in 1990, when Rampart power production was expected to be at maximum capacity, the total American population would be 100 million more than its current level.[20] On this assumption, the DRC calculated that Rampart, far from being overambitious, would supply only 0.5 percent of the total national capacity required in thirty years.[21] The report won the unanimous approval of the REAB and the data it provided became the main source for advocacy as Rampart steadily gained the support of much of the Alaskan political community and economic establishment.

Much early Rampart boosting echoed contemporary arguments for grandiose Alaskan developments (like Chariot) that emphasized Alaska's role as a beckoning frontier. The Development and Resources Corporation described Alaska as "practically the only remaining frontier." It claimed that by the end of the century Alaska would be called on to provide the service the frontier had rendered throughout American history because "people and industries will be looking for elbow room, for more land, for clean water and air." The DRC argued that the growth encouraged by Rampart would reflect the "traditional national urge to pioneer."[22] Like Chariot's supporters, Rampart advocates emphasized the appropriateness of a gigantic project to such a vast land. George Sundborg, Gruening's administrative assistant and chief Rampart aide, provided a direct connection with Grand Coulee, the last of the great river basin projects. Between 1944 and 1946, Sundborg had been an industrial analyst for Grand Coulee's builders, the Bonneville Power Administration. He was author of *Hail Columbia: The Thiry Year Struggle for Grand Coulee Dam* (1954). This account, which raised dam-building to an art, was replete with references to the "epic struggle" between man and nature and the inevitable bending of the latter to the former's "indomitable will." Sundborg compared the construction of dams such as Grand Coulee to the building of the pyramids in Egypt. Coulee, he declared, superseded the Pyramids of Gizah as the world's biggest man-made structure.[23] His brash optimism (similar to "puffing" on the nineteenth-century frontier)[24] typified the period. In 1959 Sundborg boasted that "Alaska, supplanting Texas as the biggest state . . . Now . . . comes to the fore with a proposal for construction of the biggest hydro electric project in the world—nearly three times as big as Grand Coulee, producing more energy than all of the dams of TVA combined."[25]

Rampart "pumpers" (to borrow a term from *Hail Columbia*) frequently rose to poetic heights. Gruening recruited Robert Service and enlisted frontier romance in his cause as Chariot's backers were also doing. He once quoted famous lines from "The Law of the Yukon:"

Wild and wide are my borders, stern as death is my sway;
From my ruthless throne I have ruled alone for a million
    years and a day;
Hugging my mighty treasure, waiting for man to come: . . .
and I wait for the men who will win me—and I will not
    be won in a day;

"So," Gruening commented, "the men for whom Robert Service's Yukon has waited a million years have come." It was true, he added, that the Yukon could not be "won in a day." But when it had been conquered by Rampart, he prophesied, "its mighty treasure will build 'cities leaping to stature' as it pours the tide of its 'riches in the eager lap of the world.'"[26]

This frontier-conscious mood was also reflected in the envious glances cast across the Bering Strait. Those who maintained that Rampart would rekindle the nation's flagging pioneer spirit did not find their contemporary inspiration to the south. In 1959 Gruening had visited the Soviet Union with fellow-members of the Senate Interior and Insular Affairs Committee to inspect Siberian hydroelectric schemes. His subsequent report praised the boldness that infused Soviet hydro policy and threw into relief the timid approach of the American federal government: "In contrast with Russian dreams which are now projects, the United States has not even begun to dream of the things which may be accomplished in Alaska—the area which was once Russian America."[27]

Campaigning for the presidency in Alaska in 1960, John F. Kennedy had played on the same irony. Mindful of Khrushchev's threat to "bury" the United States, he warned his audience that their Cold War competitor was already building a plant with a power output greater than that of Grand Coulee and Boulder (Hoover) dams combined. With a nod to Gruening and other Alaskans who emphasized the role of man-made as opposed to natural obstacles in thwarting Alaska's economic development, Kennedy concluded that "the tragic fact of the matter is that if Alaska still belonged to the Russians, Rampart Canyon Dam would be underway today."[28] During this visit, the future president's rhetoric pandered to the booster vision of Rampart's role. "I see another Alaska, the Alaska of the future," proclaimed Kennedy. "I see a land of over one million people. I see a giant electric grid stretching from Juneau to Anchorage and beyond. I see the greatest dam in the free world at Rampart Canyon producing twice the power of TVA to light homes and mills and cities and farms all over Alaska."

Entirely out of context, he quoted Franklin D. Roosevelt's assertion: "I, for one, do not believe that the era of the pioneer is at an end; I only believe that the area for pioneering has changed."[29] FDR had not been thinking, as Kennedy implied, about Alaska, or any other geographical frontier for that matter. In 1932 FDR had emphasized that "our last frontier has long since been reached . . . and there is practically no more free land."[30] What FDR had in mind was the "new" socioeconomic frontier that became the New Deal.

Kennedy was the only national politician to back the Rampart Dam. From its inception, the project also attracted a range of federal and private opponents, locally and nationally, who attacked it for economic, social, environmental, and other reasons. Though conservationists and Alaskan Natives figured prominently as critics, protest was not confined to these groups or to the project's implications for fish, ducks, and indigenous peoples.

The most formidable initial obstacle to Rampart's progress, ironically, was the nation's leading dam advocate, Floyd E. Dominy. The commissioner of the Bureau of Reclamation had registered the strongest objection to Rampart throughout the hearings held in Alaska in 1960. He was not unsympathetic to Alaskan power needs and acknowledged that the only existing Alaskan hydro project, at Eklutna, was inadequate. However, he questioned Rampart's logic until its advocates could prove that a market existed for the power.[31] He favored his agency's more modest Devil Canyon proposition.[32] According to Dominy, this project promised to yield a more useful quantity of power, 580,000 kilowatts as opposed to Rampart's 4,77 million, and much sooner—between 7 and 9 years rather than between 25 and 30. Moreover, this would be produced in a more accessible location; 14.5 miles east of the Alaska Railroad, at a point equidistant between Fairbanks and Anchorage— rather than 85 miles south of the Arctic Circle. Devil Canyon's cost was also more reasonable; $498 million rather than over $1 billion. It would not interfere with navigation or existing rights-of-way. Nor would it endanger any salmon run.[33]

With reference to Bloch's "cardinal principle," Dominy denied it was possible or desirable to force the pace of Alaskan development. He drew an imaginative analogy between contemporary Alaska and the early nineteenth-century West: "I consider that to think about building Rampart now, at this stage of the Alaskan economy, would be similar to have thought about building Grand Coulee in the Pacific North West when the Oregon Trail was still being traveled by covered wagons and the population was coming from east of the Missouri River. . . ."[34] The U.S. Fish and Wildlife Service also recommended Devil Canyon as a more economically viable and environmentally responsible alternative to Rampart in May 1960. This recommendation was incorporated into the Bureau of Reclamation's official report a year later.[35]

These federal critics were joined by private Alaskan groups. The Alaska Conservation Society (ACS) gave Devil Canyon its blessing soon afterwards. Environmentally, the Alaska Conservation Society judged the project essentially nondestructive as it would inundate only 68,550 acres instead of 6.946 million acres. In addition, Devil Canyon was expected to provide genuine recreational opportunities. Above all, the area was uninhabited by humans.[36]

The *Anchorage Daily News* also felt this was a more prudent proposal for the time.[37] However, the Devil Canyon project seemed to Alaskan boosters (like Joe Vogler) a fainthearted one that lacked vision.[38] Vogler denounced the attitude of Rampart's critics as a renunciation of faith in modern man, which was especially galling since the space program was making rapid progress. "I cannot believe," he wrote to William Wood, "that a civilization which is capable of sending a battery of TV cameras to the moon . . . is incapable of dealing with means for protecting and preserving ducks."[39]

Concern among organized sportsmen also revealed itself first at a local level, and at much the same time. In early 1961 an Alaska Sportsmen's Council (ASC) resolution criticized the Corps of Engineers for reducing its funding for fish and wildlife studies.[40] *Alaska Sportsman,* which enjoyed a wide circulation outside, took a formal stand against Rampart in April 1961. Emery F. Tobin, the founder, publisher and former editor, and its editor, Bob Henning, believed there were good dams, such as the Bureau of Reclamation's Glen Canyon on the Colorado, and useless dams, like Rampart.[41] Not one pro-Rampart letter appeared in the magazine's pages during the controversy. By contrast, readers from across the nation supplied it with a steady flow of protests. As well as conservationists such as Olaus Murie, critics included engineers, among them hydro experts, retired Yukon steamboat captains, and those who had suffered as a result of other Corps of Engineers projects like Garrison Dam in North Dakota and Oak Dam in Missouri.[42]

Among Alaskan state politicians, Jay S. Hammond (Republican, Naknek), the chairman of the state Senate Resources Committee, led a small group of critics. In February 1961 Hammond sponsored an amendment to a joint House and Senate resolution eulogizing Rampart that eliminated a lengthy piece of Cold War posturing and replaced it with a statement denying the existence of Alaskan unanimity regarding its benefits. The passage that offended Hammond read: "Whereas the construction of power facilities at Rampart would not only guarantee the economic development of Alaska and the comfort of its population, but [it] would act as a dynamic torch to the entire world to affirm the fact that the American democratic and free enterprise systems are the most vigorous and efficient forces existing in the world, and it would further provide other underdeveloped areas in the world with a living guide to their own progress."[43]

In the spring of 1962 the Corps of Engineers' exclusive control over Rampart was broken. For decades, the corps' rivalry with the Bureau of Reclamation had resulted in jealously, waste, and duplication of effort. In March 1962 the Department of the Interior and the U.S. Army signed an agreement to rectify this traditional situation by dividing responsibilities for water resource planning and development. This agreement (limited to Alaska and the Columbia and Missouri rivers) authorized the Bureau of Reclamation to carry out all feasibility studies, to operate projects and to market their power. The Corps of Engineers would be responsible for all design, engineering, and construction. As far as Rampart was concerned, the corps would finish the studies it had begun in 1960.[44] This reduced the significance of the Development and Resources Corporation's report (1962) since the crucial studies would now be those carried out by Interior, which began a full three-year study of all aspects of the project.

In the fall of 1962 Alaskan politicians campaigning for election to the state legislature supported the project virtually across the board. However, the project's chief advocates realized that the crucial battleground was national and were alive to the emerging strength of the opposition outside the state. Gruening had already warned a fellow-supporter that "one of the obstacles we will face is the opposition of a group of fanatical conservationists." He alleged that agents of the U.S. Fish and Wildlife Service were in the Yukon Flats bent on sabotage.[45] Moreover, advocates knew the project had no intrinsic or tangible appeal to the rest of the nation. As Mayor George Sharrock of Anchorage explained: "We will be asking the people of Massachusetts, New York, Montana and the other states to pay for a project that will directly benefit Alaska more than it will be benefit them. We cannot expect much sympathy from many senators and representatives in Congress on what Rampart can do for Alaska. We would be better off to use a different approach. It will be strictly a sales job."[46]

In the fall of 1963 Sharrock and his counterpart in Fairbanks organized a conference to form an organization to sell Rampart to Alaskans, the public in the lower forty-eight, and, not least, to Congress. The conference was attended by over a hundred prominent "Railbelt"[47] civic leaders, politicians, and businessmen.[48] Governor Egan emphasized: "We can only win if we are completely united, as we were for the statehood effort itself. All geographical areas of Alaska, all economic and other interests in Alaska, must speak with one voice." Egan alluded to the public and congressional hostility or indifference that had always confronted Alaskan efforts to secure federal assistance, which he attributed to distaste for large-scale federal intervention and the persistent grip of the icebox image.[49] Public hydro power advocates all over the United States anticipated roadblocks from private power interests, the coal industry, and Republican administrations. Conservationist opposition, however, was a relatively new proposition. Accordingly, Rampart's proponents spent much of the conference attacking conservationists, whose criticism of the dam's environmental impact was really the only form of protest that they acknowledged.

"Rampart has its enemies," warned George Sundborg, "waiting with a loaded shot gun and a red-hot mimeograph machine." He intended to expose a conspiracy against the proposal by "fanatical" conservationists, who placed the interests of ducks above human needs—a reference to the 1.5 million wildfowl of the Yukon Flats (according to USFWS figures). Adopting an unusual criterion to evaluate an environment and its culture, he told his amused listeners that the area contained "not more than ten flush toilets." Besides, since the Yukon Flats possessed no "scenic wonders," he saw no reason for conservationists to worry. He issued the following challenge: "Search the whole world over and it would be difficult to find an equivalent area with so little to be lost through flooding. . . . In fact, those who know it best say the kindest and best thing one could do for the place is put it under 400 feet of water."[50] But Sundborg saved the brunt of his criticism for the secretary of the interior. Most Alaskan politicians had been suspicious of Udall's sympathies since he had failed to revoke his predecessor's order creating ANWR. Udall's coolness towards the project had quashed their great expectations from the new administration, and Sundborg lambasted him for not attending the conference.[51] He also accused Udall of treating his job as "dealing primarily, if not exclusively, with parks and recreation."[52]

Sundborg's intimidating attitude to Udall was surpassed by Gruening's tirade again Ira N. Gabrielson, the director of the Wildlife Management Institute (WMI). The senator pinpointed the former director of the U.S. Fish and Wildlife Service (1941–46) as the source of the germ that had spread to infect the entire conservationist community. Gruening traced these troubles to the spring (1963) when Gabrielson had denounced the project as being "synonymous with resources destruction."[53] This was the start of Gruening's campaign of defamation against Gabrielson, which continued unabated for the rest of the decade. Gruening singled out the Wildlife Management Institute, the National Wildlife Federation, the Izaak Walton League, Ducks Unlimited, the National Audubon Society, Defenders of Wildlife, the Wildlife Society, and the Outdoor Writers

Association for special mention. He accused their members of infiltrating state fish and game commissions throughout the nation, the upper reaches of the Department of the Interior, and the University of Alaska's faculty.[54]

The promotional organization, Operation Rampart, which emerged from the conference, promulgated itself as the representative of the Alaskan public interest. "This is not a private lobby group," explained Ted Stevens, a trustee, "this is a group of public-spirited citizens who have donated their time."[55] In October the City of Anchorage and the Fairbanks Public Utilities Board both voted $10,000 to put Operation Rampart on its feet. This action did not meet with universal approval, even among pro-Rampart Alaskans. Robert Claus, president of the Fairbanks Businessman's Association, declared: "Taxing the people without a vote is just not good government practice. . . . I believe any such funds should be raised by public subscription. One hundred people at one hundred dollars and they would have it. From the calibre of people involved, there is no question that they're financially capable of donating the hundred dollars."[56]

Operation Rampart was renamed Yukon Power for America (YPA) in November, and C. W. Snedden was elected president. William Wood became vice-president-at-large and Irene Ryan assumed the post of executive secretary. All had served on the now defunct Rampart Economic Advisory Board, whose work was finished when it approved the DRC report in August 1962. Robert Atwood was appointed to the executive committee. Egan, Congressman Ralph Rivers, and Senators Gruening and Bartlett became ex-officio members. The first official membership list for YPA showed a total of sixty-four, predominantly from Anchorage and Fairbanks. Twenty served in an official capacity. Twenty-five represented businesses, utilities, and chambers of commerce. Nineteen were private individuals, of whom three resided out-of-state. YPA offered a special membership category for schoolchildren—"Beaverettes"—at 25 cents per annum.[57]

Yukon Power for America's motto was "Build Rampart Now," and, notwithstanding that Interior's studies were only half finished, its goal was to include the project in the Rivers and Harbors Omnibus Bill for 1964. (An omnibus bill must be passed or rejected without amendment.) Bartlett, who believed that the chances of success in any immediate sense were slim, was one of the few Alaskan politicians who preached the virtues of patience. Bartlett kept his distance throughout and did not appear at any major rally or event. Members of YPA often complained about his failure to play a more active role. As an ex-officio member of the lobby, he was occasionally required to express himself on Rampart. However, his infrequent public pronouncements were mostly perfunctory and cool. He warned Yukon Power for America of the strength of the opposition, small in Alaska but potent when allied to national groups and the federal government. "Prepare to dig in for the long pull," he advised.[58]

By the fall of 1963, a pattern of argument had been established from which neither side in the debate would depart over the next two years. The major feature of the controversy during these years was the adversaries' efforts to advance their causes at the Department of the Interior, in the nation at large, and among their opponents. Soon after the McKinley conference, George Sundborg presented the full case for Rampart to the Wilderness Society. Conservationists feared that the dam would block a salmon run of 270,000 that played an important role in the Native subsistence economy. Sundborg, however, confined himself to the observation that the run was of no commercial value. He argued that, in the unlikely event of any loss, introduced species would provide the basis for a large-scale commercial fishery. Turning to the subject of ducks, he questioned the accuracy of statistics provided by the USFWS. He claimed that the number produced annually on the Yukon Flats was closer to 500,000 than 1.5 million. The possible loss of 4,600 moose did not move Sundborg since moose were abundant elsewhere in Alaska. Moreover, he argued, they were an impediment to economic progress in areas like the Matanuska Valley. He did not think that the existing Native way of life on the Yukon Flats was worth preserving. He contended that resettlement and job opportunities created in construction, electro-process, pulp mills, and fisheries would improve the displaced residents' standard of living tremendously.[59]

Above all, Sundborg denied that Rampart's main purpose was to ease the young state off the fiscal rocks. He declared that "the unmatched contribution to the welfare of the entire U.S. which would be made by Rampart, makes any argument that this project is required only to solve financial problems of the state of Alaska absurd."[60] Nevertheless, the impression of the non-Alaskan press and public that any benefits from Rampart would be entirely local was unshakeable. Their view was confirmed by the Alaskan press's argument that Alaska was as deserving a candidate for "foreign aid" from the federal government as India or Yugoslavia.[61] Reporting on the McKinley conference, the *Washington Post* (5 October 1963) observed that "to Alaska, the Rampart dam has acquired the same mystical aura of promise as, for example, steel mills to India; it is a symbol of economic independence to a land very recently emerged from colonial status."

The 300-member Alaska Conservation Society (50 percent were "associate," i.e., non-Alaskan members) was Yukon Power for America's major local opponent. This group was the primary source of information for national conservationist critics and concentrated on the project's economic defects. The Federal Communications Commission's "fairness doctrine" gave the society an opportunity to present its case to a broader audience shortly after Mayor Brewington of Fairbanks appeared on television in Fairbanks to report on the McKinley conference. Quoting extensively from the Bureau of Reclamation's report on Devil Canyon (1961) and a recent study by Arthur D. Little for the State of Alaska,[62] Celia Hunter questioned the assumption that availability of large quantities of cheap power was in itself sufficient inducement for industry, especially electro-process, to locate in the far north. She argued that other factors, such as climate, remoteness, absence of raw materials, and high labor and living costs also needed to be taken into account. Hunter also criticized the failure to consider better sources of hydroelectric power, notably Devil Canyon, and the futility and inappropriateness of decision-making before Interior's studies were completed. Like Yukon Power for America, she emphasized the interests of humans, but, in this case, those of the region's Natives. She only referred to fish and wildlife indirectly, as the basis for a subsistence economy that the project threatened to disrupt.[63]

Rampart was winning some national support. The AFL-CIO endorsed it in December 1963. The American Public Power Association, the National Rural Electric Cooperative Association, the Inland Empire Waterways Association (covering Washington, Oregon, and Idaho), and the Pacific Northwest Public Power Association (Montana, Idaho, Washington, Oregon, British Columbia, and Alaska) followed suit. Egan informed President Johnson that construction and operation would yield an estimated 60–80,000 jobs and looked forward to a site city of 10,000 and a doubling of the state's population as a result of the growth the project would stimulate.[64] However, the cause failed to attract the president's attention, or that of other influential national interests beyond organized labor and the public power lobby.

Meanwhile, local and national hostility was rising. From early 1963 onward, letters from village councils and individuals in the affected area appeared in the *Alaska Conservation Society News Bulletin*.[65] Later that year, *Tundra Times* sent a questionnaire to all villages along the Yukon. The leading national sportsman's magazine, *Field and Stream*, announced its opposition that summer.[66] Growing numbers of sportsmen feared that inundation would destroy up to 2.4 million acres of prime duck breeding habitat. The shallow lakes, ponds, and sloughs—which are the distinctive feature of the Yukon Flats—are home to scaup, pintails, scoters, wigeons, mallards, shovelers, teal, and canvasback ducks. Canada geese, white-fronted geese, the rare trumpeter swan, loons, grebes, and sandhill cranes also frequent the area. Because of the displacement of birds from marshlands drained for farming further south in Canada's Prairie Provinces, the Yukon Flats had assumed increasing importance for waterfowl.

Sportsmen argued that the loss of nesting grounds in the Yukon Flats as a result of Rampart would mean a drastic cut in Alaska's contribution to the four international flyways. As Walter H. Pierce, president of the Tanana Valley Sportsmen's Association (TVSA, Fairbanks), explained to local members of Yukon Power for America, "the big obstacle in sportsmen's minds is: 'How do you replace ducks?'"[67] Gruening seized on such assertions of self-interest to discredit his opponents. He decried the type of conservation practiced by sportsmen who seemed interested in saving ducks in Alaska only for the opportunity to shoot them later in the lower forty-eight. Was there, he wondered, "no nobler and higher form" of conservation?[68] Warren Taylor, who represented Fairbanks in the state legislature, accused sportsmen of being "anti-people" because they opposed a measure vital to the economic well-being of Alaskans. The TVSA resented the charge that by opposing Rampart it was opposed to the interests of people. Pierce reminded Yukon Power for America that his interests were also anthropocentric: "It is a case of striking a balance between industry and recreation, both for people."[69]

In January 1964 a leading member of YPA introduced a bill in the Alaska state legislature which called for $12,000 to display and distribute promotional literature on Rampart as part of the Alaskan exhibit at the New York World Fair. The legislature voted the money. In response, the Alaska Conservation Society passed two resolutions on 11 February. The first called for equal funds to distribute information on the project's drawbacks and to investigate other power sources. The second requested that the legislature desist from any further dissemination of public funds to any private agency for promoting Rampart until all feasibility studies had been not only completed but submitted to Congress with a favorable recommendation from the secretary of the interior. That same month, eleven representatives sponsored a resolution complaining in similar vein. The most vociferous critic was still Jay Hammond, who remarked sardonically of his colleagues that "most of them would stand up and face Mecca whenever someone said Rampart."[70]

Ginny Wood of the Alaska Conservation Society had remarked of Rampart that in Alaska "to speak against it is like promoting sin or communism."[71] Gruening certainly strove zealously to quench all manifestations of apostasy. He and Sundborg scoured the press for any damaging signs of heresy, however trivial. It was rare for any criticism to stand without rebuttal—whether a *New York Times* editorial, a column in a state fish and game commission bulletin, a private letter, or an article in a national high school magazine. An example of the latter was contained in a weekly distributed widely throughout the public school system. As a result of the misinformation spread by this article, Gruening complained, his office, and those of many other congressmen, had been "flooded with letters from innocent schoolchildren who have been brainwashed."[72] YPA unsportingly insisted on allegiance or silence. For example, Ed Merdes objected to the anti-Rampart activities of an individual in the armed forces in Alaska who was a member of the Alaska Sportsmen's Council: "We believe," professed Merdes, "as a public official, he should remain neutral."[73]

In February 1964 the Bureau of Land Management (BLM) held hearings in Alaska to consider the U.S. Geological Survey's request (filed March 1963) for a 8.955-million-acre power site classification for Rampart, the largest ever applied for. This classification request was standard procedure in accordance with the Federal Power Act of 1920 and was designed to prevent land speculation. The BLM emphasized that testimony would be restricted to the question of classification.[74] Testimony in Anchorage and Fairbanks was overwhelmingly proclassification. In Anchorage, eighteen spoke in support and only two against. (Both were petroleum consultants concerned about the effects on oil leasing prospects.)

In Fairbanks, a minority—represented by the Alaska Conservation Society, Natives, and sportsmen—signified their disapproval. In the traditional parlance of Alaskan boosters, conservation measures were "lockups." On this occasion, sportsmen reversed the charge and criticized classification as single-use appropriation of the worst kind that would preclude all other forms of use and development. "Even in Alaska," explained Emery F. Tobin of *Alaska Sportsman*, "we are not so land 'rich' that we can afford to tie up an area larger than the area in several of the other 49 states."[75] The choice of *Alaska Sportsman* for a damsite in Alaska was the Taiya-Yukon in southeast Alaska, whose scale, the magazine argued, was more appropriate and would provide power close to Alaska's second major concentration of population. It also argued that the dam, near tidewater at Skagway, would be better placed to attract an aluminum industry that required imported bauxite.[76]

At Fort Yukon, on the other hand, only a representative of the USGS supported classification. Sixteen witnesses, mostly Native, denounced it as a threat to the local way of life.[77] Fort Yukon (population 700), which lay at the heart of the area to be inundated, was the largest Native village on the Yukon Flats. Trapping beaver, otter, muskrat, marten, mink, lynx, weasel, wolverine, fox, and wolf sustained both Natives and whites on the Yukon Flats, where an estimated 9 percent of Alaska's furbearers were located. Fort Yukon was the historic center of the trade. Local Native residents organized a citizens group, *Gwitchya Gwitchin Ginkye* (Yukon Flats People Speak), in preparation for the hearings.[78]

To counteract the growing evidence of Native discontent, Gruening paid a visit to the Yukon Flats in the summer of 1964, which confirmed his belief that the Athapaskans lived "miserably" and "eked out a meagre subsistence." He claimed that only one person he had met, an elderly Native, opposed the project.[79] Don Young, who has been Alaska's congressman (Republican) since 1973, was then a schoolteacher in Fort Yukon and president of the local council. Young launched his political career on an anti-Rampart platform (endorsed by Yukon Flats People Speak) when he ran for the state house in 1964 as a candidate for the local seat.[80] However, the majority of Alaskan politicians continued to support Rampart. In October YPA conducted a poll among candidates for the legislature, which posed two questions: "Do you favor the early construction of Rampart Dam?" and "Would you vote for funds to be used for promotion of Rampart Dam?" Of thirty-four politicians who responded, only four answered in the negative.[81]

In April 1964 the Rampart cause had been dealt a blow when the USFWS published the study it had undertaken as part of the Army-Interior agreement of March 1962. Its report declared that a habitat development program for waterfowl would compensate for only 20 percent of an estimated loss of 1.5 million ducks. It concluded that Rampart constituted "the greatest single threat to wildlife values of any project ever suggested in this country."[82] Warren Taylor, who was cochairman of YPA's finance committee, told Gruening that the report's authors "should be writing science fiction stories."[83] But in the report's wake the critical ranks were swelled by the Duck Hunters' Association of California, the International Wild Waterfowl Association, the Sport Fishing Institute, the Canadian Wildlife Federation, the Western Federation of Outdoor Clubs, the National Parks Association, the California and Utah state departments of fish and game, the California Cattlemen's Association, and the Conservation Federation of Missouri.

After the death of Olaus Murie in 1963, the most venerable national critic with strong Alaskan connections was Frank Dufresne, who condemned the proposal as "a monster." Dufresne quoted approvingly the comment of an old friend in Anchorage that "Alaska needs Rampart Dam like it needs another earthquake."[84] Another leading critic was George W. Rogers, the economics professor from the University of Alaska who was serving as a consultant on the economic aspects of the project for the Department of the Interior's feasibility studies. Rogers criticized Alaskans for thinking of development in terms of "promotion and gimmicks." He insisted that "Development based on knowledge is needed."[85]

By the end of 1964 the results of the *Tundra Times* survey were clear. The following Native villages, representing a total of 1200 people, which all stood to be inundated, had voted against Rampart: Birch Creek,

Beaver, Venetie, Rampart, Chalkyitsik, Stevens Village, and Fort Yukon. The downstream villages of Kokrines and Holy Cross and the delta settlements of Alakanuk and Emmonak were also opposed. One village was neutral, Nulato. Only one, Kaltag, declared itself in favor, and on condition that jobs were forthcoming.

Interior completed its field studies in January 1965. Its report did not judge the marketability of Rampart power but noted the likely impact on fish and wildlife and local Natives.[86] Udall appointed a six-member departmental task force to review its findings before making a final decision.[87] To convince Interior that a consensus impatient for action prevailed in Alaska, Rampart proponents denied the existence of any noteworthy dissent. Gruening compared the diehard Alaskan few who he admitted were opposed to those who had been against the construction of the Alaska Highway.[88] He dismissed the signatures on a petition he received in February from some Native residents of the Yukon Flats as forgeries.[89] The evidence in *Tundra Times* and the *Alaska Conservation Society News Bulletin* indicated to Gruening that Natives had been duped, like the Department of the Interior, by "fanatical" conservationists, namely, the *Tundra Times,* the Alaska Conservation Society, and the U.S. Fish and Wildlife Service.[90] He and Sundborg continued to misrepresent the attitude of some Alaskans. "It is very surprising to find any Alaskan resident opposed to the Rampart Dam proposal as most of the opposition has come from wholly uninformed persons living outside of our state," replied Sundborg to an Alaskan dissenter in 1965.[91]

A time-honored device favored by Alaskan boosters was to dismiss conservationist criticism as Outside interference. Long-term Alaskan residence was extolled as a mandatory qualification for the right to pronounce on Rampart or any other aspect of Alaskan affairs. However, YPA applied this principle selectively and paraded its own roster of outside experts; notably Gus Norwood, Ivan Bloch, and Anthony Netboy. At YPA's second annual meeting in Fairbanks on 20 June 1964, Netboy, a salmon biologist, denounced the incompetence and ignorance of Outside critics with particular virulence. Yet this was the Oregonian's first trip to Alaska.

According to Terris Moore, who had been president of the University of Alaska in the 1950s, Rampart was seen as "a threatening cloud" only by "zoologists, social scientists and wildlife enthusiasts—the latter groups a small minority in the state."[92] Sundborg estimated the ratio of Alaskans for and against Rampart variously as 99–1, 10–1, and 9–1. Nevertheless, an analysis of the correspondence Gruening received between 1963 and 1967 indicates that overall sentiment ran 3–1 against the dam and that even among his Alaskan correspondents, almost twice as many were opposed as in favor.[93]

The Canadian government's stance was already clear. Citing U.S. Fish and Wildlife Service figures, Arthur Laing, the Canadian minister of northern affairs and natural resources, expressed alarm over duck losses.[94] He told Gruening that "on that score alone the construction of Rampart would be inimical to Canada's interests."[95] Another reason for Canada's objections was the Treaty of Washington of 4 July 1871, which settled the boundary between Alaska and Canada. Article 26 stated: "The navigation of the Rivers Yukon, Porcupine, and Stikine, ascending and descending, from, to, and into the sea, shall forever remain free and open for the purposes of commerce to the subjects of Her Britannic Majesty and to the citizens of the United States."

In April 1965 Yukon Power for America and the Alaska Conservation Society (which then had 600 members) rehashed familiar arguments in a television debate in Fairbanks. Thomas K. Paskvan, a Fairbanks businessman who had recently become president of YPA, reiterated the case for the dam as the key to unlocking Alaska's natural resources. Then he turned to the subject of conservation. He accused his opponents of misusing the term, explaining that true conservation gave humans priority. He also questioned critics' understanding of natural history. He dismissed conservationist doomsaying as an insult to the intelligence of ducks as no bird was stupid enough to sit still and wait to be drowned. He explained that, like other creatures, they would shift to higher ground as the water gradually rose.

Whereas Paskvan concentrated on wildlife, Celia Hunter of the Alaska Conservation Society talked economics exclusively. She tried to build a case for the superior competitiveness of nuclear power plants, which could be located close to the demand for power.[96] YPA's second speaker, Ed Merdes, the organization's vice president, expressed astonishment that Hunter had ignored the ducks. He believed that this was the only issue which she, as a conservationist, should have addressed, besides being the only one she was qualified to discuss.[97]

An influential summary of the opposition's case was Paul Brooks's article, "The Plot to Drown Alaska," in the propreservationist *Atlantic Monthly* (May 1965).[98] Brooks denounced Rampart as speculative, an economically foolish boondoggle without recreational potential, ecologically disastrous, and an attack on the Native way of life. He preferred Devil Canyon and the nuclear alternative.[99] Characteristically, Gruening demanded, and was granted, an opportunity to reply in *Atlantic Monthly*. His retort, "The Plot to Strangle Alaska," restated a case that had not substantially changed since 1960. As ever, he maintained that the worthlessness of the Yukon Flats was an excellent justification for the project: "Scenically it is zero. In fact, it is one of the few really ugly areas in a land prodigal with sensational beauty."[100] Elsewhere, he rated its scenic and recreational potential among Alaska's lowest 10 percent.[101]

In related articles and speeches, Gruening endowed the Yukon Flats with the negative qualities historically associated with American wilderness and revived the icebox image. He called the region "nothing but a vast wasteland . . . notable chiefly for swarming clouds of mosquitos." Recently completed Glen Canyon Dam and Lake Powell, showpiece of the Bureau of Reclamation's Upper Colorado Basin Storage Project, provided inspiration for this theme. Gruening praised this project for having redeemed the "dreary" desert of Arizona. Citing Lake Powell as an example, he argued that Rampart's reservoir would provide a comparable range of popular recreational opportunities, such as picnicking and even water skiing.[102] To argue that damming improved upon nature was an old theme. Advocates of the Hetch Hetchy project had emphasized that a man-made lake would enhance the bleak Sierran canyon.

In 1965 and 1966, proponents redoubled their efforts to advertise the dam's environmental merits. Throughout the controversy, their pronouncements on Rampart's ecological effects had invariably been ill-fated and illustrate how Rampart conflicted with James Malin's principle of "natural regional adequacy." A widely held and oft-intoned fallacy was that the dam would enhance the local environment by creating a lake where there were no lakes. Grand Coulee was held up as an example of the ecological benefits of dams.[103] Privately, however, a leading purveyor of the analogy, Ivan Bloch, denied there was any real parallel between Coulee and Rampart.[104] Conservationists did not deny that Grand Coulee had created a habitat suitable for wildfowl in arid eastern Washington. But they objected to the transfer of these assumptions to the Yukon Flats, where literally thousands of small lakes already existed, providing an ideal environment for water birds. Further examples of ecological ignorance were Warren Taylor's celebrated quip in 1963, "who ever heard of a duck drowning?" and Sundborg's derisive comment, "Perhaps we should teach them to swim or even fly."[105] Boosters believed that twenty years, the time it would take the lake to fill, was long enough for wildlife to move on. This fantasy flew in the face of one of the cardinal principles of ecology—carrying capacity, which refers to the maximum wildlife population levels the available resources of a given area (food, shelter) can properly support and sustain.

Gruening exposed his poor grasp of natural science to Stewart Udall in the summer of 1965. Writing with evident irritation to inquire about the progress of the review team's assessment of the February report, he invoked the example of a constituent who had told him about the salubrious effect on wildlife of a dam he had built on his property. This dam had created a six-acre lake out of swamp.[106] The senator cited this as overwhelming proof that Rampart's benefits would be similar, only magnified. "Think of how much more Rampart could do!" he exclaimed.[107] However, as John Gottschalk, the director of the USFWS, explained, ducks spend half their time on land. Their favorite habitat is the shoreline, where they nest and feed in shallow water. The critical factor is not the quantity of water but the amount of "edge."[108] What Gruening's constituent had done was create a body of water approximating one of the 36,000 small lakes and ponds that characterized the Yukon Flats and provided an immeasurably larger amount of "edge" than one enormous lake. Ducks might not drown but they could starve.

In November 1965 the assessment of wildfowl losses in the 1964 USFWS report was judged "conservative" and the estimated mitigation possibilities "optimistic" by the National Academy of Sciences.[109] Undaunted by this setback, Gruening continued to insist that the pro-Rampart voice be heard on all occasions. A symposium entitled "Ecological Considerations of the Rampart Dam," sponsored by the Ecological Society of America, came to his attention.[110] Gruening demanded, without success, that Sundborg be included on the panel alongside distinguished wildlife biologists such as A. Starker Leopold, who would be representing the Natural Resources Council of America (NRCA). "It looks like a loaded affair," Gruening wrote to his aide, "and we should be represented."[111] On the other hand, Gruening always maintained that conservationists were not qualified to discuss the dam's economic aspects.

While the Rampart lobby waited impatiently for the decision of Udall's review team, its cause suffered further serious damage. In March 1966 a report sponsored by the Natural Resources Council of America undermined what remained of Rampart's severely eroded economic foundation. The Spurr Report's fundamental premise was Alaska's urgent need to develop power resources.[112] Its purpose was to locate the cheapest source of power for the Railbelt, the corridor along the Alaska Railroad where much of Alaska's population is located. The report's choices were natural gas from the Kenai Peninsula and Cook Inlet, coal from the Matanuska Valley, and hydroelectricity from Devil Canyon. It considered Rampart's cost prohibitive, the amount of power unrealistic (enough for 6 million whereas the state's population was only 253,000), and the prospect of exporting the surplus unlikely. The nearest market, the Pacific Northwest, was 2,000 miles away and already adequately supplied by dams like Grand Coulee. The Spurr report concluded that Rampart would be "the most expensive gamble ever suggested in hydroelectric development." The report also considered the detrimental impact on the Native population and denied that there was any recreational potential.[113] For the *New York Times* (11 April 1966) and the majority of critics, the Spurr report sealed the case against Rampart. Gruening resented its influence at the Department of the Interior and denounced it as "nothing but propaganda for a highly specialized point of view."[114]

Though Rampart's national fortunes continued to decline, the Alaska Legislature (with only minor objections) was still voting public funds to promote Rampart. In March 1966 it appropriated $50,000 by a margin of 31–7. But proponents were finding it increasingly difficult to ignore their poor prospects. Yukon Power for America attributed its lack of success to an imbalance of forces. At the lobby's annual meeting in April 1966, Austin J. Ward, who was chairman of the Fairbanks Chamber of Commerce's Rampart Power Committee and YPA's public information chairman, dismissed its image as "a powerful, well-heeled, industrial-backed monster" as a conservationist fabrication.[115] Snedden insisted that YPA was a "fledgling, volunteer-member organization, with one paid staffer and secretary," whose financial resources were no greater than "a few thousand dollars." He targeted the gun and ammunition manufacturers as the source of critics' inordinate influence in Washington, D.C., claiming that "the big money is riding on the other side."[116]

In June 1967 the Department of the Interior recommended against Rampart. Udall cited "the availability of favorable, less costly alternatives." He also explained that "the fish and wildlife losses which could result . . . are of such significance that if the other factors of concern . . . were favorable, there would be question as to the appropriateness of any recommendation."[117] He denied that there would be any recreational benefits but did not refer to any detrimental impact on the Native population of the Yukon Flats.

This decision surprised nobody. Gruening had told Udall in May that an adverse decision would not be unexpected since it had been "perfectly clear for many years that Alaskans could never expect an unbiased, objective report" from his department.[118] He summarized Udall's attitude as being, "we can't afford to antagonize the sportsmen."[119]

Interior brushed aside the inference that it had capitulated to pressure from sportsmen and other conservation "extremists." David S. Black, the Undersecretary of the Interior and vice-chairman of the Federal Power Commission, acknowledged that "the Department recommendation against it has received some criticism . . . as being short-sighted and a step intended to keep Alaska a resource storehouse and wilderness preserve." He explained that "cold analysis simply showed that Rampart is not the most economic means of providing the power needed now and in the near future to support Alaskan development. The vast expenditure required for Rampart's construction far in advance of the time its power could be utilized in Alaska would be a misplaced investment."

Black pointed to far more suitable hydro proposals, such as Devil Canyon, and the opportunity to produce electricity from Cook Inlet gas.[120] The *Washington Post* (20 June) agreed, describing Rampart as simply "beyond the fringe of economic rationality." Preservationist sympathies were not the nucleus of Interior's decision. From a strictly utilitarian conservationist standpoint, Rampart made no sense.

In 1968 the faltering cause was dealt two final blows. When Mike Gravel defeated Gruening in the Democratic primaries in the latter's bid for reelection to the Senate, the project's chief advocate lost standing and influence. More importantly, the discovery of oil at Prudhoe Bay extinguished whatever appeal the project retained for Alaskans. Some, notably Gruening, never accepted defeat. Undeterred, he sought to reassure the remaining faithful that "contrary to rumor and pessimistic reports the Rampart Project is very much alive."[121] C. W. Snedden was another diehard. Five months after the oil industry had applied for a permit to construct a pipeline across Alaska, he protested that the choice of route ignored the future location of the dam. He suggested rerouting the line because "to assume that Rampart Dam will never be built is a foolish, dangerous and short-sighted assumption."[122] However, in June 1971, the Corps of Engineers recommended conclusively that Rampart "not be undertaken at this time."[123] The corps explained that there was insufficient existing power demand to justify construction. The report also acknowledged that Rampart would not "enhance" fish and wildlife resources and that mitigation measures would be only "partially effective."[124] In the meantime, YPA had dwindled to a handful of diehards. Jack White, an Anchorage businessman and legislator who was then the group's vice president, still worried about the pipeline's location. He continued to hope that "maybe it could be re-routed."[125] By now, the lobby existed in little more than name.

The failure of Rampart can be adequately explained by its economic defects. Yet the opposition to it, even within Alaska, was not solely based on these, or even its strictly ecological consequences. To some immigrants to the north, the project symbolized everything they thought they had left behind by coming to live in Alaska. One refugee protested: "I went there to get away from dirty cities and the garish, neon-lit way of life in the states. I am not appreciative when our representatives work feverishly to industrialize a state that should remain substantially natural."[126] For those in flight from the rest of the United States, Alaska seemed to offer the return to primitive conditions that the American frontier had traditionally provided. These Alaskans also recognized that Rampart was intrinsically inconsistent with the dominant image of Alaska in the lower forty-eight. A resident of Seward tried to explain the point to Gruening: "The tourist dollar in the lower 48 is ready for harvesting in Alaska. And the Alaska the tourists wants to see, is Alaska, the Last Frontier, the land of fish and game, the last genuine wilderness in our country; they will not, dear Senator, come to see dams and artificial lakes."[127] The Tanana Valley Sportsmen's Association's resolution against the dam on 13 July 1965 likewise emphasized Alaska's supreme national significance as a recreational storehouse to supply the nation's expanding appetite for wild nature: "An industrialized Alaska of highly commercialized areas with plants that discharge air and water-polluting gases and wastes, and a landscape bound with a network of high tension power lines, will do permanent and irreparable damage to the primitive aura of Alaska's unblemished horizon, its healthful atmosphere, and natural waters—the things which tourists will travel half-way around the world to see."[128]

In spite of Gruening's claim that Rampart would enhance the environment, critics in the lower forty-eight let him know that it was fundamentally alien to their conception of Alaska. "I plan to go to Alaska some day. . . . But I won't be coming to see Rampart Dam," declared an Iowan.[129] Two New Yorkers who had just returned from a tour of Alaska assured him, "We would not have traveled a mile or spent a dollar to see a dam."[130] Others with a particularly romantic misconception of Alaska stressed the project's incompatibility with Alaska's frontier heritage and self-image. "I know Alaska has always been considered a state of rugged individualists," wrote one disillusioned Californian to William Wood, "not truck drivers looking for fat jobs."[131] Criticism from the lower forty-eight was dominated by the "last chance" theme. A representative warning from Florida characterized Rampart as the tip of an iceberg: "the destruction of wildlife and wilderness is the first step in the cancerous growth of industrialization with all its ugliness. Let's preserve Alaska from it."[132]

Rampart's supporters saw Alaska's vocation differently. According to Bloch, "This opening up of the Last Frontier is important to the Nation as a whole . . . as a location for swelling population . . . and as a traditional outlet for those who wish to pioneer." Gus Norwood, a former executive secretary of the Northwest Public Power Association, also emphasized Rampart's significance to the nation and its frontier heritage in this respect. "During the physical expansion of the U.S.," he explained, "the spirit of venture found itself mainly at the frontier. The American historian Turner described this pioneering frontier as an important element in our national make-up." "Physically," Norwood argued, "Alaska represents America's last, great frontier."[133] Like the opponents of the Arctic National Wildlife Range, Rampart's advocates believed that the integrity of American democratic values depended on the continuing availability of "free land." They also subscribed to the safety valve theory and interpreted the nation's socioeconomic ills as the product of overpopulation and land scarcity. One Alaskan supporter excoriated the dam's critics for begrudging "even one square inch for some future child who will be oppressed and suffocated by the super-industrialization of the Outside," and for refusing "to admit the necessity of industrial progress here to relieve the situations like L.A."[134]

Proponents had little patience with those who criticized Rampart on the basis of historic misdeeds. They had even less sympathy with the belief that the dam would irrevocably destroy the essence of Alaska. Gruening saw no friction between Rampart and the survival of Alaska's distinctiveness. He was convinced that Alaska could have a greater population and more prosperity as well as abundant wilderness. This desire to have it both ways produced a capacity for thought and expression that many outsiders found contradictory. Gruening illustrated this trait during an appearance on the broadcast "Washington Viewpoint" in October 1963. Asked whether he wanted Alaska's population and level of industrialization to increase to a point comparable to the rest of the nation, Gruening responded in tones befitting Rampart's severest critic: "I hope not. One of the great charms of Alaska is that there are relatively few people there, that we have a magnificent heritage of wilderness, which we do not want to see ruined as I feel California has been ruined by this tremendous influx of people, smog hanging over their cities, sub-dividers all over the lot, so that the great pristine beauty of California is disappearing."[135]

Sundborg argued that most of Alaska would remain wilderness inviolate on a perpetual basis because "the land which is above 2,000 feet in altitude is good for nothing but scenery and recreation."[136] Instead of dwelling on the project's large dimensions in their dealings with conservationists, Rampart advocates emphasized its insignificant size in terms of Alaska as a whole. One member of YPA described the lake that would be created as "little . . . the size of New Jersey."[137]

Above all, Rampart supporters found it hard to understand why conservationist sentiment was so inflamed by a project that they believed was a legitimate expression of conservation. In this respect, the Rampart controversy constitutes an illuminating episode in the continuing debate over the proper meaning of conservation and provides a striking example of its growing complexity at a time when environmentalism was stirring. The origins of conservation had been closely associated with irrigation and other forms of river control. Theodore Roosevelt and Gifford Pinchot were firm believers in the redemptive potential of reclamation in the arid West. Rampart proponents were conscious of this heritage and often referred to Pinchot as a mentor. In 1967, Gus Norwood, a founder member of YPA, was appointed chief of the newly created Alaska Power

Administration, which assumed the Bureau of Reclamation's duties for the state. In 1968 he quoted Pinchot's statement that "the coming electrical development will form the basis for a civilization safer, happier, freer and fuller of development than any the world has ever known."[138] Furthermore, engineers, civilian and military, had always been closely associated with conservation insofar as it meant efficient and scientific use of natural resources.[139]

In this tradition, Gruening explained in 1963 that "harnessing our rivers, instead of allowing them to run wastefully to the sea, is essentially a conservation function."[140] Gruening, a liberal Democrat since the early 1920s and former editor of *The Nation,* had cut his political teeth during the New Deal. He saw Rampart as a resumption of the TVA-style public works that had represented a fulfillment of the idea of multiple-purpose river basin development originally conceived by W J McGee.[141] One of TVA's main aims had been to provide electricity to remote areas and the rural electrification movment of the 1930s illustrated the reformist, socioeconomic purposes of conservation.[142] Such projects had fallen into disfavor during the Eisenhower years (though not for conservationist reasons). But John F. Kennedy had praised Rampart as a fine opportunity to return to the conservation policies of Theodore Roosevelt, which had been based on the notion of a river as "a unit from its source to the sea."[143]

Gruening portrayed Rampart as the spearhead of a campaign against deprivation in rural Alaska. He announced that it would raise living standards in the same way TVA had helped improve conditions of life in Appalachia during the depression. In April 1966, when he complained to Udall about the delay in publishing the review of Interior's 1965 report on Rampart, Gruening tried to convince the secretary that the project was essential to the success of the Great Society program's "war on poverty" in Alaska and elsewhere in the nation: "Its effects will reach beyond the state to increase opportunities for untold numbers of citizens in other areas of the nation."[144]

Preservationists had disliked dams ever since their unsuccessful campaign to prevent the drowning of Hetch Hetchy Valley to supply water to San Francisco. However, it was not until the successful fight in the early 1950s against Echo Park Dam in Dinosaur National Monument on the border between Utah and Colorado (which conservationists considered the most serious threat to the federal system of parks and monuments since Hetch Hetchy) that the private conservation community was united in condemning dams in such places.[145] This point of view, which amounted to a considerable revolution against conservation norms, provides one of the best examples of the emerging conflict between traditional utilitarian conservation and environmentalism. Samuel

P. Hays has selected the Supreme Court decision in 1967 not to allow the damming of Idaho's Snake River at Hell's Canyon as the most dramatic evidence of this watershed in the history of American conservation.[146]

The Rampart controversy, though hardly known, is an equally good, if not better example. One of the hallmarks of environmentalism was a concern with the adverse impact of industrial growth and a questioning of its value from an ecological standpoint. In addition, environmentalists were concerned not only with industrial impact on the natural world but also with its effects on people. Since Hell's Canyon was uninhabited, that controversy did not involve this vital aspect of environmentalism. Moreover, Hell's Canyon, which is deeper than the Grand Canyon, qualifies as monumental scenery par excellence, though it was not protected in 1967.

Opposition to Rampart, on the other hand, illustrated two integral aspects of environmentalism; an ecological perspective and a concern for the quality of human life and the human environment. The struggle against Rampart Dam was not a mainstream defense of a resource of recognized aesthetic, recreational, and ecological significance that can be compared to the parallel campaigns against dams in the Grand Canyon and Hell's Canyon. Rampart did not pose a threat to the integrity of a national park or monument like Yosemite, Echo Park, and the Grand Canyon, or an area like Hell's Canyon, which subsequently became a national monument, or any other more conventionally attractive area. The Yukon Flats had never been nominated for protection. (Indeed, this sudden conservationist interest in them mystified Rampart proponents like Alex Radin, of the Fairbanks Chamber of Commerce. Radin wrote to the *New York Times,* "It seems incredible to me that anyone would want to preserve in its natural state an area as wild and as bleak as Rampart Canyon.")[147] And, as Gruening stressed, by no stretch of the imagination could the Yukon Flats be considered worthy of protection on traditional grounds.

Alaskan conservationists were quick to point out that because the region was inhabited and contained no spectacular scenery it did not qualify as "wilderness" as most Americans understood the concept and valued the condition.[148] This did not mean that the region was of lesser value than the Grand Canyon or Hell's Canyon. What it meant was that its worth was more complex because cultural and human values were at stake as well. These cultural and human values were past and present, white and Native; the legacy of the gold rush era and the Athapaskan way of life based on a close association with the land and its resources.

This also meant that it was harder to mobilize national public support for the anti-Rampart cause than it was to muster forces for the defense of the Grand Canyon or Hell's Canyon. Many conservationists believed that Glen Canyon on the Colorado had been lost to reclamation in 1966 because, unlike the Grand Canyon of the Colorado itself, this stretch of the river was virtually unknown in conservation circles. The Yukon Flats, a much more obscure region than Glen Canyon for most conservationists, and whose values were far less evident, enjoyed even less of a clientele. An added difficulty was the inapplicability to the Yukon of the scarcity theory of value, which operated in favor of opponents of further dams on the Colorado. Rampart Dam was the first threat to the free-flowing Yukon, whose physical condition was comparable to that of the Colorado in the 1920s when Boulder (Hoover) Dam was proposed. In view of these circumstances, conservationist opposition to Rampart Dam is the more significant and testifies to the increasing importance of Alaska to them.

While conservation, past and present, tends to cut across the traditional American political spectrum, a distinctive ingredient of the environmentalism that emerged in the 1960s was a connection with political liberalism. By the mid-1960s, many conservationists considered political liberalism and the pursuit of Rampart Dam incompatible. The stubborn dedication with which Gruening led the cause puzzled many fellow liberals. Gruening's stance on a range of issues, not least his pioneering efforts to stop the war in Indochina (he was one of only two senators to vote against the Gulf of Tonkin Resolution), had won him many national admirers. These supporters were as confused and disappointed by his position on Rampart as he was by the criticism it was receiving from conservationists. For many of them, these apparently unrelated issues, the Vietnam War and the dam, were linked by the theme of waste and destruction. One critic wanted to know, "How can a man talk about the waste in Viet Nam, but not blink an eye when he thinks of flooding a large area full of wildlife?"[149] Another protester found it equally difficult to reconcile Gruening's brave and unpopular stance on the war with his position on Rampart, where he would "kow-tow to the home businessmen."[150]

In the end, Rampart advanced the interests of these "home businessmen" no further than Project Chariot had done.[151] Their case was weak and romantic and the chances for success were never high. In contrast, the opposition's arguments were strong and pragmatic. Private conservation organizations fought the dam in greater numbers and with more assurance than they had opposed Chariot. Again, however, conservationist criticism, federal and citizen, was far from the only factor in the project's demise. Native dissent was also a consideration in the Department of the Interior's decision-making, while the project's economic flaws were no doubt the most important factor of all.

Proponents often emphasized that the construction activity generated by the project would surpass in scale even that of the halcyon years of federal defense spending between 1950 and 1955. Critics believed that, for many Alaskans, the jobs provided during construction were the main inducement. Many Americans concurred with the *New York Times* (8 March 1966) that the ultimate economic logic was a secondary consideration. Unable to demonstrate convincing national or local benefits, Rampart boosters were forced to rely on the storehouse image. This strategy, which yielded scant success, was typified by Anthony Netboy, who had a dream that one day, "a housewife in Phoenix or L.A. will fry her eggs at breakfast with electricity generated on the far-off Yukon."[152] Rampart collapsed essentially from within. No matter how hard proponents tried, they could not sell the project to sufficient Alaskans, let alone convince the Department of the Interior and the rest of the nation of its value.

A noteworthy feature of the controversies over Project Chariot and Rampart Dam was Alaskan Natives' increasingly vociferous pursuit of their legally recognized land claims. This illustrates a major difference between the course of relations between Natives and whites in Alaska and in the contiguous United States. The Natives of the Aleutians and the Pacific coast were conquered, enslaved, and eradicated. However, the inhabitants of the interior and the Arctic, while culturally, socially, and economically affected by contact with Russian and American, were never fought, displaced, or segregated on reservations to any devastating extent. Invading cultures were primarily interested in removing resources (often marine), not in occupying the land or transforming it.

The Native Allotment Act of 1906, which granted surface right to 160 acres, did turn a small percentage of Alaskan hunter-gatherers into "homesteaders" while some widely scattered reservations were created in Alaska between 1936 and 1946. These reservations ranged in size from Venetie (1.4 million acres) to Unalakleet (870 acres). However, Alaska Natives retained comparatively undisturbed possession of their ancestral lands until the late 1950s. Project Chariot and Rampart Dam were the first major challenges to the integrity of the historic land base of Alaskan Native cultures beyond the North Slope, where the U.S. Navy's exploration of Pet 4 after the Second World War and the construction of the DEWline had marked the first disturbances of this kind.

The Treaty of Cession of Russian-America (1867), the Organic Acts of 1884 and 1912, and the Statehood Act guaranteed the aboriginal land rights of Alaska Natives on grounds of historic use and occupancy. In December 1966, Stewart Udall, who was guardian of the nation's indigenous peoples as well as steward of its natural resources, imposed a moratorium on applications for title to unappropriated federal lands in Alaska. This so-called land freeze was specifically intended to protect Native claims, which were jeopardized by ongoing state selection of lands under the terms of the Statehood Act, until congress reached comprehensive settlement of these outstanding claims.

6
Oil: The Forces Gather, 1966–1969

Great wilderness has two characteristics, remoteness and the
presence of wild animals in something like pristine variety and
numbers. Remoteness cannot be imitated in cheap materials;
and wilderness without animals is mere scenery.
—Lois Crisler, *Arctic Wild,* 1958

6.1 GOVERNOR HICKEL

In November 1966, Walter J. Hickel—the millionaire real estate developer, contractor, and hotelier, who had not previously held public office—
ran as the Republican underdog in the gubernatorial election against
the incumbent Democrat, William Egan, who had been governor since
statehood. Hickel's election by a narrow margin (80 votes) was a major
political upset, breaking the Democrats' long period of dominance in
Alaskan politics. Upon assuming office, Hickel resigned as chairman
of the Anchorage Natural Gas Company and as a director of the Anchorage Pipeline Company.[1] The latter company owned a franchise to supply
the city with oil and gas from the Kenai Peninsula, where in 1959 the
discovery of gas had followed the recent oil strike. Economic development had been Hickel's campaign theme and his campaign slogan had
expressed entrepreneurial impatience with the current rate of Alaskan
economic growth: "Let's get Alaska moving!"[2] A few years earlier, in
connection with the Rampart Dam proposal, the *Washington Post* had
remarked that "thinking little" was perhaps the worst crime in Alaska.
Hickel believed that lack of vision and timidity had been the basic faults
of previous state governments in Alaska. Governor Egan, he recalled,
was "a little bit afraid of bigness."[3]

Soon after being sworn in as governor, Hickel approved the upcoming
sale of oil leases on 37,000 acres of the North Slope, which his predecessor had postponed after protest from Natives. Natives claimed that
Hickel's decision violated the land freeze recently imposed by Stewart
Udall, the secretary of the interior. To date, the State of Alaska had
selected less than a quarter of its 104-million-acre entitlement under

the Statehood Act and Hickel stressed that nothing should be allowed to hinder oil development.[4] On 10 February 1967 he filed suit in the U.S. District Court for Alaska, seeking to nullify Udall's freeze as a violation of the Statehood Act.[5]

In his inaugural address in January 1967, Hickel explained that Alaskan economic development was fifty years behind the rest of America. He established the "opening" of the Arctic as a major priority and advocated an extension of the Alaska Railroad to Nome and the Arctic Slope to facilitate arctic mineral exploitation. He drew attention to arctic oil as the economy's brightest prospect.[6] To accelerate the pace of economic growth, Hickel set up the Northern Operations of Rail Transportation and Highways (NORTH) Commission to explore the feasibility of extending the Alaska Railroad to Nome and Kobuk and building a winter road to the North Slope by way of Anaktuvuk Pass.[7]

As Rampart Dam's boosters had done, Hickel stressed the advances made by America's Soviet neighbors and competitors. Addressing the Anchorage Chamber of Commerce on 19 July, a week before the NORTH Commission's first meeting in Fairbanks, he declared "we are seventy years behind the Russians and their opening up of the Asiatic Arctic." Then he summarized the commission's aim: "We intend to telescope time—to open up our Arctic within the next ten years." He explained that by encouraging the development of northern resources, the commission would "forever dispel and banish the myth that the Alaskan Arctic is a land locked in ice and snow."[8] In the fall Hickel stirred memories of Project Chariot when he announced his support for the establishment of a permanent Atomic Energy Commission (AEC) testing site on the North Slope near Point Lay, an *Inupiat* village that was just beyond the western boundary of Pet 4. Hickel believed, as Edward Teller had in the late 1950s, that AEC tests would provide jobs for local Natives. More importantly, he felt that a nuclear testing site would stimulate interest in an extension of the Alaska Railroad.[9]

In the vanguard of the effort to "open up" Alaska in 1967 was the oil industry. After the strike of 1957 on the moose range, the center of its attention was the Kenai Peninsula. Following statehood, the focus of exploration shifted. One new area was Cook Inlet, which is adjacent to the Kenai. In 1958 there had been no oil production in Alaska but by 1963 oil and gas accounted for 77 percent of all mineral production.[10] By 1965, largely due to brisk oil and gas production from Cook Inlet, Alaska's income from all forms of natural resource extraction exceeded that from federal military expenditures for the first time in Alaskan history.[11] In 1967 oil displaced fisheries in Alaska as the leading source of state income.[12]

In the meantime, the emphasis of exploration activity had moved northwards. In 1961 the Sinclair Oil Corporation ran a widely distributed advertisement that showed two "restless" petroleum geologists alighting from a helicopter onto the snow-covered tundra. The caption, "Go and look behind the ranges . . ." was a quotation from Rudyard Kipling. Sinclair explained that the nation's "insatiable demand for energy" was pushing its "roving" "oil seekers" into far-flung and inhospitable corners of the globe, such as the Alaskan Arctic, into which private oil companies began to move soon after statehood. In 1964, the State of Alaska, claiming that the land was free of Native use and occupancy, had chosen two million acres of potentially oil-rich terrain on the Arctic coast. The state sold leases in 1964, 1965, and 1967, by which time a total of 900,000 acres had been leased in the Arctic. In January 1968 Atlantic Richfield repeated its pioneering success on the Kenai with the first commercial strike on the state-leased lands in Arctic. Subsequent exploration confirmed that the field contained an estimated 9.6 billion barrels, making it the largest ever discovered in North America.

But Hickel saw no reason why the NORTH Commission's attention and arctic natural resource development should be restricted to oil: "We wouldn't build that railroad into the Arctic just for oil or any one industry. . . . We need that railroad to get people in there to bring out copper, iron ore, coal and timber, and bring in the millions of tons of supplies needed to develop this oilfield. We would bring this wealth down from the Arctic, through the middle of Alaska, and open up this country at last."[13]

Drawing attention to how Russia had opened up its Siberian frontier in the 1870s with the Trans-Siberian railroad, he believed an extension of the Alaska Railroad would accomplish the same for the American Arctic. He attacked the icebox image ("this is not the harsh kind of region people think it is") and stressed the Arctic's habitability. He envisaged, within his lifetime, "self-sustaining communities," like those in the Soviet Arctic, close to major mineral deposits such as coal (west of Barrow) and copper (on the Seward Peninsula). He also foresaw the emergence of a petrochemical industry at Pacific tidewater and targeted Japan and other far eastern nations as a market for arctic minerals. He also reported that the Rampart Dam project was attracting fresh interest because the development of new long-range transmission lines was bringing Arizona and California into the market for Alaskan hydroelectricity.[14]

While the NORTH Commission was making plans, Hickel acted to forge an immediate overland connection with the newly discovered oilfields. Hitherto, the needs of the oil industry had been served mostly by Hercules air transports, but also by barge traffic from Seattle, via the Bering Strait, during the brief summer months of ice-free ocean transit. For what can only have been the benefit of the Alaskan trucking industry, the bulldozers of the state Department of Highways blazed a 400-mile winter road north from Livengood, 60 miles beyond Fairbanks, where public roads in Alaska ended. Construction lasted from November 1968 to mid-March 1969. The State of Alaska had intended to build the road as far north as Bettles itself and its road crews had reached the Yukon River by the end of November. As it transpired, the state Department of Highways also constructed the section beyond Bettles, apparently because the bids it received from private contractors were too high. Work began briefly on the northern section in late December but had to be suspended because of an extreme cold spell, during which the average temperature was minus 63 degrees Fahrenheit. Work resumed in mid-January 1969 when the temperature rose to a balmy minus 50 degrees.[15] The road crossed the Yukon at Stevens Village and proceeded up the John River from Bettles. It traversed the crest of the Brooks Range at Anaktuvuk Pass and terminated at Sagwon, some 90 miles south of Prudhoe Bay. By the time the road was completed, Hickel had left Alaska to become President Nixon's secretary of the interior. His former secretary of state and successor in Juneau, Keith H. Miller, promptly named the impulsive project the Walter J. Hickel Highway.[16]

This undertaking was not an economic success. During its single month of operation in 1969 (from mid-March to mid-April), the road carried 7,464 tons of freight.[17] The cost of trucking freight to Sagwon was $240 per ton, the same as by air. The total tonnage carried by road could have been delivered by Hercules air transports in three trips.[18] Conservationists also saw it as an instant ecological fiasco. Those who had built seasonal, unimproved arctic and sub-arctic roads had learned, usually the hard way, that successful winter road construction in these regions demanded special methods. To lay an effective winter road, a crew needed to accumulate snow, compress and compact it into a firm, raised bed, and then pour water pumped from rivers over the hardened snow to create ice. This technique avoided disturbing the permafrost underlying the tundra, taiga, and muskeg. Ignoring these lessons of the past, the blades of the bulldozers of the state Department of Highways, directed by veteran road builder, Woodrow Johansen, scraped away the protective mat of vegetation together with the snow.[19] The road deteriorated into a ditch once the permafrost was exposed to thaw during spring melt. The Department of Highways used the same crude construction methods the following winter (1969–70), when it rerouted severely eroded sections in like style at considerable expense, which produced parallel ditches come break-up.[20]

The Hickel Highway, which some conservationists described as a *debacle,* precipitated the first major protest against oil-related activities in

Alaska since the beginning of Alaska's petroleum era in 1944. (There had already been some disquiet over discharge from pipeline breaks and tankers in Cook Inlet.)[21] Road construction had begun shortly before Udall imposed a "superfreeze" on land use in Alaska in January 1969, but conservationists claimed that nevertheless it was illegal. George Marshall, who had campaigned tirelessly for a "Gates of the Arctic" park in the central Brooks Range since 1963, was particularly upset.[22] In his introduction to a timely reissue of his brother's account of adventures in these mountains, George Marshall lamented that the winter road, which penetrated the heart of the country explored by his brother, represented "the first violent change, the first major intrusion of the modern industrial world," which split the range in half and destroyed its "unity with past ages."[23]

Another preservationist who believed that the road's implications were savage and profound was the Reverend Samuel A. Wright, a professor of human ecology from Berkeley, California. While Hickel's winter road was being pushed through the Brooks Range in the winter of 1968–69, he was spending a sabbatical year living in a cabin built and formerly used by gold miners. It was located at Big Lake, near the hamlet of Wiseman, which had been Bob Marshall's base in the area. Wright's purpose during this first arctic winter was to study the human ecology of the Eskimos of the interior, the Nunamiut. Bob Marshall was his mentor and he often trod deliberately in his footsteps. (For example, Wright intended to monitor the progress of the experimental seedlings Marshall had planted forty years ago during his studies of tree growth at northern timberline.) "The effect of his book on my own life has been dramatic," he commented of *Alaska Wilderness,* explaining that "the earlier edition [1956] was a major factor in my decision to make this last great wilderness my home."[24] Wright was to become a participant in the Trans-Alaska Pipeline debate through his contributions to various journals and testimony at hearings held in Alaska.[25] From his cabin at Big Lake he published in rudimentary fashion a quarterly "Journal of Human Ecology" called *A View From the Top.*[26]

Wright agreed with George Marshall that the "Hickel Canal," as conservationists dubbed it, marked the crucial break with the past in northern Alaska. As vehicles were driven north over the Brooks Range for the first time, he reflected ruefully on the explosive "end of thousands of years of solitude."[27] After the road's second winter in operation (1969–70), the nation's oldest and biggest-selling sportsman's magazine, *Field and Stream,* described it as "only one small, ugly symptom of the fever that has Alaska by the throat."[28] The State of Alaska was unmoved by conservationist indignation and anger. Governor Miller claimed that Hickel's action was authorized by a law of 1866 that permitted road construction by states across the unreserved public domain. Despite his new national office, and efforts to appease conservationists, an uncontrite Hickel retorted in the fall of 1969: "So they've scarred the tundra. That's one road, twelve feet wide, in an area as big as the state of California."[29] In Hickel's view, the problem was that conservationists in the lower forty-eight did not appreciate Alaska's size: "Can you imagine the state of California without a single road?"[30]

Hickel's nomination as secretary of the interior provoked an outcry matched in American history only by the reaction against James Watt in 1981. Hickel's hostility to preservationism was well-known and his national image was not improved when he denounced "conservation for conservation's sake" at a press conference for cabinet designees in mid-December 1968.[31] In the weeks preceding his confirmation hearings, the press speculated gleefully over Hickel's alleged links with "big oil." Accusations focused on certain appointments Hickel had made when he was governor of Alaska. His commissioner of natural resources, for example, had been Tom Kelly, a former geologist with Alaska Oil, who was the stepson of the owner of Halbuty Oil. The press also made much of the role of Atlantic Richfield's president, Robert O. Anderson, a close friend of the president-elect, in Hickel's selection for national office.[32] Nearly all leading newspapers outside Alaska, and the national and Alaskan conservationist communities without exception opposed Hickel's nomination. The critics also included Alaskan chapters of national sportsmen's groups, such as the Tanana Valley Sportsmen's Association (TVSA) and the Alaska chapter of the Izaak Walton League.[33]

Whereas the national conservationist community complained that there was no designated wilderness in Alaska, Hickel, at his confirmation hearings, made the familiar booster observation that the state was afflicted by a surfeit of wilderness. He argued that any formal designation would be superfluous and retrogressive because Alaska was "practically all wilderness."[34] The opinions he expressed at these hearings on a range of other issues did nothing to redeem him in conservationist eyes. His predecessor Udall's disapproval of Rampart and the dam's unpopularity among Americans did not discourage Hickel, who called it "a great, great engineering project. I think it would do so much for so many." However, he did admit that it was unlikely to be built in his lifetime.[35] He also argued that oil development on the Kenai National Moose Range had improved conditions for moose. While he did not advocate abolition of the range, Hickel did favor reducing it to the higher montane regions where there were no oil reserves.[36]

Of more immediate concern to conservationists was his attitude toward arctic oil development. Hickel wanted to open the Arctic National Wildlife Range (ANWR) to oil exploration and to transfer jurisdiction over National Petroleum Reserve Number 4 from the U.S. Navy to the Department of the Interior, to facilitate leasing to private interests.[37] These stances prompted allusions by Senators Jackson (Democrat, Washington) and Metcalf (Democrat, Montana) to Albert B. Fall, the interior secretary who had achieved notoriety in the 1920s for his role in the Teapot Dome oil leasing scandal.[38] Debate over Hickel's oil interests dominated the hearings. Investigations by the director of the state Division of Lands for Alaska revealed that between 1953 and 1963 Hickel had owned a 250,000-acre lease through the Yakutak Development Company in southcentral Alaska and one lease on the North Slope from 1962 to 1965.[39] Shortly before leaving office, Udall had extended his freeze of December 1966 by withdrawing all unreserved public lands in Alaska from any kind of transaction until 31 December 1971. Hickel had to promise the Senate Interior Committee that he would consult it prior to taking any action to modify the freeze, before the committee eventually confirmed his nomination by 14 to 3.

6.2 OIL DEVELOPMENT AND ARCTIC WILDERNESS: IMPACT AND RESPONSE

On 10 February 1969 the Trans-Alaska Pipeline System (TAPS) announced plans to build a pipeline south from Prudhoe Bay. The Sierra Club's Eleventh Biennial Wilderness Conference, held in San Francisco a month later, was devoted exclusively to the subject of wilderness under threat in Alaska. At the previous conference in 1967, protection of Alaskan wilderness had ranked sixth in the hierarchy of club priorities. (Topping the list was the campaign for Redwood National Park; second was protecting the Grand Canyon from dams; and third was the struggle for a national park in the North Cascades.) In San Francisco in 1969, George W. Rogers, a prominent Alaskan economist, criticized what he called "the last frontier syndrome." With thinly veiled references to the attitude and style of former Governor Hickel and his successor, Keith Miller, Rogers explained that "many Alaskans, including political leaders, think of themselves as heroes in a TV western. In the northern version of the last frontier, as in the western version, wilderness is to be despoiled and destroyed as valueless, a nuisance, or a threat."[41]

Brock Evans, the Sierra Club official with responsibility for Alaska, reflected on what he saw as the rapid recent loss of wilderness in northern Alaska. He offered a gloomy prediction: "What has taken one hundred and fifty years to happen in Ohio, one hundred years in California or fifty to my adopted state of Washington will probably take only a generation to accomplish in Alaska, if present trends, attitudes, and policies continue."[42] Between 3,000 and 5,000 men had been living and working on the North Slope that winter assembling drilling rigs and accommodation units, constructing air strips, roads and drilling pads, digging gravel, and producing garbage and sewage. Robert B. Weeden, speaking as Alaska Conservation Representative for the Wilderness Society and the Sierra Club, estimated that as a result of such activity the wilderness character of an area the size of Massachusetts had been lost since 1968.[43]

At the time of the Klondike-Alaska gold rushes, John Muir did not fear that this flurry of activity would be followed by permanent development and settlement along the lines of the West. "Fortunately," he reflected, "nature has a few big places beyond man's power to spoil," among which he included "the two icy ends of the globe."[44] Muir was correct in his prediction that Alaskan wilderness (especially in the Arctic) was not threatened by the form of agrarian colonization that had prevailed east of the Mississippi, or the destructive logging and grazing that he had witnessed in his beloved Sierra Nevada. However, since Muir could not foresee the emergence of other agents of assault, he was mistaken in thinking that the absence of agriculture and timber harvest would leave Alaskan wilderness unviolated.

At the time the Trans-Alaska Pipeline was proposed, certain Alaskans familiar with the Arctic were convinced that large-scale, irrevocable destruction of wilderness conditions and atmosphere had already occurred there over the past quarter-century. An Alaskan particularly well-acquainted with the post-war history of arctic oil exploration, arguably the most damaging and without a doubt the least regulable phase of oil activity, was Sam O. White. White was a veteran bush pilot and a former colleague of Frank Dufresne on the Alaska Game Commission. A week before the Sierra Club's San Francisco conference (March 1969), he provided its officials with a description of activities on the North Slope. White, who had characterized the post-war exploration of Pet 4 as "killing and burning,"[45] claimed that the private oil companies were "even going the Navy one better" at Prudhoe Bay. He wondered if the caribou would go the way of the buffalo and the passenger pigeon.[46]

Another critic was Harmon "Bud" Helmericks. Helmericks and his wife Constance had lived on the coast in the Colville Delta since 1952 and were authors of a number of popular books about life in the Arctic. They were the sole inhabitants, white or Native, of this area, which lay some fifty miles west of the Prudhoe Bay operations. The twenty years he had spent operating a guiding service for hunting polar bear along the coast and other big game in the Brooks range overlapped with the later phase of Pet 4 exploration and the period when the first commercial companies began to work on the North Slope. While he had no doubts about the desirability of oil development, Helmericks was appalled by the carelessness of those searching for oil in the 1960s. He referred to "bandido operations" that "tore up the countryside fantastically." He denounced indiscriminate, out-of-season hunting by survey crews operating from "cancerous camps" and condemned harassment of wildlife by helicopters, airplanes, and snowmobiles. As a result, Helmericks claimed, large predators had been virtually eliminated from the area where he lived. Since the Prudhoe Bay strike, he estimated that numbers of grizzly, wolf, and wolverine had declined there by 90 percent.[47]

Helmericks stressed, as Bob Marshall, Olaus Murie, Frank Dufresne, Lowell Sumner and George Collins had done in earlier decades, that the character of arctic wilderness was extremely fragile. For Helmericks, visual pollution ruined the wild character of such a sparsely vegetated region, which he called "prairie." He identified the chief culprits as discarded 55-gallon fuel drums. He once counted 280 of these offending objects along five miles of lakeshore near his home.[48] This and other kinds of debris, left mainly by seismic crews, particularly galled him: "There was absolutely no attempt made in the first wells to pick up anything from empty coke cans to boxes of high explosives or empty drums—and wire was left all over the prairie to tangle up the caribou."[49]

It was seen in connection with the campaign for ANWR that conservationists valued wilderness (among other things) as a direct physical source of the distinctive qualities associated with Americans by Frederick Jackson Turner. In May 1969, in the context of oil exploration and the pipeline proposal, John P. Milton of the Conservation Foundation published an account of a 300-mile hike he had undertaken with two companions through ANWR in the summer of 1967. The trio had walked from the southern foothills of the Brooks Range across the Arctic Divide, proceeding down onto the Arctic Slope and finishing at the Barter Island DEWline station.[50] For contemporary pioneers like Milton, who claimed to be the first humans (white or Native) to travel along this route, the Brooks Range offered wilderness on a scale comparable to that which the mountain men (fur trappers) confronted in the Rockies in the mid-nineteenth century. Milton explained that "when we go into wilderness we are consciously reversing a historical process."[51] The mountain man, who, according to Ray Allen Billington, reverted to the primitive, being "conquered" by the wilderness rather than subduing it, as other types of pioneers did, was such conservationists' favorite type of frontiersman.[52]

Kenneth Brower's book, *Earth and the Great Weather: The Brooks Range* (1971), based on the same trek, included an introduction in which Milton discussed the Trans-Alaska Pipeline's threat to arctic wilderness. In this account, Kenneth Brower, the son of David Brower, the former executive director of the Sierra Club, referred to the "Lewis and Clark

Syndrome" of the third member of the trip, Steve Pearson: "We were a lot like Lewis and Clark, he said as he contemplated the Okpilak River bluffs, which reminded him of the bluffs Lewis and Clark saw on the Missouri. We had seen the same grizzlies and the same great plains, he said, except that here the plains were tundra instead of buffalo grass and the herds were caribou instead of buffalo."[53]

This type of "fancy" again illustrated what William Tucker, an arch-foe of wilderness preservation and environmentalism in general, has more recently described (perceptively, but not kindly) as "the attempt psychologically to reopen the American frontier."[54]

6.3 THE NATIONAL CLIMATE

The influence of arctic oil development in elevating the fate of Alaskan wilderness to such prominence among conservationists so suddenly was complemented by larger national factors. In the late 1960s an increasing number of Americans perceived an urgent "environmental crisis." In late 1969 and early 1970 many leading magazines devoted special issues to the problems of ecology and environment.[55] One social scientist contended that there was only one precedent in American history for such a rapid and deep-seated shift in public attitude, namely; "the flip-flop from isolationism to interventionism occasioned by Pearl Harbor."[56]

This crisis was symbolized by the Santa Barbara oil spill. Hickel's first test as interior secretary came at the end of January 1969, a few days after he was confirmed, when a "blow-out" occurred at a Union Oil drilling platform in the Santa Barbara Channel.[57] One historian has argued that this catastrophic spill boosted the emerging environmental movement in the same way that brutal police treatment of black demonstrators in Birmingham, Alabama, in April 1963, galvanized public support for the nascent civil rights movement.[58] Hickel himself has selected this event as the beginning of the environmental movement.[59] For many, the Santa Barbara catastrophe cast the oil industry unmistakeably as the bête noire of the environmental cause.

There was a good deal of overlap between environmentalism and traditional conservation. The wilderness preservation and wildlife protection traditions associated with the Sierra Club and the Boone and Crockett Club respectively became planks—though not necessarily central—in a wider platform.[60] However, environmentalism was also shaped by postwar socioeconomic forces that nurtured the civil rights, consumer, and antiwar movements.[61] Environmentalist beliefs were sometimes part of a broader outlook characterized by a more or less critical stance toward many prevailing American values, institutions, and practices. The 1960s brought a revival of the Progressive muckraking tradition and produced a new breed of muckraker; the investigative journalist who tackled the old corporate enemies, typified by the "fourth branch of government"— "big oil"—from a comparatively new, environmentalist angle. Environmentalism was frequently sceptical of orthodox, laissez-faire economics and sympathetic toward technological modesty and restraint. Concerned about threats to humans as well as the dangers confronting the natural environment, all disturbing and damaging manifestations of later twentieth-century urban-industrial culture came under its purview. Substances environmentalists opposed, such as toxic chemicals, pesticides, and food additives, and policies they advocated, like birth control, recycling, mass transit, and energy conservation, were issues of equal (sometimes greater) importance to some environmentalists as the protection of wilderness and endangered species.

Environmentalism's coming of age in the late 1960s affected the American conservationist community in two main ways. Some of the new wine found its way into old bottles. Established organizations underwent an unprecedented growth as new recruits swelled existing ranks. Contemplating this trend, a veteran member of the Izaak Walton League of America (IWLA), founded in 1922, remarked, "where once we seemed no more than a corporal's guard, now there's an army."[62] In his capacity as executive director between 1952 and 1969, David Brower transformed the Sierra Club from a genteel coterie of trail companions into a popular force. Between 1960 and 1971, its membership increased from 15,000 to 135,000.[63] In 1966 the club lost its tax-deductible status over the fight to keep the Grand Canyon free of dams. This resulted from its growing aggressiveness and effectiveness in the political arena.

However, there was sometimes a limit to the capacity of traditional organizations for assimilating fresh contents. While many conservationists also became environmentalists, others remained uneasy with the new issues and disliked the increasing politicization of their organizations. Some sportsmen conservationists suspected the new urban recruits of being antihunting and antibusiness and disliked their left-of-center, confrontationalist politics. A number of conservationists yearned for the days when it had seemed possible to campaign for wilderness preservation without "burning the flag." The expulsion of David Brower from the Sierra Club leadership in 1969 was in some ways a reflection of these tensions.

Brower transferred his beliefs and the radicalism the Sierra Club's establishment had rejected to an organization of his own making, the John Muir Institute for Environmental Studies, which was soon superseded by a group called Friends of the Earth (FOE). FOE was an international body that attracted 27,000 American members within a year of its establishment in 1970. In many ways, it typified the new generation. Yet in other respects, not least due to the continuity provided by Brower, FOE shared much in common with the Sierra Club. Wilderness preservation retained its hold as a special cause despite the influx of new concerns.

The most innovative of the groups born of the new era was the Environmental Defense Fund (EDF), which pioneered in using the lawsuit to advance the environmentalist cause. The EDF, founded in 1967 in New York City, grew out of a suit filed in 1966 by a thirty-one-year-old lawyer, Victor J. Yannacone, in Suffolk County, Long Island, against the spraying of DDT to control mosquitos.[64] The suit alleged that DDT, the pesticide brought sharply to public attention by Rachel Carson's book, *Silent Spring*,[65] was killing fish in local lakes. Yannacone, whose motto was "Sue the Bastards," won a temporary injunction for a year. Effectively, however, the application ("dumping") of DDT in Suffolk County became a thing of the past. The EDF followed this action up with suits against a number of towns in Michigan that were using DDT to control Dutch elm disease. The EDF, whose membership consisted principally of lawyers and scientists, allied legal skills with scientific expertise. The group came to national attention in 1967 when it took legal action against officials of the agriculture departments of the federal government and the State of Michigan to prevent them spraying pesticides—in this case, dieldrin—in the Lake Michigan watershed.[66]

Like the older conservation organizations, the EDF was concerned with protecting wildlife, especially birds. However, its outlook was more comprehensive and its philosophy was more integrated. The EDF stressed that pesticides affected all forms of wildlife, whether they were formally "protected" in a refuge, wilderness area, or national park or not. Like radioactive fallout, pesticides are mobile and persistent. They consist of substances that spread to great effect far beyond the original place of application and become increasingly concentrated as they rise up the food chains. Here was another object lesson in ecology; all forms of life are linked somehow and anything toxic introduced into the natural system will travel throughout the network, becoming more dangerous in the process. Accordingly, the EDF advocated the right of all forms of life, human and nonhuman, animate and inanimate, to a pollution-free environment. Given existing concepts of law in the United States, or in any other nation for that matter, this was an entirely revolutionary demand.

The EDF's promotional literature, which depicted human crowds as often as natural scenes, illustrated its concern with the total environment. According to one brochure: "The environment is more than birds and fish and forests. People are the most important natural resource in it." The EDF maintained a more holistic perspective on environmental problems than most older groups. Its principal method of operation, the lawsuit, indicated a growing impatience with the standard techniques of traditional conservation organizations. The EDF was cynical about the efficacy of available political, administrative, and educational channels as ways to halt and reverse the onslaught of environmental deterioration. Towards the end of the 1960s, the emerging ecological consciousness and the growing awareness that environmental problems existed throughout the nation was unmistakeable. Equally apparent was the increasingly combative mood of the new conservationists and some of the older ones. In view of the unrestrained activities of the oil industry in the Alaskan Arctic and the exuberant approval of the Alaskan state government, a major clash could not be far off.

7
The Trans-Alaska Pipeline Controversy: 1, 1969–1971

We have spent many years trying to roll back the wilderness
in Alaska. Don't worry about it, it doesn't roll easily.
—Ernest Wolff, president of the Alaska Miners Association, to the
Senate Interior and Insular Affairs Committee, 26 February 1971

The rise and fall of obstacles to authorization provide a natural organizational framework for this analysis of the first two years of the controversy. To start with, the project needed to win the approval of the Senate Committee on Interior Affairs and the secretary of the interior. During 1970 Native land claims and the requirements of the National Environmental Policy Act (NEPA) became the immediate barriers facing an increasingly defensive project. Public hearings on the draft environmental impact statement issued for the pipeline project in accordance with NEPA illustrated the variety of critics and criticism that the proposal continued to face in the spring of 1971, when proponents' fortunes were at their lowest and those of the opposition were at their zenith. The chapter concludes with a reflection on how the adversaries in the TAPS contest used historical analogy and frontier images for the purposes of advocacy.

The strategy of Alaskan pipeline proponents involved the propagation of three half-truths concerning opposition to the proposal. Firstly, they argued that the delay in authorization was entirely due to the objections of private conservation organizations. Secondly, they maintained that the conservationist position was monolithic and monotonic. Thirdly, they claimed that this opposition resided entirely within the "lower forty-eight." This chapter, and the next, which deals with the third, fourth, and fifth years of the controversy, pose these questions: Were there other reasons for the delay? How diverse was criticism? Did it have an Alaskan dimension?

Investigation reveals that the reasons for delay were multiple and that criticism was not confined to private conservationist groups. Administrative and legal requirements, technological difficulties, and, not least, Native claims, were impediments as well as concerns within and outside the federal government regarding the project's ecological consequences. Moreover, study of the TAPS controversy—as did analysis of the campaigns for ANWR and against Chariot and Rampart—challenges the Alaskan booster tendency to juxtapose the interests of Alaskans and those of the national conservationist community.

7.1 First Reactions and Responses to the TAPS Proposal, April 1969–February 1970

Following confirmation in the summer of 1968 of the size of the oilfield discovered in January, the oil companies involved began to consider the question of transportation. They conducted aerial reconnaissance and ground surveys of a pipeline route between Prudhoe Bay and the Pacific coast. In October, the Trans-Alaska Pipeline System (TAPS), a loose and unincorporated consortium of the "big three" operating on the North Slope, was formed.[1] At this stage, however, the oil companies gave some attention to other routes. A task force concluded that an all-land route through Canada was also a possibility; politically, technologically, economically, and environmentally.[2] In February 1969 TAPS announced its choice; a 798-miles, sub-surface, hot oil pipeline from Prudhoe Bay to Valdez, the northermost ice-free port in the United States.[3]

The 95 percent of Alaska still federally owned in 1969 represented 46 percent of all American public lands and 552 miles of the proposed pipeline crossed the public domain. In April a North Slope task force was set up within the Department of the Interior under the leadership of the undersecretary, Russell E. Train. Its purpose was to ensure that oil exploration and development were consistent with the proper use of public lands, the protection of the environment, and the rights of Alaska Natives. On 9 May President Nixon expanded it into an interdepartmental Federal North Slope Task Force on Alaskan Oil Development. In its final form, the task force consisted of, in addition to Train[4] and his science adviser, William T. Pecora, chief of the U.S. Geological Survey (USGS), and the directors of the Bureau of Land Management (BLM), Fish and Wildlife Service (USFWS), Federal Water Pollution Control Administration (FWPCA), Bureau of Indian Affairs (BIA), and Bureau of Commercial Fisheries.[5] The task force was completed by an ad hoc Conservation-Industry Committee cochaired by John L. Hall, assistant executive director of the Wilderness Society, and Richard G. Dulaney, the chairman of TAPS.[6] Later in 1969 the so-called Menlo Park Working Group was formed. Based at the Alaska branch of the U.S. Geological Survey, Menlo Park, California, this group operated

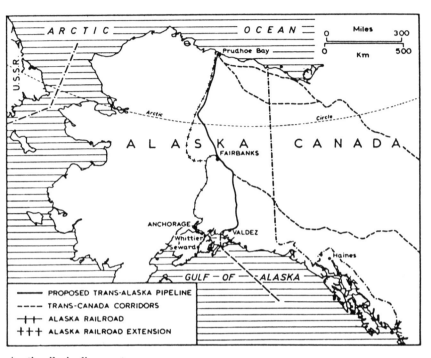

Arctic oil pipeline routes

as a subcommittee of the federal task force. Its purpose was to examine the technical aspects of the proposed pipeline.[7]

Federal concern over the engineering problems posed by Alaska's climate and physical environment soon became evident. On 25 April, the assistant Resources Director of the Bureau of Land Management in Fairbanks, Robert C. Krumm, warned the agency's director that "a pipeline break in the wrong place at the wrong time could be devastating to a broad spectrum of the ecology of a significant area." Krumm condemned the litter disfiguring the North Slope and drew attention to the problems for fish associated with removing gravel from streambeds for construction purposes. He also pointed out that any elevated pipe could impede caribou migration.[8] After an inspection of the proposed pipeline route in May, William T. Pecora and his assistant, Max C. Brewer, wondered how much of the pipe could actually be buried in view of widespread permafrost conditions. According to Brewer, the director of the Naval Arctic Research Laboratory at Point Barrow, there was "not one chance in a hundred that any pipes would be buried north of the Brooks Range."[9]

On 6 June 1969 TAPS applied to the Department of the Interior's

BLM for a pipeline right-of-way of 100 feet. TAPS also requested two additional rights-of-way for ingress and egress to the primary right-of-way and to eleven pumping station sites.[10] In addition, the consortium asked for a transportation corridor of 100 feet to build an adjacent service (haul) road from Livengood, the northernmost point on the Alaskan road system, to Prudhoe Bay. A right-of-way 200–500 feet wide was requested for construction camps at river crossings. TAPS desired approval by the end of June to enable construction to begin in September. The consortium had already signed contracts with three Japanese steel companies for 800 miles of 48-inch pipe.[11] Undersecretary Train responded to the application with seventy-nine questions, "to which satisfactory answers will be required before permits can be given for the use of public lands."[12] Train emphasized that it would be in TAPS's interest to supply these answers within a month, "in our anticipation of industry's timetables." Only two questions inquired whether TAPS had considered any other pipeline routes (oil or gas), such as one through Canada. The reply from the chairman of TAPs on 19 June sought to reassure Train that "good pipeline design dictates . . . procedures that will cause a minimum of disturbance of the natural environment." Dulaney included a twenty-page summary of TAPS's research to date. For TAPS, "sound conservation principles" meant a buried line of the same kind as the industry habitually laid in Texas and Oklahoma. Hickel's reply (27 June) confirmed his eagerness to dispense the necessary permits "as expeditiously as possible," but indicated that levels of information currently provided by TAPS concerning roads and the construction workpad as well as the pipeline were insufficient.[13] On 18 July the North Slope Task Force on Alaskan Oil Development issued its first set of stipulations to govern the construction and operation of a trans-Alaska pipeline. These required the revegetation of disturbed terrain and the protection of stream beds and anadromous fish spawning areas. They banned the use of DDT and other harmful pesticides. They also stipulated measures that would ensure freedom of movement for wildlife across the pipeline corridor.

In August the House and Senate interior committees approved Hickel's request for a modification of Udall's land freeze to clear the way for construction of a 53-mile highway from Livengood, a gold mining community, to the Yukon River.[14] This highway would parallel the first roadless stretch of the pipeline route, and without it workers, equipment, and materials could not be moved north. TAPS built the road that summer according to an agreement with the State of Alaska that it would meet state secondary highway standards and would be handed over to the state when the project was finished. However, Alaskan pipeline proponents were already growing impatient. Governor Miller interpreted the delay in issuing the permit for the project as a whole as another example of the historic neglect of Alaskan economic needs and Alaska's mistreatment by the federal government in league with conservationists.[15]

During the summer months, oil companies vied frantically to gather information in preparation for competitive (sealed) bidding on 450,000 acres of land adjacent to Prudhoe Bay, which were up for lease in September. In contrast to the $12 million gained from the three sales it had previously conducted, the State of Alaska netted $900 million from this auction, enough to cover all state government expenditures, at the current budget level, for four-and-a-half years. Governor Miller described the sale as "a rendezvous with our dreams."[16]

A report on the sale in *Time* magazine's business section set out to establish a longer frame of reference for the current situation in Alaska. It opened with a lesser-known quotation (1914) from Frederick Jackson Turner that went against the grain of the historian's earlier apocalyptical announcement of the death of the American frontier: "Already Alaska beckons on the north, and pointing to her wealth of natural resources asks the nation on what new terms the new age will deal with her."[17] Most of Alaska, the reporter argued, was "as raw and untouched as the Great Plains and Rockies were 150 years ago."[18] The scramble of activity on the North Slope that summer prompted many popular comparisons with the Klondike-Alaska gold rushes. Conservationists, however, felt the differences were more revealing and overshadowed any similarities. The National Parks Association (NPA) offered this sober reflection on the pioneering qualities, frontier flavor, and romance of arctic oil development: "The new oil rush differs from the Gold Rush of 1898. The old stampede was a clamor of sourdoughs, lone men out for themselves, enduring hardship and danger, substance of a chaotic but free society. The new development is the work of great corporations within which roughnecks and engineers alike are cogs. . . ."[19]

A few weeks before the sale, the Twentieth Alaska Science Conference had been held in Fairbanks in an atmosphere of increasingly adverse publicity and growing public concern over oil activity in the Arctic. Its proceedings were devoted to discussion of the changes oil development would bring to Alaska, environmentally, politically, economically, and socially. Over a thousand scientists, scholars, politicians, conservationists, and representatives of industry gathered to discuss the theme "Change in the North: Petroleum, Environment, and People."[20] The conferees included the political scientist, Robert Engler, a renowned critic of "big oil."[21] His chief adversary was Frank Ikard, president of the American Petroleum Institute. Among the conservationists who attended were Robert Weeden of the Alaska Conservation Society, Edgar Wayburn, vice president of the Sierra Club, and Frank Fraser Darling, the vice president of the Conservation Foundation. Alaskan pipeline advocates reacted angrily to the discussion of Alaska's interests and future by a roster of what U.S. Senator for Alaska, Ted Stevens (Republican) derided as "outside experts." "I am up to here with people who tell us how to develop our country," Stevens burst out at the end of the conference. The selective nature of the Alaskan booster assault on the icebox image (illustrated by the advocates of Chariot and Rampart) again became apparent during the course of these remarks, when Stevens made the ill-fated and widely reported assertion that "there are no living organisms on the North Slope."[22] This insinuation that there was nothing of ecological value in the Arctic—and, by implication, nothing to damage—aggravated inestimably the tender conservationist sensibilities he ridiculed.

A few days later, Undersecretary Train conducted public hearings in Fairbanks to consider the draft environmental stipulations drawn up for the project by the federal task force in July.[23] To fully appreciate the anxiety expressed on this occasion by federal geologists, some basic information about permafrost in Alaska is essential. Permafrost is any rock or soil material that has remained below 32 degrees Fahrenheit continuously for two or more years.[24] In parts of the earth where the average annual temperature of the air is sufficiently low to maintain a continuous average surface temperature of below 32 degrees Fahrenheit, the depth to which the ground freezes in winter exceeds the amount of summer thawing and a layer of frozen ground develops. The cold moves downwards into the earth until it is balanced by heat coming up from the interior. Excluded from the definition of permafrost is the surface layer of ground (the "active layer" between 1.5 and 6 feet thick) that freezes and thaws alternately each year. Permafrost is found in Antarctica but is most widespread in land surrounding the Arctic Ocean. In the Northern Hemisphere, two major zones can be distinguished: a northern zone where permafrost forms a continuous layer close to the surface of the ground and a southern zone of discontinuous permafrost that contains areas of permafrost-free ground. A third, much less distinct belt of "sporadic" permafrost can be found beyond the southern limits of the discontinuous zone.

Permafrost underlies approximately 80 percent of Alaska. The whole of the North Slope and a good deal of the western half of the state are zones of "continuous" permafrost. The seasonally frozen active layer which tops the permafrost is generally no more than a foot thick in these regions. In places like Prudhoe Bay, permafrost can be 2,000 feet thick. Most of the rest of Alaska lies in a zone of "discontinuous" permafrost. The permafrost areas become more scattered and less extensive

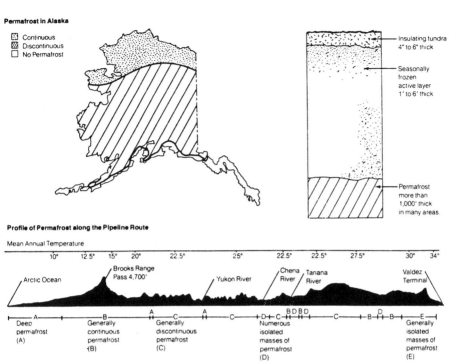

Permafrost in Alaska

☒ Continuous
▨ Discontinuous
☐ No Permafrost

Insulating tundra
4″ to 6″ thick

Seasonally
frozen
active layer
1′ to 6′ thick

Permafrost
more than
1,000′ thick
in many areas

Profile of Permafrost along the Pipeline Route

Mean Annual Temperature

| 10° | 12.5° | 15° | 20° | 22.5° | 25° | 22.5° | 22.5° | 27.5° | 30° | 34° |

Arctic Ocean
Brooks Range Pass 4,700′
Yukon River
Chena River
Tanana River
Valdez Terminal

A · C · A · C · D · C · B D B D · C · B · B · D · E

Deep permafrost (A)
Generally continuous permafrost (B)
Generally discontinuous permafrost (C)
Numerous isolated masses of permafrost (D)
Generally isolated masses of permafrost (E)

Courtesy of Alyeska Pipeline Service Company.

as one moves southwards. At its southernmost margins, permafrost constitutes only small, isolated islands, less than a foot thick.

From the soils engineer's point of view, the most important properties of permafrost are ice content and soil particle size and texture. These two factors largely determine whether permafrost is "thaw-stable" or "thaw-unstable," in terms of the soil's capacity to provide a strong foundation for construction. Permafrost soils that are rich in ice may be strong and hard when frozen but like thin mud when thawed. Large amounts of ice are usually present in permafrost when the active layer is thin, drainage poor, and the soil consists of fine-grained sediments, such as silts and clays. Sometimes, permanently frozen soil is in fact virtually ice. Soil that does contain large amounts of moisture can literally flow when the thermal balance is upset. On the other hand, permafrost in bedrock and properly drained, coarse-grained sediments such as glacial outwash gravel, and a number of sand and gravel combinations, can be practically water-free and is thaw-stable. In short, the strength of these soils is much the same in a frozen and a thawed condition.

Another significant property of permafrost is its temperature, how close it remains to 32 degrees Fahrenheit. In the continuous zone, permafrost usually stays below 30 degrees and may be as low as 10 degrees.

In the latter case, considerable heat can be applied without the permafrost melting. However, in the discontinuous zone much of the permafrost is just below 32 degrees and, in some spots, the addition of as little as half a degree will thaw it. This type is called "warm" to distinguish it from the "cold" permafrost found north of the Brooks Range. Further engineering problems result from the process of ice segregation that takes place during freezing where there is a large amount of moisture in the ground. This moisture accumulates in lens-shaped pockets "ice lenses"), layers, veins, and other masses. Ice segregation also occurs where the active layer freezes. As ice forms, it pushed the surface of the ground upwards. This process (discussed in chapter 1 in connection with the WAMCATS telegraph project at the turn of the century) is known as "heaving" and occurs outside permafrost zones as well. Unless a structure is built to resist this kind of movement, it will be pushed toward the surface during freezing and does not normally return to its former position when the active layer thaws again. This net upward movement is called "jacking."

The construction plans for conventional burial of the pipe, which TAPS officials presented at the hearings in Fairbanks in the fall of 1969, confirmed the worst fears of Interior's geologists. As Henry W. Coulter, the USGS official responsible for monitoring the survey's investigations

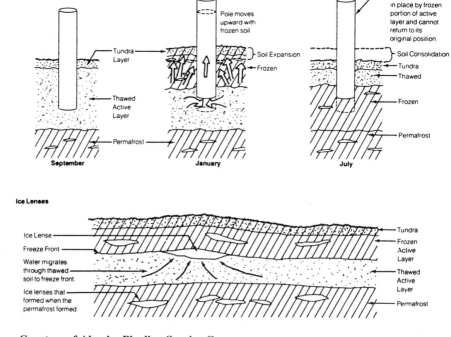

Courtesy of Alyeska Pipeline Service Company.

in Alaska, has recalled (1981): "TAPS simply planned to dig a ditch from one end of Alaska to the other and bury the pipeline in it, notwithstanding the existence of permafrost."[25] Undersecretary Train and USGS officials responded to TAPS's plans with information based on research by Arthur H. Lachenbruch of the geological survey's Menlo Park Working Group. According to Lachenbruch, "substantial amounts of thawing can be expected promptly in typical permafrost media." As a result, the pipe would "float" to the surface. Max Brewer exclaimed that in areas of ice-rich permafrost, "we could expect to see buried pipe swinging in the air."[26] Moreover, they expained that solifluction, the gravity controlled mass movement of thawed, water-saturated sediments, could occur on slopes as low as three degrees. Since many sections of the pipe would be on slopes, they could twist and break. According to Coulter, Sam M. Corns, chief engineer for TAPS, "appeared unable to comprehend the problem."[27]

In early September the Department of the Interior published a revised set of stipulations requiring the oil companies to accept absolute liability for cleaning-up in the event of any spills.[28] Train commented a month later: "To my knowledge, no private construction project has ever been asked to accept such strong constraints or such continuing strong direct control by the federal government."[29]

Boosters in Alaska saw these guidelines as a further addition to the crippling burden of regulations that had long strangled Alaskan development.[30] Sections of the Alaskan press attributed such "obstructionism" to the persistence of the icebox image. Albro Gregory, who had been a fervent supporter of Chariot and Rampart, alleged, in reaction to a critical *New York Times* editorial, that the writer of the piece in question "probably thinks, like many others, that Alaska is one large chunk of ice."[31] Why "outsiders" should care so much about the fate of an icy wasteland was not explained. Like many Alaskan pipeline proponents, Gregory seemed unaware of the emergence of a national environmental consciousness, especially with regard to Alaska, and the growing political power of the environmentalist lobby.

While still stubborn in spite of the serious engineering problems federal geologists had brought to its attention, the oil industry began to respond to other public concerns. In the summer of 1969 ARCO hired Angus Gavin, a former vice president of Ducks Unlimited for Canada, to study the ecology of the North Slope. Soon afterwards, ARCO signed contracts with the University of Alaska for revegetation experiments. The oil industry was primarily interested in revegetation to control erosion around the pipeline. Ralph F. Cox, ARCO's manager at Prudhoe Bay, claimed that the new grasses would be hardier and faster growing than native flora and would provide more abundant and better feed for herbivores.[32]

On 15 September the North Slope Task Force submitted a preliminary report to the president. Though convinced that "oil development and environmental protection are not incompatible," the task force listed numerous problems that remained unsolved. The report stressed that no American pipeline had ever faced problems comparable to those presented by Alaska's terrain and climate. On the North Slope, winter temperatures sometimes fell to minus 80 degrees Fahrenheit (with the windchill factor) but precipitation was low and the climate desert-like in some respects. Eight-hundred miles south at Valdez, temperatures were much milder, but snowfalls were heavy and total annual precipitation as high as 300 inches.[33] The task force's major concern was the effect of a hot oil pipeline buried in permafrost.[34] Sources of gravel were another important consideration because millions of cubic yards would be needed to insulate the permafrost around those sections of pipe that could be buried, to protect the permafrost under the construction work pad, and to lay the bed of the haul road. The report drew attention to the problems of human waste disposal and water pollution from land spills and marine discharge of oil. It also referred to seismic dangers (a big worry given that the lower two-thirds of the proposed route traversed a prime earthquake zone), and the potential disruption of wildlife activities north of the Yukon, especially caribou migration, by human and mechanical intrusion into a hitherto largely undisturbed area. In addition, there were difficulties stemming from Section 28 of the Mineral Leasing Act of 1920, which specified a maximum of 50 feet for rights-of-way, quite apart from Native land claims.[35]

Native claims constituted a restraining factor independent of reservations relating to the technological and environmental features of the project. The Senate hearings in the fall were dominated by concern over the project's environmental aspects. Yet they had been required primarily as a result of the land freeze Udall imposed to protect Native claims. Hickel had promised at his confirmation hearings that he would consult the Senate Interior Committee before acting to modify the freeze.[36] Senator Henry M. Jackson (Democrat, Washington), the chairman of the Senate Interior committee, had explained by way of introduction of the first part of the subsequent hearings on 9 September that "the direct interest and involvement of the Committee in the proposed pipeline . . . relates to the protection of the rights of the Alaskan Native people pending a legislative settlement of the Alaska Native land claims issue. . . ."[37]

On 30 September Hickel asked the Senate Interior Committee to approve a second modification of the land freeze to accommodate TAPS's right-of-way request.[38] He wanted to grant permits section-by-section as TAPS delivered the requisite information. He hoped this would avoid the delay that might result should the permit be made conditional upon approval of the project as a whole. His policy also applied to the haul road, which the Senate interior committee, to the chagrin of TAPS and the State of Alaska, insisted on treating as a separate matter. TAPS, however, was not prepared to accept Hickel's piecemeal approach.

Aware of his committee's responsibilities with regard to Native claims, Jackson deferred a decision on Hickel's request until public hearings had been held. Moreover, the committee was under mounting pressure from conservation organizations that saw Udall's freeze as the only defense against a wholesale disposal of Alaskan public lands on a nineteenth-century "first-come-first-served" basis. On 7 October, Senator Lee H. Metcalf (Democrat, Montana) received a telegram expressing alarm from Joseph Penfold, the director of the Izaak Walton League; Charles Callison, the executive vice president of the National Audubon Society; William E. Towell, the executive vice president of the American Forestry Association; Stewart Brandborg, the executive director of the Wilderness Society; Richard Stroud, the executive vice president of the Sport Fishing Institute; Edgar Wayburn of the Sierra Club; Thomas Kimball, the director of the National Wildlife Federation; R. A. Kotels, Washington representative of Trout Unlimited; and Fred Evenden of the Wildlife Society.[39]

The varied nature of their criticism was displayed in Washington, D.C., on 16 October. A prominent theme heard on this occasion was one that had been voiced with increasing regularity since the campaign for ANWR in the late 1950s; Alaska's status as the nation's last, large, undeveloped frontier region. Almost invariably mentioned in the same breath was the opportunity this gave Americans, in the words of David Brower, "not to repeat the mistakes we made in the south 48."[40] Whether utilitarian or preservationist, conservationist critics, without exception, preached extreme caution. Echoing federal officials, they all deplored the inadequacy of existing understanding of arctic ecology and the equally deficient state of technological competence in arctic regions that the proposal had brought to public attention. John C. Reed, who had been in charge of the U.S. Navy's postwar exploration of Pet 4, warned that the lessons of arctic petroleum history had not been learned. Now senior scientist at the Arctic Institute of North America (AINA), Reed emphasized that tracked vehicles scarred the tundra badly.[41] Mindful of the recent disaster at Santa Barbara, Robert Weeden advocated an immediate moratorium on further oil development and delivery plans to permit further study of the environmental problems involved and to formulate a long-term land use "master plan" that would restrain the private sector and protect public (notably wilderness) values.[42]

From Livengood to Valdez the pipeline route paralleled sections of the Alaska Highway and then followed the Richardson Highway. The Marine Terminal at Valdez was planned for a site originally occupied by Fort Liscum, the WAMCATS service camp. After the fort was abandoned when WAMCATS was superseded by the wireless telegraph in the 1920s, the site had been used as a sawmill and then as a cannery (Dayville). North of Livengood, however, there was no existing right-of-way except the Hickel Highway. Accordingly, some witnesses were most concerned about the project's impact on the Brooks Range, which was universally perceived by American conservationists as North America's, if not the world's, last great wilderness frontier. Tom J. Cade, a zoologist with extensive research experience on the North Slope, who testified on behalf of the Wilderness Society, lamented: "What I knew as wilderness twenty years ago is already a thing of the past in northern Alaska." He expressed doubt that one could find, "except in some of the mountain fastnesses of the Brooks Range . . . a one hundred square mile plot of ground east of the Colville River—including the Arctic National Wildlife Range—that does not show some irreparable sign of man's activities."[43]

The TAPS proposal made arctic wilderness protection an urgent subject of popular and public policy discussion. Senator Gaylord Nelson (Democrat, Wisconsin), a member of the Interior Committee who was in the vanguard of congressional support for environmentalist causes, approached arctic oil development as the crucial testing ground for the nation's nascent ecological conscience. Nelson's colleague, Senator Metcalf, realized that once the committee had assented to Hickel's request, it relinquished any further control over the project.[44]

Meanwhile, as the first sections of Japanese pipe began to arrive at Valdez, Alaskan proponents were growing increasingly restive. Deemphasizing TAPS's novelty and the environmental threats its posed, Senator Gravel compared it to the Haines Pipeline (part of the Canol Project built during the Second World War) and contrasted the absence of citizen protest against this earlier line with the intensive federal scrutiny of TAPS and conservationist uproar over the proposal.[45] C. W. Snedden, who had been an enthusiastic advocate of Chariot and Rampart, saw the continuing delay as another example of arbitrary interference by an unsympathetic federal government and conservationists, with a decision vital to the state's future once again being made in Washington, D.C. Boosters had hoped that Hickel's accession to national power would lead to a sharp reversal in federal attitudes toward Alaskan economic development. In view of Hickel's background, ideology, business interests, and gubernatorial policies, Snedden felt that Alaskans had been betrayed.[46] Governor Miller, returning from a trip to Japan, denounced the promise that Jackson's committee had extracted from Hickel at his confirmation hearings (to consult it prior to seeking a modification of the freeze) as "illegal"—an example of the legislative branch overriding the power of the executive. Miller reported that the Japanese were extremely interested in buying North Slope oil to reduce their dependency on Middle Eastern supplies.[47]

On 23 October Senator Jackson sent Hickel a list of questions reflecting the predominantly critical tone of the recent hearings, when only TAPS officials and the representatives of the State of Alaska had advocated immediate authorization of the project. Jackson indicated that satisfactory answers were required before the modification could be approved. The questions covered every aspect of the project but concentrated on Native claims and permafrost problems, especially in the Copper River basin north of the Valdez coastal mountains, a discontinuous zone of warm permafrost. They also emphasized Hickel's failure to disclose the contents of the North Slope task force's presidential report (15 September) and the problems of effective oversight by the BLM (funding, staffing, and stipulation enforcement).[48]

Replying almost a month later, Hickel asserted that there was no conflict between granting a permit and the just settlement of Native claims; that no Native village or association had "reacted unfavorably" to the project; and that the Mineral Leasing Act authorized the right-of-way. He also denied that Interior had an official policy for "opening up" Alaska north of the Yukon. On the other hand, he admitted that engineering problems, especially with regard to permafrost, remained considerable. Then he identified the danger of oil spillage as the project's major public risk. (The task force had noted that each mile of pipeline would carry half a million gallons of oil, twice the amount discharged into the Santa Barbara Channel in January.)[50] Hickel described the main public benefits as being national (oil delivery to the West Coast, where a grave shortage existed) and local (jobs for Alaskan Natives).[50]

On 3 December Hickel informed the Senate Interior Committee that his department would not sanction the project until TAPS had solved its problems with permafrost. Apparently reassured, the committee notified Hickel on 11 December that it no longer objected to him lifting the freeze for the road, provided the project did not hinder the settlement of Native claims and that the North Slope Task Force's stipulations were properly enforced. The committee's decision was influenced by the imminent passage of the National Environmental Policy Act (NEPA). Jackson pointed out to Hickel (11 December) that NEPA, which he had co-sponsored, was in its final form and ready to be submitted for a final vote.

Prior to NEPA, there was no mechanism within the American political structure at any level—federal or state—to compel consideration of environmental factors in the policy-making process. Instead of government simply reacting to environmental problems and crises once they had occurred, Jackson believed that institutions and procedures should be set up to anticipate them. At a practical level, Jackson hoped to achieve this by influencing the activities of the federal agencies and departments responsible for approving and administering projects. The bill had been making its way through the various legislative stages unobtrusively. It is doubtful that many of the overwhelming majority of congressmen who supported NEPA were aware of its extensive implications for a variety of projects, large and small.[51]

Jackson advised Hickel to pay close attention to the implications of NEPA for the TAPS proposal because section 102(2)(C) of the act required a detailed public statement of the environmental impact of any project that was federally financed or that involved federal lands. However, NEPA did not indicate whether it was sufficient for the agency in question to just identify and describe the environmental impacts it anticipated from a given project or proposal and then to go ahead and authorize it. The act provided no criteria for deciding what constituted an intolerable environmental impact or whether a project should be stopped if an agency discovered unacceptable environmental impacts. It would obviously be left to the judiciary to decide whether an agency had complied with the demands of NEPA.

The House Interior and Insular Affairs Committee voted in favor of Hickel's action (16 December) by fifteen to seven. Hickel now modified the freeze to enable him to grant the road permit. Though empowered to do so, and under enormous pressure from Alaskans, he did not issue the permit immediately. His restraint reflected three considerations: his serious doubts about the project's feasibility at this point, his awareness of the opposition's political influence, and, not least, his need to improve his standing in the environmental community. Since he became interior secretary, frontier-conquering rhetoric and sentiments had been conspicuously absent from his pronouncements. Conservation policies were urgently needed in the United States, he explained (without referring to Alaska) because "now there is no more frontier."[52]

The BLM began to reclassify 5 million acres along the proposed road and pipeline route as a transportation corridor. Taking note of the Mineral Leasing Act and the North Slope Task Force's objections, TAPS refiled its application in December, requesting the 54 feet maximum legal right-of-way and temporary Special Land Use Permits (SLUPs) for additional footage of 200 feet along the entire road and up to 500 feet in places (river crossings). On 7 January 1970 Hickel signed PLO 4760, which enabled him to grant a permit for the 390-mile road from the Yukon to Prudhoe Bay. Anticipating the permit soon and eager to start construction in the spring, TAPS began issuing letters of intent to contractors who wanted to move their equipment north over the Hickel Highway to strategic locations before the spring thaw when the ice bridge over the Yukon would become unusable. In mid-January the road had been reopened from Livengood to Bettles at a cost of $433,000. In early March, the Alaska State Legislature appropriated a further $250,000 to re-open the road between Bettles and Sagwon. Since much of the original road was useless, construction began on a new one parallel to it.

Against the background of these activities, William Pecora, now chief of the North Slope Task Force, had addressed the State Senate in Juneau in early February. Pecora frowned on the impatience which seemed to be "the order of the day" in Alaska and warned that it was extremely unlikely that construction would begin on the pipeline itself that year because "we are still in the stage of groping for answers."[53] The Department of the Interior, especially its Menlo Park Working Group, had had no choice, in view of the pipeline consortium's inadequacies, but to take a more active, interventionist role than it originally intended.

7.2 THE PROJECT ENCOUNTERS NEW OBSTACLES, MARCH 1970–JANUARY 1971

On 5 March, Hickel informed the president of his intention to issue the road permit. To comply with NEPA (effective 1 January 1970), Interior produced an eight-page environmental impact statement that found no environmental dangers arising from the project.[54] In spite of his efforts to reconcile conservationists in other respects, Hickel remained steadfast in his desire to see northern Alaska "opened up." Nothing had affected his conviction that the state had a right to a permanent road link with the Arctic Ocean.[55] He was poised to issue the road permit when the entire project was halted. On 9 March five Native villages north of Fairbanks filed suit against Hickel's proposed action in the federal district court in Washington, D.C. The suit claimed that TAPS had failed to honor a promise, made the previous summer, to hire Native contractors and workers for the project. The plaintiffs explained that only on the basis of this understanding, which TAPS denied having entered into, had the Tanana Chiefs Conference (the regional Native political organization) agreed to waive its claims to the right-of-way.[56]

On 26 March the Wilderness Society, Friends of the Earth (FOE), and the Environmental Defense Fund (EDF) brought their own action against Interior in the same court. Their suit claimed that the TAPS proposal violated both the Mineral Leasing Act and section 102(2)(D) of NEPA, which required the study of alternatives to the proposed action.[57] On 3 April, Judge George L. Hart, Jr., issued a temporary restraining order that enjoined Interior from issuing a construction permit for the road across 19.8 miles of the proposed route that traversed land claimed by the sixty-six residents of Stevens Village, a settlement close to where the pipeline would cross the Yukon. On 13 April Judge Hart also ruled that since the road was an "inseparable part" of the pipeline project, an impact statement covering the entire proposal had to be prepared according to section 102(2)(C) of NEPA before Hickel could decide to approval all or any part of it. Accordingly, Hart issued a preliminary injunction against the entire project.[58] In this way, the pipeline became one of the first major tests of NEPA's effectiveness as a law for advancing the environmentalist cause.[59] In addition, Hart ruled that the project violated the maximum right-of-way provisions contained in the Mineral Leasing Act.

The Native litigants emphasized that their injunction was different from the environmentalist suit. "We are talking about property, not conservation," declared David P. Wolf, attorney for Stevens Village, lest an identity of interest and outlook should be assumed: "This is similar to practice everywhere. If the state or some government unit takes your land for public purpose you expect to be reimbursed for that land."[60] The preliminary injunction went into effect on 23 April. As a result, road-building equipment sat idle at ten camps north of the Yukon. For Alaskan contractors who had invested heavily in anticipation in springtime construction, the additional delay was disastrous.

It was left, unenviably but not inappropriately, to Hickel to spell out what many Alaskan proponents could not understand or accept. The secretary of the interior delivered the major speech at the University of Alaska's celebrations to mark Earth Day (22 April), a series of nationwide environmental observances at over 2,000 colleges and universities, organized by Environmental Teach-In, a student body set up on the initiative of Senator Gaylord Nelson.[61] Hickel explained that delay was inherent in the project's complex nature: "I am not convinced that we have encountered any delay that would not have been encountered by the pipeline people in any event . . . even if they had been granted [the permit] the very day they applied for it." He emphasized that any pipeline crossing public lands in the lower forty-eight would be subject to a comparable delay. He stressed that if anyone was to blame for the delay it was TAPS, which had underestimated the formidable nature of the undertaking and had given Interior no reason to believe it was capable of doing the job properly. Hickel reminded Alaskans that it was not the federal government's responsibility to design the pipeline. Nevertheless, he was confident that TAPS could supply the necessary information and chose this occasion to assure Alaskan advocates that he would issue the permit eventually: "There is no longer a question of 'if.' . . . And I have announced that I will issue the permit for the pipeline right-of-way."[62]

However, Alaskan supporters remained unconvinced. Some felt that other, more sinister, factors were at work. The date chosen for Earth Day, 22 April, was also Arbor Day, but Joe Vogler, a miner and property developer from Fairbanks, drew attention to the fact that this date was Lenin's birthday[63]—as did a number of ultraconservative congressmen from the South and organizations like the Daughters of the American Revolution. Three days after Hickel's speech, an Alaskan task force led by Governor Miller arrived in Washington, D.C., to lobby for the immediate issue of a permit for the entire project. They hoped to achieve this by explaining its economic importance, not only to Alaska but to the nation as a whole, and by convincing the Department of the Interior and national opinion that Alaskans were deeply committed to environmental protection. (A briefing prepared for the task force warned its members: "Don't talk strictly economics.")[64] The delegation of 120, composed of businessmen, community leaders, and labor officials, included Robert Atwood, George Sullivan (the mayor of Anchorage), C. W. Snedden, H. A. "Red" Boucher (acting mayor of Fairbanks), and Hickel's brother and business partner, Vernon Hickel. These men viewed the pipeline situation as a return to the days of Alaska's prestatehood subservience and pondered the irony of their plight. Shunned and misperceived for a century, they complained, Alaska and its future had suddenly become of tremendous interest to other Americans, but, unfortunately, not in a way that served their own cause.[65]

While in the capital, Miller announced his plans to approach his legislature for $120 million so that the state could build the road to Prudhoe Bay itself, thereby putting the inactive equipment north of the Yukon to work. Miller claimed that a statute dating from 1866, which granted to states rights-of-way across unreserved public lands, empowered him to authorize the road. The *Fairbanks Daily News-Miner* (27 April) praised his proposal as being "very much in the Hickel tradition" and most of the Alaskan press agreed that a road was part of the state's untransacted but clearly manifest destiny. The veteran journalist, Tom Snapp, reflected editorially two months later (3 July) in the first issue of his newly founded *All-Alaska Weekly* that "for conservationists to attempt to block a single road being built to open up an area of Alaska roughly the size of Texas makes as much sense to us as an attempt to block a Daniel Boone from cutting a trail through a Cumberland Gap." The Alaska Legislature proceeded to pass a bill authorizing construction, but only if TAPS reimbursed the full cost. This was met with consternation by the consortium, which informed Governor Miller that it preferred to proceed through the regular political channels.

The stubborn failure of Miller's task force to grasp the true nature of the delay astounded members of the federal North Slope task force. William Pecora, who was then undersecretary at Interior as well as task force chief, repeated what he had said in Juneau nearly three months earlier. He explained to the Alaskan lobbyists, "We have not seen any design plan for the pipeline yet—below or above ground."[66] Pecora stressed that the Russians, who had pioneered in building pipelines (for gas) in permafrost regions, had elevated all their lines. He reminded them that "our pipeline people never have built a line in permafrost."[67] However, the oil industry sought to downgrade the problems associated with construction and downplayed the project's novelty and potential for environmental damage. One tactic was to draw attention to "epic" feats of engineering from Alaskan history, such as the Davidson Ditch, which the industry portrayed as a reassuring precedent. *Oil and Gas Journal* asserted, with reference to the ditch: "The problems encountered and the solutions rendered by the Fairbanks Exploration Company compose a basis for the assurance that the engineering of the proposed Trans-Alaska Pipeline System, a similar project, will also meet success. . . . The problems have been solved and are being solved again."[68]

A few more astute Alaskan commentators frowned upon this complacency and rejected the booster explanation. In April, Robert Zelnick, a journalist, addressed the question of responsibility for the delay. He dismissed the viewpoint that conversationist opposition, notably in the form of the injunction, was the chief obstacle. Instead, the blamed TAPS (a "non-entity") and praised Hickel's "tough" stance.[69] Robert Knox, the editor of *Alaska Industry* and no foe of oil development, regretted the delay as much as TAPS and Miller's task force. Soon after the injunctions went into force, he also assigned primary responsibility to TAPS's dubious technological competence and total inexperience in arctic engineering.[70]

Another perceptive observer was Arlon R. Tussing, a economist with the University of Alaska's Institute of Social, Economic and Government Research (ISEGR). In July 1970 he delivered a paper to the Anchorage chapter of the Alaska Press Club in which he argued that the media and the political and business community in Alaska had failed to communicate the real reasons for the delay. As a result, he claimed, the Alaskan public had been misled and an atmosphere of naive hopefulness had been created. The genuine sources of delay that Tussing identified were five-fold: Firstly, he agreed with Pecora and Hickel that TAPS was still incapable of building its pipeline. The second important factor was the ineptitude of TAPS as an organization. NEPA was the third reason he gave and the objections of private environmentalism were the fourth. However, the crucial obstacle in Tussing's mind was Native land claims, and he was surprised that the oil industry had neglected this issue. He predicted that it would probably be between four and six years before the pipeline was authorized.[71]

The formation in August 1970 of Alyeska Pipeline Service Company, a formal corporate entity, was a great relief for all federal officials who had dealings with the pipeline concern. Comments on the TAPS consortium's defects flowed freely from those who had contact with it. Hickel himself wryly described the difficulties of dealing with an organization that operated on a committee basis without a single point of contact and an effective executive organ. TAPS, he recalled two years after its replacement by Alyeska, had "more consultants than a psychiatric ward, more colonels than Kentucky, and not a general in the place." What is more, he continued, "by the time all the principals could be telephoned in Texas, New York, California and London, it could take weeks just to get an agreement on what color to paint the toilets in the construction camps."[72]

Alyeska began to respond to certain criticisms made of TAPS and to tackle some of the obstacles its predecessor had faced. Above all, Alyeska awoke to the importance of settling Native claims and spent the rest of the year and much of 1971 lobbying for their despatch. BP took the initiative here, hiring the two former aides to the late Senator Bartlett, Hugh C. Gallagher and William C. Foster, to lead its efforts on Capitol Hill.[73] Yet even at this stage, the oil industry failed to act positively with respect to Tussing's first point. At late as August, Hickel's press secretary, Josef Holbert, announced that Interior had still not seen "a single sheet of design specifications."[74] Because Alyeska was slow supplying the necessary data, Interior could not begin to prepare an environmental impact statement until October 1970. Once Alyeska had delivered its design and engineering information, Interior completed the statement within a month.

Alyeska did try to improve the oil industry's environmental credibility among the public. The Arctic, qualifying as a cold desert by virtue of its aridity, is characterized by a rate of biodegradation as slow as the Grand Canyon's. It had become commonplace to remark that there was no other part of the world, with the possible exception of the hot deserts, so badly designed for withstanding and concealing the traces of humankind and its technology. The federal task force noted in its report to the president (15 September 1969) that "orange peelings last months, paper lasts years, wood scraps last decades and metal and plastic are almost immortal. . . ."[75] In 1970 film wrappers and debris from plastic apple trees carelessly discarded during oil company public relations shooting on the ice of the Beaufort Sea were enough to provoke an outcry from easily offended conservationists.[76]

A notorious example of "arctic graffiti" fifty miles east of Prudhoe Bay was frequently cited in the press and conservationist bulletins. A bulldozer crew had engraved the initials of an exploration company, Geophysical Services Inc. (GSI), in letters 150 yards high. This damage on the west bank of the Canning River, which formed the western boundary of ANWR, had been inflicted in 1965. By 1970 the letters had eroded into 10-foot-deep channels. "Horror" pictures of the tundra criss-crossed with the scars of "cat tracks" were rife in conservationist journals.[77]

These publications concentrated on the aesthetic damage caused by so-called scarification. Rarely if ever mentioned, let alone discussed, were its geomorphological and ecological consequences. The North Slope around Prudhoe Bay is dotted with small lakes and dominated by a polygonal pattern caused by a process known as the thaw lake cycle. These landscape features have been shaped by the natural process of freeze and thaw. Artificial disturbance of the terrain aggravates natural processes considerably. Damage to the insulating mat of vegetation above the active layer induces permafrost thaw. A pool of standing water soon becomes a shallow pond and eventually a thaw lake forms. These lakes are rapidly enlarged as the underlying permafrost continues to thaw and the banks are eroded by waves produced by the high winds characteristic of the region. The lakes grow quickest along the axis perpendicular to the wind and become know as oriented thaw lakes. As a result, existing drainage patterns are significantly altered. When it captures a stream, a lake drains. The exact nature of the ecological injury and long-term biotic repercussions of scarification and the alteration of drainage patterns for the complex arctic ecosystem are far from clear. However, small mammals (shrews, lemmings, voles, and mink) cannot easily cross the eroded ruts originally caused by vehicles. The formation of thaw lakes reduces the availability of food and shelter in an environment already characterized by scarcity. Changes in drainage patterns can flood the dens of small animals. Moreover, heavy loads of silt resulting from erosive activity in thaw lakes can be carried into streams and interfere with fish spawning.[78]

Public indignation ran high and protest was loud over the trash strewn across the Arctic Slope, especially over the host of empty 55-gallon fuel drums. As part of a concerted effort to improve its public standing, the TAPS consortium had launched a clean-up drive, which Alyeska continued. Much of the trash the oil companies were obliged to remove— which consisted to a large extent of these fuel drums—represented the combined legacy of the U.S. Navy's postwar Pet 4 explorations and the construction and operation of the DEWline. As two prominent Alaskan wildlife biologists pointed out, the worst scarification on the North Slope was inflicted by "cat" trains during the exploration of Pet 4.[79] Many of the problems associated with arctic oil exploration, which caused so much consternation in 1970, existed two decades before they were discovered by large numbers of conservationists.

The industry's appreciation of the need to respond to damaging publicity was also reflected in its efforts to incorporate the new environmental awareness into its public image and policies. This strategy to win over the opposition was reflected in an advertising campaign begun in the summer of 1970. In the fall ARCO ran an advertisement concerning environmental enhancement through the revegetation of disturbed tundra (to prevent erosion). This bore the caption: "How to help keep America beautiful . . . plant a lawn in a deep freeze."[80] ARCO explained that good husbandry such as this was designed to make "the world we live in . . . just a little bit better than when we started."[81]

The oil industry had also grown more cautious than Alaskan politicians and businessmen in its use of frontier rhetoric. Aware that publicly uttered frontier-busting sentiments had lately become a liability, the industry borrowed words such as "delicate" and "fragile" from the conservationist lexicon. At pains to counter its "hit-and-run" image, it emphasized how careful it was not to disturb the existing life on the North Slope. British Petroleum advertised its stewardship with a picture of its clean-up crew on the North Slope under the caption "the Neighborhood Improvement Association."[82] Another advertisement ("ah, wilderness!") showed a discarded oil drum dominating an otherwise featureless white desert. BP gravely called its presence there "one of man's most enduring monuments." BP explained that since "the frozen tundra refuses to accept it for burial," clean-up crews were painstakingly collecting drums and flying them out at great expense. This was being done because "when B.P. Alaska touches the wilderness, we try to touch it gently."[83]

This policy contrasted with the traditional advertising symbols still used by Alaskan construction firms and contractors, in which the overriding image of a merciless, hostile land survived. In mid-winter at Prudhoe Bay, the sun stays below the horizon for forty-nine consecutive days. Mean temperatures for January, February, and March are minus 30 degrees Fahrenheit and are usually driven lower by the windchill factor. Insurance Incorporated, understandably, evoked in its advertisements the scenario of "mile after mile of frozen tundra. Perma-frost infested deep in the earth. A harsh wilderness, a relentless climate," and added, "but this is where you work."[84] The Burgess Construction Company (Fairbanks), which had built the road from Livengood to the Yukon River in the summer of 1969, was dedicated to "Opening Up the Alaskan Wilds" for "a better Alaska."[85] ERA Helicopters spoke proudly of "providing the hard-working men and machines to help tame Alaska's wilderness."[86] Even the oil industry was reluctant to relinquish the frontier tradition in its entirety. While divesting itself of the discredited aspects of the pioneer ethos, it still hoped to capitalize on some of the frontier heritage. Environmentally sensitive language notwithstanding, the apparently indelible old image appeared alongside the new one. Prudhoe Bay, 200 miles above the "ARCO Circle," remained a dark, frozen, and rugged place at "High noon" in winter.[87]

7.3 THE DRAFT IMPACT STATEMENT AND THE PUBLIC HEARINGS: THE RANGE OF CRITICS AND CRITICISM

There could be no major developments before the Department of the Interior released its draft environmental impact statement for the project in January 1971. The statement recognized that there would be some oil spillage, but denied that ecological damage would be significant. It argued that most of the disturbances would be temporary and that caribou were "adaptable" and unlikely to be restricted in their movements by the pipeline.[88] However, the authors freely conceded the drastic impact of a transportation corridor that would split the unity of arctic Alaska in half: "It is clearly recognized that no stipulation can alter the fundamental change that development would bring to this area. . . . for those to whom unbroken wilderness is most important, the entire project is adverse . . . because the original character of this corridor area in northern Alaska would be lost forever."[89] The authors accepted that a pipeline across Alaska would be the least environmentally destructive and most economically advantageous method of transporting the oil deemed essential to meet the nation's (particularly the West Coast's) rapidly expanding energy requirements.

The statement's treatment of other sources of energy was curt, referring to increased coal production, tar shale extraction, and nuclear power as other, much less satisfactory ways to meet energy needs. Within the realm of oil, increased production from fields in the lower forty-eight and greater importation were briefly considered. Discussion of other pipeline routes and other modes of transportation was also cursory. Two pages were devoted to a possible Canadian route via the Mackenzie Valley. But the authors argued that this would "shift the location of ecological problems rather than cure them." They also rejected a Canadian line because it would be twice as long as the Trans-Alaska line, because of the complexities of negotiating with a foreign government, and, not least, due to the two-to-four year delay in oil delivery that they envisaged.[90]

The statement touched upon the alternative forms of transit that had been considered before TAPS chose a pipeline. During the summer of 1969, Humble Oil undertook a $50 million experiment involving the tanker S.S. *Manhattan*. At 115,000 tons deadweight, this ship was the largest merchant flag in the American fleet. In preparation for the 10,000-mile round trip from Philadelphia to Prudhoe Bay it was cut into sections and distributed to various shipyards for reinforcement and conversion into an icebreaker. Having been welded back together, the S.S. *Manhattan* embarked upon a historic three-month voyage through the Northwest Passage, succeeding as the first commercial vessel to do so, but only with the help of Canadian and American icebreakers. These ships were often called upon to disengage the beleaguered tanker from the pack ice. The S.S. *Manhattan* suffered considerable damage before it finally reached Prudhoe Bay, where it picked up a symbolic barrel of oil before returning to the East Coast.[91]

Its voyage was under way at the time of the first pipeline hearings (1969), when Dulaney's reply to an inquiry about the ship's progress confirmed the consortium's dedication to a pipeline. "The trans-Alaska Pipeline must be built, regardless of the success or failure of the *Manhattan*," declared the chairman of TAPS.[92] There were other ambitious modes that were less seriously considered than icebreaking supertankers, and soon dismissed as too expensive, too dangerous, too difficult, and too time-consuming to develop. These included Boeing jumbo jet tankers and General Dynamics Corporation's proposal to build 170,000 ton, 900-foot-long nuclear-powered submarines that would cruise under the ice pack. The chief disadvantage of the proposal to move oil out to the Pacific by submarine were the precariously shallow waters of the Bering Sea, which, the authors of the impact statement explained, would not provide sufficient clearance for a submarine with a hull 85 feet deep. The impact statement also considered the waters of the Beaufort Sea insufficiently deep for submarines headed west to terminals in Greenland or Newfoundland. Any form of marine transportation would also have to contend with the near-impossibility of constructing a tanker-loading terminal in the shallow waters off Prudhoe Bay, which those in charge of Pet 4 explorations had noted in the 1920s and 1940s. Adequate clearance for supertankers such as the S.S. *Manhattan* (at least 90 feet of water) is not available within twenty miles of the shoreline at Prudhoe Bay. In addition, the impact statement stressed that it would be difficult to build a loading terminal strong enough to withstand the assault of the shifting ice pack.[93]

Another suggested mode was an old booster favorite; an extension of the Alaska Railroad. Quoting the statistic that in 1968, 76 percent of crude oil in the United States was moved by pipeline, 23 percent by ship and barge, and a mere 1 percent by railroad tank cars, the statement's authors concluded that a pipeline was by far the cheapest and safest method available. They also claimed that a pipeline was the soundest method ecologically. A railroad, they argued, would require far more gravel and trains would pose a continual hazard to wildlife. They claimed that the number of trains required to move the same amount of oil as the pipeline would make a railroad "the busiest . . . in the world."[94]

Some conservationists also supported a railroad–but through Canada, not Alaska. David Brower of Friends of the Earth preferred nonexploitation of oil, but, assuming that extraction was unavoidable, was willing to support a railroad down the Mackenzie, connecting with the Great Slave Lake railhead. He claimed that a railroad would be "spillproof," and "earthquake-proof," and would avoid the marine pollution threat and the dangerous seismic zone in southcentral Alaska.[95] From an early date, Canadian conservationists saw a contradiction in such a policy. In May 1970 Terry Simmons of the Sierra Club of British Columbia informed the club's national director, Michael McCloskey, that "the construction of a pipeline in the Yukon would leave the conservationists of North America with a hollow victory."[96] George Marshall confided to Samuel Wright that any pipeline or railroad that crossed the Arctic National Wildlife Range would lead "from frying pan into the fire."[97]

A few days after the impact statement was released, Rogers C. B. Morton replaced Hickel as secretary of the interior.[98] Morton's appointment did not improve the project's chances of swift approval. The former congressman from Maryland stressed that he was not subject to the same pressures from the oil industry and Alaskan businessmen that Hickel had been. At Morton's Senate confirmation hearings at the end of January, which were dominated by discussion of the pipeline, Alaska's Senator Stevens asked the secretary-designate whether construction could begin that year. Morton responded with cautious criticism of the draft impact statement, referring particularly to its failure to address engineering problems.[99] However, Morton reaffirmed the department's fundamental support for the pipeline, which, he explained, "has been considered the method most efficient and least detrimental to the environment."[100] Nevertheless, the following month at Senate Appropriations Subcommittee hearings on funding for the Department of the Interior, Morton stated that he was "a long way" from approving TAPS.[101]

The impact statement released in January 1971 was not intended to be definitive, a point which Morton stressed. The introduction contained the caveat: "It should be understood that any of the statements or conclusions contained herein is subject to change and that it is expected that the final environmental statement may differ in significant respects from this draft." NEPA had established a three-member Council on Environmental Quality (CEQ). The council's guidelines for implementing NEPA required that the draft statement be circulated to the public for comment. Accordingly, the Senate Interior Committee held public hearings in February.

The hearings indicated how widespread opposition to TAPS continued to be at both federal and private levels. The 246-page document was roundly denounced in Washington, D.C., between 16 and 18 February and in Anchorage from 24 February until 1 March. Due to an unexpectedly high level of public interest, the hearings ran overtime in both cities. By the end of March, more than 12,000 pages of direct testimony, appendixes, and comments had been amassed. The complete record consisted of thirty-seven volumes. Its distinctive feature was the large contribution from those without conservationist, scientific, or academic backgrounds, in addition to the professionals in these fields who had dominated the hearings in the fall of 1969. Again, however, the great majority of witnesses were critical.

A prominent "expert" critic was Ira L. Wiggins, a professor of biology at Stanford University who had been Scientific Director of the Naval Arctic Research Laboratory at Point Barrow (1947) from 1949 until 1956. In this capacity, Wiggins had been closely involved with the U.S. Navy's postwar exploration of Pet 4. Wiggins found nothing in the statement to indicate that the environmental lessons of this earlier experience had been taken into account.[102]

Critics of Alyeska's state of unpreparedness placed great emphasis on a report by a prominent member of the USGS's Menlo Park pipeline technical task force. This report, by Arthur H. Lachenbruch, represented the culmination of federal concern over the project's engineering features. The so-called Lachenbruch Report of December 1970 (to which the draft environmental impact statement briefly referred) used heat conduction theory to describe the "bulb" thaw effect in permafrost of a pipe containing oil at a temperature of 158–176 degrees Fahrenheit as a result of the initial temperature of the oil (coming from depths greater than 2,000 meters) and frictional heating caused by movement. Lachenbruch estimated that a 48-inch diameter pipe—the size ordered by TAPS—when buried 6 feet below the surface, would thaw a cylindrical region around itself 20–30 feet in diameter within a few years in typical permafrost conditions. Lachenbruch warned that after twenty years of pipeline operation, the thawed area could become enlarged to 50 feet in southern areas of discontinuous permafrost and to between 35 and 40 feet in northern Alaska where the permafrost is colder. He explained that this dissolution of the permafrost would produce large quantities of slurry which could lead to severe slumping as a result of solifluction. Accordingly, the pipeline might rupture.[103]

TAPS was not the first engineering project to face the problem of permafrost. But the oil industry seemed unaware of numerous historical warnings. Permafrost had caused difficulties for those who built the Trans-Siberian Railroad, early Alaskan wagon roads, the Alaska Railroad, the Copper River and Northwestern Railroad, the Alaska Highway, and Canol. Most pertinently and more recently, Soviet experience with gas pipelines through permafrost regions in Siberia also placed TAPS's problems in a wider context. The erosion caused by vehicular disturbance of the mat of insulating vegetation, the dangers of burial, and the need for raised and supported sections of pipe were noted in a Soviet report consulted by USGS officials.[104] The American steel industry certainly acknowledged the lessons to be learned from Soviet pipelining. *Iron Age,* the industry's trade journal, commented that, in view of the Soviet experiences, conservationists who were wary of TAPS were not "necessarily the alarmists that they're sometimes labelled."[105]

The impact statement's most distinguished lay critic was the former interior secretary, Stewart Udall, who represented the Wilderness Society at the hearings. According to Udall, the statement brought "no credit on the department" and he professed himself "deeply disturbed by its glaring omissions." He argued that the present TAPS study compared unfavorably with the Atomic Energy Commission's thorough environmental studies for Project Chariot in the early 1960s, as a result of which, he claimed, the atomic blast at Cape Thompson had been canceled. According to Udall, "There was no predetermined conclusion for that impact study."[106] Concerning omissions, criticism focused on two areas. The first was the danger of tanker spills in Prince William Sound. Like other critics, Udall argued that an adequate and credible statement demanded detailed consideration of all aspects of the delivery system. The second area of concern was the statement's violation of NEPA because it failed to broadly consider all alternatives, including the possibility of no action ("deferral").

The TAPS controversy had already become the *cause célèbre* of "the Age of Ecology," a label coined by journalists for the new decade in 1970.[107] A good deal of the criticism heard at the hearings in Washington, D.C., and Anchorage and submitted for the record did not relate to any specific environmental or engineering problem. In this sense, the tone of the hearings reflected the influence of the new ecological perspective, which, in its extreme form, resembled abolitionism and prohibitionism in the strength of its moral ardor. Some critics viewed the proposal as a symptom of a sick, "gluttonous," and sordidly materialistic society that was overdependent on technology and built precariously on a profligate and highly polluting economy as much as they saw it as an insult to the last great American wilderness.[108] They were dissenters from what they saw as the national secular religion of unbridled economic growth. For them, the project served as a springboard for diatribes against a number of manifestations of a chronic malaise in American culture, including electric toothbrushes, smog, and the Vietnam "war machine."[109] There was a powerful antitechnological streak in the makeup of radical environmentalism in the late 1960s and early 1970s. The TAPS project became a prime target for the assault on so-called technocratic chauvinism and runaway science, which, many believed, bore a heavy responsibility for the ecological crisis.

A further ideological component of TAPS criticism characteristic of the new environmental ethos was biocentricity ("deep ecology").[110] At root, applied ecological science was a subversive force, as suggested by the title of a book by Paul Shepard and Daniel McKinley, *The Subversive Science: Essays Toward an Ecology of Man*, 1969. For the philosophical objective of the advocates of "biotic rights" and the champions of "ecological equalitarianism" was to dislodge humankind from its dominant position in the earth community.[111] In Anchorage, Bob Marshall's protégé, Samuel Wright, assumed responsibility for what Friends of the Earth (FOE) called "the job of representing Alaska's disenfranchised electorate."[112] Liberationists such as Wright sought to free from human overlordship the long-established yet "voiceless" faunal and floral residents of the Brooks Range wilderness. In Wright's opinion, the caribou, "who have called this their home for thousands of years," were Alaska's first citizens. Among the plants for whom he claimed aboriginal rights were the dwarf spruce at northernmost treeline. Wright emphasized that these arctic trees, though still only a few feet tall, had already been seedlings "when George Washington was inaugurated."[113] Wright concluded that "all the oil in Alaska is not worth the loss of this last great wilderness."[114]

On the other hand, most Alaskan pipeline proponents with less daily distance from wilderness than many of its admirers saw nothing benign or uplifting about it. The feelings toward wilderness of Ernest Wolff, the president of the Alaska Miners Association, were reminiscent of the attitude of the first Europeans in the New World. Wolff saw wilderness as hostile, harsh, and ultimately life-threatening. Referring to the cabin at Big Lake, near Wiseman, which Wright had reoccupied, he declared: "My partner froze in 1957 trying to reach that cabin. He froze to death. That's the wilderness."[115] Wright had spoken of the need to defend the Brooks Range "from within" this wilderness, of which he described himself as "a resident," alongside the caribou and the dwarf spruce he had championed.[116] However, Wolff argued that Wright was living "in a little highland of civilization established by miners," which contradicted the wilderness concept.[117]

The oil industry and its Alaskan supporters did not lament the loss of wilderness. On the other hand, they did not necessarily condone environmental degradation. From their point of view, the best use of the land was the broadest and so the land management principle that they favored was "multiple use," a 1960s term for the brand of conservation originally associated with Gifford Pinchot. Some conservationists agreed. Charles S. Collins, the Pacific Coast representative of the Izaak Walton League of America, thought the oilfields at Prudhoe Bay were clean and orderly in contrast to the scruffy condition of other settled parts of the state (especially Native villages).[118]

Pipeline advocates also emphasized the modest, even innocuous nature, of the project. Ted Borden, the deputy director of the State of Alaska's Industrial Development Division, stressed that the pipeline corridor would only occupy sixty square miles; one-hundredth of 1 percent of the total area of Alaska.[119] Vide Bartlett, the widow of the late Senator Bartlett, strained her imagination to convey this relative insignificance. "If you took a daily newspaper," she explained, "laid it out to one full double page and stretched across it a piece of ordinary black thread, you'd have some idea of just how much the pipeline is going to take."[110] In one advertisement, Alyeska compared its role in Alaska to that of the Matanuska colonists in the 1930s, who, Alyeska explained, brought agriculture to Alaska. Alyeska stressed that "the land to be used for construction of our pipeline would fit comfortably into one small corner of the Matanuska Valley."[121]

For preservationists, however, such proportions and statistics were irrelevant since the project was alien in principle to their image of northern Alaska. In Anchorage, one Alaskan supporter dismissed the pipeline as no more than "a scratch on the back of a hand,"[122] to which an opponent, drawing on Shakespeare, had replied: "Somehow I am reminded of the scene from Romeo and Juliet, when Mercutio had been run through with his sword, and Romeo said to him, 'Courage man, the hurt cannot be much.' And Mercutio replies, 'No, 'tis not so deep as a well, nor so wide as a church-door but 'tis enough, 'twill serve'."[123]

The pipeline's potential impact on specific aspects of the natural environment and its biota (for example, the effect on fish spawning of siltation caused by removing gravel from streams) were identifiable, to some extent quantifiable, and, with careful planning and construction, mitigable in part. What could not be measured and was not amenable to any form of redress was the inscrutable spiritual, emotional, psychological, philosophical, and symbolic injury perpetrated by *any* pipeline. One exasperated oilman remarked that there was nothing the oil industry could do "if a pipeline five miles away, something he can't see or hear or smell or otherwise sense, is going to ruin a man's experience."[124] Of all the reasons for opposing the pipeline, this one, which was cultural and emotional as well as ecological, was perhaps the most significant in terms of the intellectual history of environmentalism.

The authors of the draft impact statement had stressed that the additional development the pipeline service road could facilitate might be more environmentally damaging than the pipeline itself.[125] Some conservationists agreed that while a pipeline across Alaska was objectionable enough, this other aspect of the project in fact posed a greater long-term threat. A pipeline could be removed when the oil ran out—indeed, there was provision for this eventuality in the stipulations included in the environmental impact statement. But the road was likely to become "as permanent as the pyramids," in the words of James W. Moorman, executive director of the Sierra Club Legal Defense Fund.[126] Sydney Howe, the president of the Conservation Foundation, warned: "The oil might last for fifty years. A road would remain forever."[127] In contrast, an Alaskan proponent, Carl Heflinger of Fairbanks, hoped the road would "open up new country for future generations to move to, thus relieving the strain in congested areas like New York and San Francisco."[128]

The testimony of Julian G. Rice, the mayor of Fairbanks, illustrated the booster attempt to deny the validity of pipeline criticism and the failure to acknowledge its diversity. Rice quoted approvingly a remark that William R. Wood, president of the University of Alaska, had recently made to him. According to Wood, pipeline opponents were "anti-God, anti-Man and anti-mind."[129] Rice explained that Wood had accused them of being "anti-God" because their attitude toward the pipeline violated man's biblical injunction to subdue the earth and use the resources placed there by "a supreme force."[130] Rice explained that Wood's claim that critics were "anti-Man" applied to their rejection of the anthropocentric tradition (based on man's status as "the highest order of being"). The final part of the accusation conveyed Wood's conviction that the stance of pipeline critics entailed an abdication of reason and denial of mankind's ability to improve his material condition. In short, criticism of the pipeline heralded the triumph of emotion over rationality. What preservationists really wanted, Rice argued, was "to limit Alaska to a pioneer society, living a frontier existence."[131] In the same vein, Les Spickler, representing the Juneau Chamber of Commerce, cited a placard that he had seen displayed in the window of a motel in the state capital. The sign had proclaimed, "[the] conservation way is reason, [the] preservation way is treason."[132] Former Rampart promoters Irene Ryan, Ed Merdes, and C. W. Snedden also gave the impression that most of the opposition to TAPS was preservationist in character.[133]

To reinforce their case, pipeline proponents frequently invoked the icebox image. Dismissing fears that oil development would ruin the Arctic, Rice depicted the state north of the Yukon as being "already an arctic wasteland . . . covered by ice and snow much of the time and . . . with swamps and pools of water the balance of the year."[134] The absence of anything to spoil in the Arctic was also one of Walter Hickel's favorite themes at this time. Hickel contrasted the bleakness of the North Slope with the mountain beauty of the Grand Tetons of Wyoming and the Sierra Nevada of California, and was relieved that oil had been discovered in the Arctic rather than in these splendid regions.[135]

Rice attributed the intensity of criticism facing the pipeline project largely to the mood of the time. He believed that the project had been chosen for a "grand apologia."[136] Mike Gravel, Alaska's Democrat in the Senate, felt the same way: "Many are opposed to the pipeline not because of problems related to the pipeline itself but rather because of general overriding concern for the environment."[137] John Havelock, Alaska's attorney general, agreed that much of the opposition was only incidentally concerned with genuine conservation issues; the project provided "a convenient symbolic focus for the nation's anxiety."[138]

Other proponents believed that objections of a conservationist nature were merely a facade. They denounced the Department of the Interior's handling of the permit application as antithetical to the time-honored American way of doing business and evidence of a socialistic thrust (while emphasizing that the Soviet Union was not plagued by environmentalists). Organizations like the National Rifle Association (NRA) had taken an official stand against the pipeline.[139] But John E. Clark, a Fairbanks resident and former gold miner who had lived in Alaska for thirty-five years, believed opposition to it was communist sabotage. According to Clark, "ecology and environment have very little to do with this controversy; this is an ideological contest between the free enterprise system and some totalitarian concept."[140] Like Rice and Hickel, Clark exploited the icebox image to undercut conservationist concerns. He sought to reassure a leading critic of the pipeline proposal, the *Christian Science Monitor:* "This pipeline will run through a bleak, barren, hostile, unpopulated country. If there is anyplace in America where a pipeline would not damage anyone or anything, this must be the place."[141] Edward J. Fortier, the editor-proprietor of *Alaska* magazine (formerly *Alaska Sportsman*), adopted the same tactics. During a television debate he dismissed the North Slope as "the most desolate area in the world. If I never see it again, its fine with me."[142]

Adverse reaction to the draft impact statement was not confined to the public. Federal representatives repeated many of their criticisms. NEPA required that the secretary of the interior consult with and obtain comments from any federal agency that had jurisdiction or special expertise with respect to any environmental impact before issuing a final statement. Foremost among interested federal agencies that showed their dissatisfaction with the draft impact statement was the Environmental Protection Agency (EPA), which was established as part of NEPA. EPA's director, William D. Ruckelshaus, drew attention to the insufficient design data, the inadequate discussion of the marine leg of the transportation system, the absence of a fair appraisal of alternatives, notably a trans-Canadian pipeline, and the dubious efficacy of federal task force stipulations. Ruckelshaus urged thorough further study.[143] In a critical assessment of the impact statement, a representative of the National Park Service sought to draw attention to the AEC's bioenvironmental studies at Cape Thompson for Project Chariot (*Environment of the Cape Thompson Region, Alaska,* 1966), which the official described flatteringly as "in effect . . . an environmental impact statement." The official commented that "while it may not be feasible because of lack of time to use the Cape Thompson study as the model for environmental impact statements, we believe an effort should be made [with regard to the Trans-Alaska Pipeline] to treat biological resources as something more than game animals and timber."[144]

The Department of Transportation's Office of Pipeline Safety wanted a good deal more information about how Alyeska proposed to detect, monitor, and deal with oil spills on land and sea. In fact, the transportation secretary, John Volpe, preferred a railroad to a pipeline.[145] The Alaska Branch of the U.S. Army Corps of Engineers surprised some conservationists by emerging as another powerful critic. The corps' interest in the project stemmed from its regulatory responsibilities under the Rivers and Harbors Act. Its highly critical report on the statement was leaked to Jack Anderson of the *Washington Post* (28 February). This, too, condemned the impact statement for noncompliance with NEPA in its limited provision of design data, for "unsupported opinions" in lieu of genuine analysis of environmental effects, and for its "unconvincing" consideration of alternatives, one of which ought to have been "deferral of construction or non-development."[146]

A further source of objections was the Department of Defense, whose environmental expert focused on the federal North Slope task force stipulations and technical regulations, which were thought "too general to support the positive assurances given throughout the report that adverse ecological changes and pollution potential will be eliminated or minimized." What bothered the official in question was the general dearth of information in the statement concerning the dangers associated with permafrost, oil spills, and the impact on fisheries (salmon, herring, crab) in the Valdez area. The report recommended "a more definitive discussion" of alternative pipeline routes and modes of transportation to answer the serious questions that the project faced.[147]

Like the Department of Defense, the secretary of commerce, Maurice

H. Stans, stressed the need for swift oil development in the Arctic and repeated the economic and security arguments for TAPS.[148] However, the department's Assistant Secretary for Environmental Affairs, Sidney Galler, indicated at length that the impact statement had paid insufficient attention to the threat of pollution in Prince William Sound from spills, tanker collisions, and ballast discharge. Unladen oil tankers carry water ballast in their tanks for stability. Since it is usually carried in the oil tanks, this water becomes contaminated with dregs of oil from the previous shipment. Galler addressed in particular the threat to Alaska's biggest employer, the fishing industry: "Conflict between tanker operations and commercial and sport fishing activities is inevitable and these uses of Port Valdez and approaches may be mutually exclusive."[149] The Department of Health, Education, and Welfare (HEW) was concerned about the project's potentially adverse impact upon people, especially Natives and the residents of Fairbanks and Valdez.[150]

In the spring of 1971, it was clear that many Americans, including a number of prominent federal officials, especially the USGS Menlo Park Working Group and Technical Advisory Board, remained unconvinced that Alyeska could build a pipeline capable of withstanding Alaska's environmental hazards and that would not hurt the environment unduly. This conditional criticism was reinforced by the absolute opposition of those who believed that the wrong pipeline route had been chosen, and the even harder stance of those who, for various reasons, wanted North Slope oil left in the ground. Before discussing this spectrum of critics and proponents' redoubled efforts to counter their objections, it is appropriate to stand back and look at the role of frontier themes in the controversy.

7.4 THE USE OF HISTORY

A notable ingredient of public testimony in February 1971 was historical consciousness. The deployment of historical images and analogies for partisan purposes had been a widespread feature of the controversy since 1969 and would continue to figure prominently. From the outset, critics drew upon the history of the West to dramatize the project's dangers. In June 1969, Lloyd Tupling, the Sierra Club's representative in Washington, D.C., had commented that TAPS's permit application reminded him of "the day a hundred years ago when the Golden Spike was driven at Promontory, Utah, linking the first transcontinental railroad lines." Tupling argued that the pipeline, also linking two oceans, would be an equally powerful agent of frontier conquest.[151] He was also reminded of the generous land grants the federal government had bestowed upon the railroad companies during the Gilded Age: railroad companies received 49 million of the 130 million acres of public land that the federal government disposed of to private interests between 1850 and 1871.[152]

In 1972, a journalist with the Hearst newspaper group, a leading pipeline advocate among the press in the lower forty-eight, tried to envisage government stipulations and environmental restrictions similar to those confronting TAPS facing acclaimed American construction projects of the past. Imagining that travel by time machine was possible, the journalist conjectured that the United States was "unwilling to finish" the Panama Canal "rather than suppress disease-bearing insects." Casting his net wider, he drew further analogies with two crucial agents of westward expansion; the Erie Canal, which opened up Indiana, Illinois, and Ohio to settlement, and the transcontinental railroads. He tried to imagine work being suspended on the construction of the Erie Canal in the 1820s "to keep intact the vast wilderness of New York." In the context of the transcontinental railroads, he brought up the favorite conservationist analogy between the buffalo of the West and Alaska's caribou. The oil industry was under pressure to ensure that the pipeline would not impede caribou movements. The journalist imagined that the railroad's builders waited for "detailed surveys of the migration routes of buffalo herds."[153]

The pipeline project's potential impact on caribou provided a symbolic focus for much of the public anxiety and emotion it engendered. Conservationists regarded the well-being of the northern Alaskan caribou—an animal that they believed was "wilderness-dependent"—as the litmus test of the Arctic's ecological health. They typically described these herds as "America's last remnant of the great grazing assemblages that once roamed the continent."[154] In a typewritten leaflet he distributed to motorists entering Alaska at Tok Junction on the Alaska Highway in the summer of 1971, Chip Thoma recalled that the transcontinental railroads had divided buffalo herds numbering 60 million into northern and southern parts and that the creatures were killed in huge quantities by trophy hunters, market hunters, and railroad passengers.[155] In 1973 TAPS critics in southern California cited the case of the antelope north of what is now Los Angeles. They contended that these animals, whose numbers had stood at 30,000, were reduced by half through starvation between 1882 and 1885; the antelope refused to cross the Southern Pacific Railroad, finished in 1876, which cut across their migration route from east to west.[156] Pipeline critics pointed out that, in view of the short summer season in northern Alaska, the unpredictability of arctic weather even in summer, and the need for cows to calve before mosquitoes and other insects became too abundant, the timing of the arrival of caribou herds at the coastal calving grounds was crucial. For cows arrived there emaciated and exhausted after their long trek from the winter ranges in the mountains to the south. There is little margin for error in the Arctic, and conservationists believed that anything that interfered with the caribou's normal activities might be disastrous.

The fear that the improved access provided by the pipeline corridor would increase hunting pressure, especially illegal, was most famously expressed in a controversial poster released by FOE in the summer of 1973. It depicted six caribou carcasses, including a fawn, hanging from the forklift of a truck. The picture had been taken three years earlier at the state-owned Deadhorse airstrip, which served the North Slope oilfields. The caption read: "There is more than one way to get caribou across the Alaska Pipeline"—a reference to the features which Alyeska had promised to incorporate into the pipeline's design to facilitate game crossing.[157] FOE alleged that the caribou had been shot by off-duty oil workers in contravention of the hunting ban at the oilfields. John Ratterman, Alyeska's public affairs spokesman, retorted that they were legally shot by employees of Wien Air.[158]

Habitual references to the transcontinental railroads and the memory of the buffalo had reached a haunting zenith at the hearings in 1971. Witnesses prophesied that the pipeline project would disrupt the ancient process of caribou migration in the same way and with the same disastrous consequences.[159] Native Alaskans—such as Eben Hopson, the executive director of the Arctic Slope Native Association, and those whites sympathetic to their culture—worried about the project's consequences for the Native's way of life. For, in northern Alaska, they argued, this depended as much on the Central Arctic and Porcupine caribou herds as the Plains Indians culture had on buffalo.[160] Another caribou herd, the Nelchina, was located further south, in the Copper River basin. The Nelchina migrates twice annually across the pipeline corridor en route to eastern wintering grounds and western calving and summering grounds. Whereas the Central Arctic and the Porcupine herds are mostly hunted by Natives for subsistence, the Nelchina is heavily hunted by sportsmen from Fairbanks and Anchorage.

The character of the buffalo-caribou analogy needs to be considered more closely. The precise nature of the railroads's contribution to the near-extermination of the buffalo is difficult to assess. David Klein, a leading Alaskan wildlife biologist, argues that it would be impossible to isolate the direct impact of the railroads on buffalo migration patterns as a factor in their demise since hunting (facilitated by the railroads, of course) eradicated the herds so swiftly.[161] In Alaska in the 1970s, state game laws and oil company regulations, which prohibited hunting within five miles of the pipeline corridor, precluded anything like a repetition of the direct physical slaughter of the buffalo that occurred in the West. Mass hunting aside, Klein has stressed that the ultimate fate of the buffalo was linked to the usurpation of its habitat by the large-scale colonization and livestock grazing that railroads facilitated. There was no danger of the Arctic being settled and used in this way. Nor, it turns out, does the contention that TAPS would severely disrupt the migration patterns of all Alaska's arctic caribou have an entirely sound empirical basis. Alaska's largest herd, the Western Arctic, migrates through Anaktuvuk Pass, west of the proposed pipeline corridor.

The strong historical awareness and tactical application of historical evidence that characterized the TAPS debate was also apparent in the search of Alaskan pipeline supporters for a usable past to assist their cause. Whereas critics drew upon the history of the West, proponents tapped the reservoir of Alaskan history. Firstly, they dwelt on the frustrations of earlier generations who had fought to develop Alaska's natural resources and extend local self-determination. They saw a strong parallel between the detrimental effects on Alaska of the current protest against the pipeline and the influence of measures enacted during the first epoch of conservation.

A leading purveyor of historical analogies was Alaskan-born Evangeline Atwood, who introduced herself as "a housewife and student of Alaskan history."[162] She interpreted opposition to the pipeline as the culmination of a long period of resistance to the legitimate desires of Alaskans to exploit their abundant natural wealth. At the hearings in Anchorage, she delivered a diatribe against conservation as a pernicious "political pollutant" that had "hung over" Alaska "like a pall for 100 long years." Invoking John Hellenthal's *The Alaskan Melodrama*, the anticonservationist history of Alaska (1936), she alluded bitterly to the coal and timber "lock-ups." She decried Richard Ballinger's "vilification," and celebrated the famous protests against the withdrawal of coal lands that took place at Cordova and Katalla.[163] John Hellenthal, the nephew of the author of *The Alaskan Melodrama*, also drew attention to the incident at Cordova during the hearings. He described the new ecological concern and its overbearing focus on the pipeline as "even fiercer" and "more messianic" than the first era of conservation had been.[164] Evangeline Atwood regarded conservationist preoccupation with the pipeline as a strident and crippling manifestation of persisting colonialism that made a mockery of statehood.[165] In a classic exposition of what William H. Wilson in 1970 called the traditionalist interpretation of Alaskan history,[166] she cast pipeline critics in the same mould, and assigned to them the same role as the fish trusts, the transportation companies, and the mining corporations.[167]

However, TAPS proponents did not only portray Alaskan history as a dreary and pitiful chronicle of thwarted ambitions devoid of achievements in the realm of efforts to control nature and extract its wealth for the benefit of Alaskans. They also made use of its technological triumphs. By teaching their opponents some of the more encouraging and glorious aspects of Alaskan history, advocates hoped to dispel the image of Alaska as a fragile wilderness and untouched, virgin land on the verge of sacreligious despoliation. They marshaled evidence from the frequently overlooked record of white pioneer activity in Alaska, not to indicate that "the last frontier" had already been conquered, but as proof of the country's resilience and ultimate indomitability. A related purpose was to de-emphasize the technological novelty of TAPS and the environmental dangers that it posed.

One native-born white Alaskan who took this approach was Steve McCutcheon, who was under contract to Alyeska as the official pipeline project photographer. Born in Cordova in 1911, McCutcheon had been a leading statehood activist and a delegate to the Constitutional Convention of 1955. His father, Herbert H. McCutcheon, an immigrant from California, had worked on the construction crew that laid the first ties on the Alaska Railroad at Anchorage and had been a territorial legislator in the 1940s and 1950s.[168] The main lesson that he tried to teach was the history of the Davidson Ditch, which had included Alaska's first significant pipeline of any description. McCutcheon set out to reassure the sceptics by minimizing the differences between early twentieth-century hydraulic engineering and modern petroleum technology.

Critics feared that an oil pipeline raised above the ground in areas where permafrost made conventional burial too hazardous would interfere with caribou migration. According to McCutcheon, caribou would have no trouble crossing the elevated sections. His confidence was based on personal observation of the Davidson Ditch, which carried water to the Fairbanks Exploration Company's dredging operations. He recalled having seen in 1943, "a bull caribou being harried by either large dogs or wolves clear jump a syphon" (52 inches spanning a creek and valley on trestles) "which, according to . . . later measurements, was 81 inches from tundra to top. . . ." He also denied that it had left a permanent scar on the land, pointing out that trees 4 inches in diameter were now growing in the ditches that fed water to the dredging operations.[169] Another Alaskan who referred to the Davidson Ditch in a similar way was Alfred J. Loman, a mining engineer and third generation Alaskan, who was vice president of Yukon Equipment (Fairbanks).[170]

Other aspects of the mining history of the interior were also summoned to counteract the emphasis pipeline critics placed on Alaska's "delicate" ecology. Like her former husband, the late Alaskan senator, Vide Bartlett had been raised in and around Fairbanks. At the session of the hearings in Washington, D.C., she told the story of Cleary Creek, a gold mining settlement some twenty miles north of Fairbanks, where she grew up near her father's claim. In 1905 the town had boasted a population of 20,000.[171] Now there was virtually no trace of its existence. She explained that abandoned wooden buildings had rotted back into the earth. She claimed that once-sterile mine tailings from the dredging operations of the Fairbanks Exploration Company, such as the one at Cleary Creek, now bore a mixture of willow, alder, birch, and scrub in contrast to the black spruce that had formerly dominated the area. For Vide Bartlett, this was not only proof of how quickly the imprint of man and his activities faded in Alaska. It was also a prime example of environmental enrichment through human intervention. Ironically, while emphasizing the temporality of earlier technological intrusions in Alaska, with regard to the Trans-Alaska Pipeline she exclaimed, "may it survive as long as the Roman aqueducts."[172]

Another purveyor of history was James W. Dalton. His Oklahoman father came to Alaska in 1882 and later developed the lucrative Dalton Trail, a toll road over which cattle and other livestock could be driven from Lynn Canal to the Yukon mining camps. The son was a pioneer in arctic engineering who had supervised DEWline construction and been involved in the early years of arctic oil development by commercial interests as a civil engineer and consultant. The example chosen by James Dalton, who began his career as a gold mining engineer, was the narrow gauge (3 feet) Tanana Valley Railroad, which ran for forty-five miles between Fairbanks and the Chatanika mining camps during the heyday of the gold mining era.[173] Construction of this line began in 1905, when Fairbanks had a population of around 5,000. Since the line was wood-powered, the hills around Ester and Chena became denuded of trees. The railroad ran from Chena to Gilmore via Fairbanks, and was later extended to Chatanika. Its fortunes mirrored those of the mining industry. Privately funded and operated, it carried 15,809 tons of freight in 1909 and moved 53,248 passengers the following year. In 1917, when mining was in decline, the management of the line was taken over by the federally owned Alaska Railroad. Three years later, the population of Fairbanks had dropped to less than 1,200 people. In 1930 the line was discontinued and the rails were salvaged. The wooden trestles have since crumbled and the embankments have become overgrown. According to Duane Koenig, "only a few ties and trestles remain."[174] Pipeline advocates were convinced that this kind of evidence discredited the unshakeable preservationist notion that wilderness, like virginity, cannot be regained. These defenders of the frontier tradition claimed that instead of being rolled back irrevocably, the Alaskan frontier had in fact rebounded.

To underline the long-term harmlessness of past pioneer practices and the robustness of a far-from-fragile land, Everett Patton (an Alaskan resident since 1913) related another part of the interior's transportation history. The average steamboat burned twenty-five cords of wood daily.[175] Woodyard operations had sprung up by 1900 because steamers were having to stop every six hours to cut wood. (The same practice had flourished illegally on the public domain along the Mississippi and western rivers prior to the Civil War.)[176] During fifty years of steamboat operation, the forest along the Yukon had been eventually cut back up to two miles. During the later period, Caterpillar tractors allowed woodchoppers to exploit areas further from the riverbank. At the hearings, Patton described how fifty years of steamboat operation had stripped the trees along the Yukon.[177] However, Patton also explained that the trees had since regrown and that, in general, there was less evidence of human activity along the Yukon than there had been almost three-quarters of a century earlier.

At least two historians agree that frontier and wilderness conditions have been "recreated" along the Yukon. Declining by the late 1930s, steamboat operations ceased in the mid-1950s. This was due less to competition from the Alaska Railroad (which operated its own steamers from Nenana) as it was to competition from cargo planes and the result of the decline of mining along the Yukon. In view of the passing of the steamboat and sourdough mining era and the demise of the wood-yards, fish camps, trading posts, and roadhouses, David Wharton has concluded that the Yukon "reverted to a true primitive area."[178]

These interpretations should be assessed in the light of botanical evidence indicating that full ecological "recovery" among the white spruce—which dominate the strip of needleleaf evergreen forest along the banks of the sub-Arctic Yukon and its tributaries—takes much longer than seventy-five years. According to one source, a mature white spruce reaches 40 meters in height and attains a girth of a meter; spruce of this stature along the Yukon are at least a century old.[179] Other, more conservative botanical data points out that white spruce between one and three inches in diameter can be seventy-five years old and that it takes two centuries for a white spruce to reach a diameter of two feet.[180]

The contrast between the speed with which the Alaska Highway and Canol projects were planned and built, and the sluggish rate at which pipeline plans—beleaguered by technological, political, and legal obstacles—were proceeding, was not lost on proponents who testified in 1971. Dalton was just one of many whites born in Alaska or long-term residents of the state, who, at the height of the pipeline imbroglio, recalled the free hand enjoyed by the armed forces thirty years before the onset of "the Age of Ecology" and the demands of the National Environmental Policy Act. Granted, TAPS was proposed in peacetime by private enterprise and the Alaska Highway and Canol Project were wartime, federal initiatives. Despite this basic difference, it is hard to disagree with Dalton's opinion: "the construction and operation of the proposed Trans-Alaska Pipeline System would have been unattended by the controversy and strife currently being generated had occasion for its existence occurred twenty-five years ago . . . at the time of the construction of the Alaska Highway and the Canol Project."[181] Likewise, Steve McCutcheon drew attention to the uncontested nature of the DEWline to highlight the victimization of TAPS and what he regarded as the novelty and irrationality of national conservationist concern for the Alaskan environment.[182]

Evangeline Atwood observed prophetically that finer conservationist sensibilities would have been brushed aside and the current intricate debate over the merits of the pipeline abruptly terminated if there was a national emergency or a wartime situation comparable to that which precipitated the Alaska Highway's construction. "We who lived here when World War Two became a reality know what happened then," she explained. "The earth was torn asunder ruthlessly as bulldozers worked around the clock making ready for the arrival of defense forces."[183]

The Alaska Highway provided proponents of immediate pipeline construction with more than just an example of a project that went ahead without legal hindrance and public harassment. They also selected the highway as a prime example of an engineering project which, despite the rudimentary techniques employed in its construction, could serve as a reliable guide for future action. According to these Alaskans, who were scornful of what they called "the delicate ecology pitch," the Alaska Highway exemplified the indestructibility of the natural environment they claimed to know and understand far better than pipeline opponents, most of whom had never visited Alaska, let alone lived there. Bruce Campbell, commissioner of the Alaska State Department of Highways, was confident of the land's capacity for self-renewal. For though the road had been "rushed through 1500 miles of virgin country in record time . . . with little regard for environmental or aesthetic considerations," he claimed that there had been no more than a temporary disturbance of the natural order: "The scars of that wartime experience have largely disappeared, the land has healed itself."[184]

Pipeline supporters drew further support for this contention from the various components of the four-unit pipeline system known as the Canol Project. None of them disputed that Canol had been an unsophisticated piece of engineering. The pipe, according to a 1982 oil industry retrospective, "frequently broke and thousands of barrels of oil were lost" during its brief operating history of fourteen months.[185] What Alyeska supporters did deny was that these various mishaps had left anything but the merest of vestiges. Edward L. Martin, Director of Public Works for the city of Anchorage, had worked as an engineer on Canol Four. Martin stressed that it was now "impossible" to locate it either from the air or from the ground, since the pipe itself and the project's other accoutrements had been removed after its closure.[186]

Repeating Dalton's point about the Alaska Highway, Martin stressed that Canol's construction and operation had not been burdened by government regulation, public inquiry, or environmentalist objection. Moreover, it had been built without the benefit of any of the safety features, such as automatic shut-off valves, weld X rays, and anticorrosion devices, which were available to Alyeska's designers and engineers. Nonetheless, Martin argued, Canol had encountered the same hazards posed by climate and terrain, as well as problems such as wildlife migration, all of which its builders had handled without malign consequences. Martin rode roughshod over the differences between Canol, a subarctic line of 4-inch diameter, and TAPS, measuring 48 inches, of which 300 miles traversed terrain north of the Arctic Circle. Like those pipeline supporters who cited the parallel between TAPS and the Davidson Ditch, Martin concluded that the hue and cry over TAPS was unjustifiable.[187]

Likewise, Alaskan proponents claimed that the Haines pipeline, which had replaced Canol Four, had proved to be environmentally harmless, despite having received little attention, during construction or afterwards, from conservationists within or outside the federal government. Taking seismic threats into consideration, the Haines line was "snaked"—as the Alyeska line was to be—to allow for lateral motion.[188] However, at the hearings in 1971, the U.S. Fish and Wildlife Service, which contributed to the initial federal studies of TAPS in 1969, denied any comparability between the two lines.[189] As a military report on the Haines line stated in 1971, there was no statutory requirement to report environmental damage caused by spills. Hence, "few official records or observations were made which reflect this area of concern so prevalent today." Nonetheless, the report did provide some damage information. Before it came online in 1956, a de-icing operation proved necessary because of water left in the pipe after hydrostatic testing in the fall and winter of 1955. To free the line, twenty-eight cuts were made at different locations, resulting in the loss of "significant quantities" of petroleum, the effects of which, "unfortunately, because of lack of emphasis on environmental control at this early period," were not monitored.[190] Because of a rupture caused by external corrosion of the buried pipe near Haines, 3,000 feet of line had to be replaced between 1964 and 1968. In 1968 the worst spill in the line's fifteen-year history was recorded near Dezadeash lake in Canada's Yukon Territory. As a result of corrosion, an estimated 4,000 barrels of diesel fuel were lost, much of which flowed into the lake. No conservation organization in Canada or the United States reacted to this event, after which twelve miles of pipe were replaced.[191] Conservation organizations made little use during the TAPS controversy of either Canol One as an example of ill-advised action to be avoided in the future, or of this evidence of spills along the Haines line. At the hearings, only Frederick C. Dean of the Alaska Conservation Society referred to the Dezadeash spill.[192]

The most recent historical experiences that Alaskan pipeline advocates sought to capitalize on was Alaska's previous oil boom in the late 1950s. Turning their attention from the pipeline itself to the broader subject of oil development's environmental impact, supporters held up the operations of Atlantic Richfield Oil Corporation on the Kenai National Moose Range as a vindication of the multiple use philosophy of land management (as opposed to "single use" for wilderness) favored by Alaskan boosters. At the hearings in 1971, Julian Rice, the mayor of Fairbanks, praised these activities as an example of the peaceful coexistence between nature and technology that was possible in Alaska and that could serve as a model for conduct at Prudhoe Bay.[193] James Dye, a former mayor of Kenai, and Harold E. Pomeroy, a former chairman of the Kenai Peninsula Borough, also referred to Atlantic Richfield's Kenai operations. Pomeroy described them as "a good bargain with nature." Dye, a freight owner, had a direct personal interest in the authorization of TAPS. He said he had $5 million invested in equipment sitting idle along the proposed route.[194]

None of the conservationist organizations (such as the National Wildlife Federation) who had opposed oil leasing on the Kenai Moose Range in the 1950s seriously quarreled with the verdict that any loss of moose habitat through oil development within the range had been inconsequential.[195] Leading Alaskan conservationists also found no cause for major grievance. They agreed that, for the most part, the oil companies had complied with the federal regulations governing their activities.[196] However, as conservationists also pointed out, there was no sound basis, economic or ecological, for taking the Kenai case as a guide for Prudhoe Bay. Firstly, the difference in scale was enormous. In 1970 Kenai production stood at 13,000 barrels per day. In contrast, the daily flow from the North Slope fields was expected to reach a maximum level of 2 million barrels. Secondly, most wildlife biologists seem sceptical of comparisons between the behavior of moose and caribou. Moose appear more tolerant of human presence and activity than caribou, which are known to shy away even from telegraph wires.[197] In addition, moose thrive on the vegetation that succeeds "disturbances" such as fire, trails, and mining. Caribou, on the other hand, as the fires on the Kenai Peninsula in the late nineteenth century demonstrated, are dependent on stable ("climax") vegetation communities that provide plenty of lichens.[198]

The final argument heard from Alaskan pipeline advocates was the protest that, despite having a record innocent of environmental abuse, contemporary Alaskans were being punished for the "crimes" against nature committed by Americans in temperate regions both now and in earlier times. To highlight their sense of persecution, Trans-Alaska Pipeline proponents compared the many thousand miles of oil pipeline already lacing the continental United States with the 800 miles proposed for Alaska. One editorial on this subject, submitted as part of the hearings record by Irene Ryan, once a leading supporter of Rampart Dam, and now commissioner of the Alaska Department of Economic Development, was entitled "217,000 versus 800." The author of this editorial from the *Ketchikan Daily News* of 18 February 1971 complained that "we have not heard of oil pipelines blamed for killing wildlife or stopping animal or people migration [in the lower forty-eight]." (No critic, incidentally, ever claimed that the pipeline would obstruct human movement.)

What the *Ketchikan Daily News* did not point out was that these pipeline miles in the contiguous states were all subsurface. Nor did the Alaskan newspaper admit the slightest significant difference of climate, terrain, or ecology between the North Slope on the one hand and Texas or Pennsylvania on the other. The emphasis that this Alaskan pipeline proponent placed upon Alaska's physical and ecological conformity with the rest of the nation on this occasion was inconsistent with the usually fierce booster insistence on Alaska's environmental distinctiveness, which, they argued, made them the only Americans qualified to comment on the pipeline project. This editorial exemplified the dubious applicability of most historical analogies during the pipeline debate. It also illustrates the difficulty of making legitimate comparisons between ecologically disparate regions.

8
The Trans-Alaska Pipeline Controversy: 2, 1971–1974

8.1 THE OPPOSITION ORGANIZES—NATIONALLY AND LOCALLY

Coherent bids to protect the natural environment in the United States, whether successful or ultimately defeated, have generally depended on the creation and mobilization of a nationwide constitutency to overcome local inertia and the hostility of prodevelopment interests. In 1913, during the crucial Senate debate on the bill to authorize the damming of Hetch Hetchy Valley, Senator James A. Reed of Missouri, a proponent of the dam, expressed incredulity at the scale of the furor over the proposed inundation of two square miles of Yosemite National Park. He asserted: "The degree of opposition increases in direct proportion with the distance the objector lives from the ground to be taken. When we get as far east as New England the opposition has become a frenzy."[1] In some respects, the TAPS controversy conformed to this traditional alignment of forces.

During 1971 the opposition to TAPS crystallized and became organized. After the hearings in February 1971, the Alaska Public Interest Coalition (APIC), an umbrella organization, was formed to coordinate and intensify the campaign against the pipeline. The impending congressional settlement of Alaska Native claims provided the immediate spur. Conservationists feared that Alaska's wilderness would be vulnerable to exploitation when the land freeze was lifted upon the settlement of these claims. So they formed a group known as the Alaska Coalition to support an amendment introduced by Congressman John P. Saylor (Republican, Pennsylvania), which sought to impose a five-year moratorium on land development in Alaska to allow for the formulation of a general land use plan for the state. The Alaska Coalition, one of APIC's forerunnners, consisted of the Sierra Club, the Wilderness Society, the National Wildlife Federation, the Wildlife Management Institute,

217

FOE, Defenders of Wildlife, Trout Unlimited, the National Rifle Association, Zero Population Growth (ZPG), Environmental Action, the Citizens Committee on Natural Resources, and the Alaska Action Committee.

APIC's chairman was Brock Evans, recently appointed the Sierra Club's representative in Washington, D.C. FOE's legislative director, George Alderson, became responsible for coordination. APIC had a number of antecedents in addition to the Alaska Coalition, such as the Coalition Against The SST. Worried about the hazards of sonic booms and the danger to the ozone layer, the Sierra Club, the Wilderness Society, FOE, Environmental Action, Common Cause, and Zero Population Growth formed this group in April 1970 to fight the supersonic air transport scheme. The House of Representatives vote in March 1971 to cut off federal funds for SST was the most noteworthy environmentalist victory during the Nixon administration and provided APIC with some of its inspiration, political momentum, and organizational expertise. Other recent, successful campaigns from which APIC could draw encouragement were those against the Everglades Jetport (proposed to serve Miami and to be located just north of Everglades National Park)[2] and the Cross-Florida Barge Canal. Nixon canceled the jetport in January 1970. A year later, he stopped the Army Corps of Engineers' canal, which was already one-third complete.[3]

Like these predecessors, APIC was sustained by the involvement of a diverse blend of old and new. This illustrated the catholicity of the new environmentalism and testified to the broad range of issues raised by the pipeline proposal. Some of APIC's seventeen members were active political lobbyists. Others, whose status was tax-exempt, assumed a more passive role. The three recently established organizations, all registered lobbies, whose direct interest in Alaska was not as strong as that of other coalition members, were Common Cause, the Consumer Federation of America, and Zero Population Growth. Common Cause was a citizen pressure group located in the capital with 100,000 members, which wanted American troops withdrawn from Vietnam and pursued a variety of domestic reforms. Formerly known as the Urban Coalition Action Council (1968), Common Cause listed its legislative interests as "employment, education, health, consumer protection, environment, family planning, law enforcement and the administration of justice. . . ."[4] The Consumer Federation of America (1967), also located in Washington, D.C., provided an umbrella for 180 local, state, and national organizations that represented a total of 30 million Americans. The federation was involved in campaigns to ensure the safe packaging of hazardous household materials and to impose reductions on automobile exhaust emissions. For Common Cause and the Consumer Federation, the pipeline raised general questions about corporate behavior and government accountability. The third lobby, Zero Population Growth (ZPG), had been formed in 1967 by the Stanford University biologist and noted conservationist, Paul R. Ehrlich, author of *The Population Bomb* (1968). The purpose of this 20,000-member group, based in Los Altos, California, was "to promote legislation to: ensure the ready availability of all methods of contraception; establish tax and other incentives to the smaller family; provide government support for research, study and implementation of new approaches to the population problem."[5]

The other coalition member only incidentally concerned with conservation in Alaska was the United Automobile Workers (UAW). The participation of the UAW (Detroit, Michigan, 1937) is especially interesting in view of its 1.5 million members' stake in higher vehicle production levels (and therefore greater gasoline consumption). UAW's involvement with APIC reflected the union's general concern with a variety of environmental hazards. By the end of the 1960s, articles in *Solidarity*, the UAW organ, on topics such as the Santa Barbara oil spill and the moribund state of eutrophied Lake Erie—probably the most potent symbols of the national environmental crisis—began to supplement the staple diet of material on wages and working conditions.[6] *Solidarity* reasoned that industrial workers could no longer afford to occupy themselves exclusively with issues of immediate self-interest such as factory safety. The new age, it explained, required an expanded consciousness recognizing the interrelated nature of all issues, whether part of the human or natural environment ("we are all workers in a factory without walls").[7] At its twenty-second constitutional convention in May 1970, the UAW became the first American union to adopt pollution as a collective bargaining issue and to call for the enactment of an environmental bill of rights.[8] UAW's interest in conservation was strongly influenced by the personal commitment of its leader, Walter P. Reuther.[9] Still, one should not assume that TAPS was a priority issue for the union. *Solidarity* carried no article or news item on the Alaska Pipeline during the controversy.

The conservation organizations that were members of APIC can be divided according to age, size, outlook, and level of participation. Most of the established groups have been introduced in earlier chapters; the Sierra Club (1892), with 140,000 members; the National Audubon Society (1905), based in New York City, with 75,000 members; the National Parks and Conservation Association (1919) of Washington, D.C., with 55,000; and the National Wildlife Federation (1935), also based in the capital, with 700,000 members. Two older, regionally based groups, emphasized recreation. These were the Izaak Walton League of America, founded in 1922 in Glenview, Illinois (the headquarters of this 50,000-member group were relocated to the capital in 1971) and the Western Federation of Outdoor Clubs (1932), of Bozeman, Montana, which represented 48,000. A more recently established regional sportsmen's club that formed part of APIC was Trout Unlimited (1959), of Denver, Colorado. The purpose of this 15,000-member lobby was to promote the preservation of trout habitats. Another older group, primarily concerned with wildlife protection, was Defenders of Wildlife (1925), of Washington, D.C., which had 30,000 members. Fund for Animals (based in New York City and founded in 1967) described itself as "a combination Red Cross and Ford Foundation for animals."[10] The young environmentalist generation was represented by the Environmental Defense Fund (EDF) (37,000 members) and its fellow litigant in the antipipeline suit, Friends of the Earth. The only Alaskan member of the coalition was the Cordova District Fisheries Union (CDFU), which filed the third lawsuit against TAPS in March 1971.

The strength of APIC lay in the large population centers of the lower forty-eight. Accordingly, in June 1971, the State of Alaska and Alaska's congressional delegation encourged the Justice Department to move the conservationists' and the fishermen's suits to Alaska. They hoped this would handicap the efforts of organizations based in the East and remove the case from the public eye. They also sought to take advantage of the more favorable political climate in Alaska.[11] Mitchell Melich, a Department of the Interior solicitor who was sympathetic to the move, declared that "these cases ought to be tried in the areas where the problems have arisen."[12] He argued that the Alaskan court possessed greater knowledge of the issues and had more time available. However, as the *Washington Post* (23 July 1971) pointed out, the likely result would be "the discontinuance of the suit by plaintiffs who lack the resources in terms of money and manpower—to litigate the case in Alaska." Resisting this pressure, Judge Hart denied the move. He did so because the Justice Department had waited fourteen months before seeking a move and on account of the amount of work already done by the plaintiffs in the capital. At the end of August, the State of Alaska and Alyeska intervened in the case as defendants. The plaintiffs challenged the latter's right to do so but on 10 September Hart upheld the right of the State of Alaska and Alyeska to intervene.

Nevertheless, the dominant national thrust of pipeline criticism was complemented and reinforced by an Alaskan dimension with its own distinctive flavor—just as a faction within the San-Francisco-based Sierra Club, led by John Muir, had been in the vanguard of opposition to the damming of Hetch Hetchy. For many Alaskans, the greatest dangers were those associated with the marine leg of the oil delivery system. When he visited Alaska for the first time in the fall of 1970, David Brower, perhaps in deference to his audience, selected tanker pollution as "the greatest threat of all."[13] The Cordova District Fisheries Union

(CDFU) was at the center of the dispute over the location of the pipeline terminal and tanker berths across the bay from the town of Valdez on a fjord in the northeastern corner of Prince William Sound. In Cordova (population 1,100) the fishing and fish processing industry accounted for 50 percent of total employment. Many others were employed in related services. Valdez, forty air miles away, was not a fishing community as such.[14]

The CDFU had alerted state politicians to the threat of pollution in Prince William Sound even before the pipeline proposal was announced.[15] The draft impact statement, which admitted that transportation by tanker was more dangerous to the environment than movement by pipeline, but devoted only two pages to the threat of marine pollution, galvanized the fishermen's protest. In February 1971 the fishermen supported a resolution introduced by state Senators Jay Hammond (Naknek) and Bob Palmer (Ninilchik), both fishermen, which directed Alaska's state departments of Fish and Game, Economic Develoment, and Natural Resources to compare the long-term economic and environmental impacts of a Canadian route with TAPS. Shortly afterwards, the union held a special public meeting. According to a report in *The Thunder Mug* (4 May 1971), a newsletter the union published between April and October 1971, more than half the adult population of Cordova attended. Subsequent testimony at the hearings on the draft environmental impact statement in Anchorage suggested that the most neglected aspect of the project was the marine leg. The gravity of the global environmental crisis was brought home to Cordovans through the implications of TAPS. "When something threatens your bacon and beans," explained one fisherman, "it sort of dumps part of a world-wide problem right in your own lap. . . ."[16]

These fishermen realized the need to protect and manage a valuable, renewable resource on a long-term sustained yield basis. From historical experience, Cordovans were wary of the boom and bust cycle associated with the activities of outside natural resource exploiters such as the fish trusts and mining corporations. Ironically, rival promoters in Cordova and Valdez had vied at the beginning of the century for the terminus of the Copper River and Northwestern Railroad. Cordova had won and became the port from which copper mined at Chitina by Kennecott was moved south. By the late 1930s, the ore was exhausted. Many modern Cordovans opposed the transportation of another nonrenewable resource through their region, knowing that fisheries offered an enduring alternative. At the hearings, Alaskan TAPS proponents either ignored the CDFU or trivialized its concerns. Julian Rice dismissed the fishermen as "a small but vocal group of Alaskans" representing a "limited segment" of the Alaskan fishing industry.[17]

On 7 March the Cordova fishermen voted $10,000 to fund a legal protest. In April, at hearings before a committee of the Alaskan legislature, they opposed a bill sponsored by Governor Egan to spend $500,000 to correct misconceptions in the lower forty-eight concerning all forms of Alaskan natural resource development; they believed it was simply a lobby for Alyeska.[18] The fishermen initiated their suit against the secretaries of interior and agriculture in Judge Hart's court on 28 April. It was to be financed by self-assessment—a one cent levy on each fish caught. Their complaint claimed that the pipeline would "cause substantial damage to the marine environment . . . essential to the livelihood and economic well-being" of their members.[19] The suit charged that Interior had failed to meet its obligations under NEPA and that the Department of Agriculture had illegally issued a special use permit for 802 acres of land within the Chugach National Forest since the legal limit for "industrial purposes" was 80 acres.

The CDFU claimed that of the 476 members who were balloted only seven opposed the suit.[20] Bob Reeve, the founder of Reeve Aleutian Airways—who was chairman of Speak Up Alaska!, a private pipeline lobby formed that spring—denounced them as traitors. (Reeve also denounced the NRA for its critical stance on the pipeline and called Interior Secretary Rogers C. B. Morton "public enemy number one.")[22] According to the *Fairbanks Daily News-Miner* (14 May 1971) those who did approve were mostly absentee members from Washington State and local high school students. Robert Atwood also denied that their legal action enjoyed the support of the majority of the union's membership.[22] Prominent figures such as Ross Mullins, the leader of the CDFU's pollution action committee, were defamed as "cheechakos." This term (the opposite of "sourdough") is used in Alaska to denote a newcomer/greenhorn.[23] The Alaskan press also emphasized the support for a Valdez terminal from the mayor of Cordova, Arthur P. Knight, who was the local distributor for Standard Oil of California, which supplied fuel to the fishing community, as well as that of the editor-proprietor of the *Cordova Times*, Harold E. Bonser.[24]

Allegations that the CDFU enjoyed no local support were belied by the cooperative stance of the Cordova town council and chamber of commerce. The CDFU had also received the official backing of the Copper River and Prince William Sound Cannery Workers Union. In addition, the CDFU was supported by the Homer-based North Pacific Fisheries Association, the Seward Chamber of Commerce, Point Chekalis Packers (a Cordova salmon and crab processing plant), St. Elias Ocean Products, and Bristol Bay Resident Cannery Workers, and, later, the United Fishermen of Alaska. Outside supporters included Whitney-Fidalgo Seafoods of Seattle.

Dell Goeres, the wife of a fisherman, was the CDFU's executive secretary and the editor of *The Thunder Mug*. She explained to Brock Evans of APIC that by ignoring the fishermen's protest, the State of Alaska and the oil industry implied "by omission that the big obstacle [to the proposed pipeline route] comes from environmentalists, ecologists and conservationists." While acknowledging the integrity of the latter's criticisms, she stressed that they could be "passed off as coming from birdwatchers, wolf-pamperers and ivory tower idealists."[25] The national conservation organizations whose suit was joined with the fishermen's tended to regard them as recent converts to the environmentalist cause. After meeting with a CDFU delegation in December 1971, Samuel Wright wrote to George Marshall that "they are rugged individualists who are becoming ecologists overnight because of this issue."[26] This assessment ignored the substantial areas of disagreement between the allies in the lawsuit. Most fishermen in southcentral Alaska were not opposed to the development of North Slope oil reserves nor to its transportation by pipeline. All they objected to was the Valdez terminal.[27]

Indeed, their conviction that the dangers in Prince William Sound surpassed those associated with the overland section[28] was often accompanied by indifference and insensitivity toward the ecological integrity of the Arctic. According to Tom Parker, an air taxi service operator: "Tundra ain't good for much anyway. But why ruin this?" He suggested that the oil industry use Cook Inlet instead since "they've ruined that already."[29] In August 1971, after an inquiry from Alyeska concerning the use of the Cordova's city dock facilities for unloading its pipe from Japan, to which the CDFU did not object, Dell Goeres declared: "It is no concern of ours what they do with their pipe. They can boil it up for spaghetti . . . just as long as they don't run oil thru' it from Prudhoe Bay to Valdez. . . ."[30]

Conservationists were delighted to accept the CDFU as an ally and FOE Alaska cultivated close links with the fishermen. James Kowalsky, FOE's Arctic representative, based in Fairbanks, strove to publicize the marine pollution issue as much as his counterparts in the lower forty-eight emphasized the threat to the "last great wilderness." But the fishermen shrank from too close an identification with the environmentalist cause. While it strengthened their case in Washington, D.C., the fishermen realized that the association was a liability in Alaska. Preservation of their autonomy was seen as a wise defense against Alaskan pipeline proponents who tried to dismiss them as lackeys of the Sierra Club.[31]

The CDFU's action, though joined with his conservationists' suit, was based on conventional legal standing that could not be challenged.[32] However, Alyeska did request evidence of Alaskan connections from the other litigants. In 1970 the Sierra Club had trouble establishing its standing in a lawsuit to protect Mineral King in the Sierra Nevada from ski resort development by the Walt Disney corporation. George Alderson, FOE's legislative director, was afraid that the defendants in the action against the pipeline would also challenge the conservationists' standing.[33] Accordingly, James Kowalsky requested affidavits from FOE's Alaskan members stating how they, as individuals, would be affected by TAPS.

Those who responded saw the pipeline as fundamentally inconsistent with their way of life. A resident of Anchorage explained: "We came to Alaska to avoid the 'boomers,' not to make big money and leave. . . . this damn oil has come along to bring [my children] back to the very type of life we left. . . ."[34] A dentist from Wasilla was even more direct, protesting: "We moved to Alaska solely because of her raw beauty and unspoiled wilderness. All economically-oriented development . . . bleeds our state of its most treasured asset—uncrowded spans of land."[35] In the same vein, Ernst W. Mueller, the president of the Alaska Conservation Society who became Commissioner of the Alaska State Department of Environmental Conservation in 1975, attacked the "mad rush towards development."[36] Other FOE members who felt directly affected were sportsmen who hunted in the Brooks Range, professional wildlife photographers such as Charlie Ott,[37] and property owners like Orlando W. Miller. Miller, a professor of history at the University of Alaska, Fairbanks, and a member of the Sierra Club, owned a cabin on the Salcha River (southeast of Fairbanks) and objected to the pipeline being routed there. There were 180 vacation homes along a 40-mile stretch of the river, and Miller was one of the many owners who formed the Salcha River Homeowners Association in 1972 to fight for a rerouting of the pipeline in that area. These cabins were accessible by riverboat and floatplane. Their owners objected to the road access the pipeline would provide.[38]

According to the draft environmental impact statement for the pipeline project, "the only identifiable negative cultural influences that could be associated with the implementation of the project would be a reduction in remnant hunting and fishing cultures that still characterize some Native groups."[39] Despite this blithe assertion, FOE members in villages north of Fairbanks regarded TAPS as a threat to their chosen rural way of life. One white resident of Allakaket, who was married to a Koyukon Native, reported that local people were shocked by the omissions in the draft impact statement. Because permafrost lies so close to the surface in northern Alaska, any surface liquid moves sideways and spreads quickly. The proposed pipeline route crossed the Koyukuk River three times and villagers were afraid that spills would soon find their way into its waters.[40] Ray Bane, a white resident of the Athapaskan Indian village of Hughes, also on the Koyukuk, explained that he relied on wildlife and fish to feed himself and his sled dogs, and on trapping for clothing and income. Bane was particularly concerned about the effect that removing millions of cubic yards of gravel for various construction purposes would have on fish spawning.[41] Removing gravel from river bars, flood plains, and braided streams can destroy spawning beds directly and indirectly affect fish reproductive capabilities by producing silt. Silt falls over the spawning gravels obstructing water flow through the gravels, thus the eggs die from lack of oxygen. Moreover, many small streams with gentle gradients do not flush silt away effectively. For once gravel is removed, surfaces are left unprotected and soil is easily loosened by the current. This turbidity can also kill plant life.

None of the twenty-five (ten permanent) residents of Wiseman, the village where Bob Marshall had stayed at various times between 1929 and 1939, were members of FOE or any other conservation organization. Yet one, Joe Strunka, a thirty-three-year-old gold miner, considered himself and fellow residents the people most directly affected by the project because the pipeline corridor passed within two miles of the village. He complained to the BLM that the impact statement did not address the consequences for them.[42] These disaffected individuals felt that a social impact statement that discussed the human ecology of northern Alaska was needed as well as an statement dealing strictly with environmental impacts.

8.2 THE FINAL IMPACT STATEMENT AND THE RESPONSE, MARCH 1971–JULY 1972

After the February 1971 hearings, Interior Secretary Rogers C. B. Morton set up a "102 Statement Task Force" under the chairmanship of David A. Brew of the U.S. Geological Survey to prepare the final document and imposed a deadline of 1 June.[43] In May, William A. Vogely, director of Interior's Office of Economic Analysis, began to study the economics of other methods of transporting North Slope oil, the national security implications of an Alaskan and a Canadian route, the impact of an Alaska pipeline on the U.S. balance of payments and the revenues of the State of Alaska, and future American energy supplies and demand. Vogely solicited contributions from the departments of State, Commerce, and Defense, the Treasury, and the Office of Economic Preparedness. An indispensable part of the task force preparation for the final statement was a comprehensive Project Description from Alyeska. However, this was not forthcoming and the deadline for the final statement was set back first to 1 July, then to 15 September 1971. Morton (like Hickel in 1970) blamed the oil industry for the continuing delay. He told the Anchorage Chamber of Commerce: "We're moving as fast as the information comes to us. . . . The monkey is on the oil companies' back, not mine. I've been two-weeked to death, of commitments for delivery data made by ALPS."[44] Alyeska was ready to submit its twenty-nine-volume Project Description in late July and early August. The Project Description provided the USGS's Menlo Park Working Group and the Technical Advisory Board with its first maps of the proposed pipeline route. It also indicated which parts of the pipeline would be buried and which would be raised. The Menlo Park group was still not happy and requested additional submissions which provided mile-by-mile design data. Morton now set 15 November as the fourth target date for the final environmental impact statement.

In deference to criticism that other pipeline routes had not been adequately considered, Morton approached Alyeska for information regarding Canadian attitudes. On 22 September, Harry Shooshan, the director of Interior's Office of International Affairs, explained that though the Canadian government had not taken an official position, its attitude to a Canadian pipeline was positive. He indicted, however, that a Canadian line would involve a one-to-two year delay. For Shooshan, the critical factor was that the oil industry was not really interested in this alternative. Meanwhile, Morton established a new deadline of 15 January 1971 for the final statement.

In general, the preparations of Brew's task force had proceeded on the assumption that NEPA did not require discussion of anything beyond what Alyeska had submitted. Then, on 13 January 1972, the U.S. Court of Appeals in Washington, D.C., affirming a District Court judgment the previous month in a litigation brought against Interior by the Natural Resources Defense Council (NRDC), ruled that NEPA required that the impact statement consider alternatives. This case concerned Outer Continental Shelf leasing off Louisiana. The alternative suggested on this occasion by the Court of Appeals was reduction or elimination of oil import quotas.[45] Consequently, Morton set up a special task force to write a section for the final impact statement on alternatives that had not hitherto been fully weighed. Richard D. Nehring, who worked at the Department of the Interior's Office of Economic Analysis, was assigned to study the economics of a joint oil and gas pipeline corridor through Canada.

By this time, Native claims had long been settled. The Alaska Native Claims Settlement Act (ANCSA) of December 1971 extinguished further Native land claims. In return, it gave Alaskan Natives outright ownership of 44 million acres of Alaska and a cash settlement of almost $1 billion.[46]

ANCSA specified that if the Secretary of the Interior wanted to set aside a pipeline transportation and utility corridor, neither the State of Alaska or Natives could select lands within it. Half the cash settlement was to be paid from royalties on oil production. This gave the Native community an interest in the pipeline's rapid authorization. An Alaska Federation of Natives press release put it simply: "The rejection or substantial delay of the Trans-Alaska Pipeline could easily undermine and violate the intent and terms of ANCSA."[47] Following the settlement of Native claims, NEPA and the Mineral Leasing Act now became the major legal obstacles to the project.

The *Final Environmental Impact Statement on the Proposed Trans-Alaska Pipeline* was released to the president's Council on Environmental Quality and to the public on 20 March 1972. The nine volumes included three on the economic and security aspects of the proposal and amounted to a total of 3,500 pages, prepared at a cost of $12.7 million. The statement was full of warnings about potential damage to the environment. It was constantly emphasized that insufficient information was currently available upon which to base accurate estimates of the actual extent of injury and interference that would result from the project. Concerning the movement of big game, the statement commented that the "effect of the above-ground portions of pipeline on large mammal movement cannot be conclusively predicted. Knowledge of the behavioral reactions of large mammals to obstructions is as yet quite limited."[48] Discussing the problem of gravel removal and siltation, the statement remarked that "while measures presented in the Stipulations are designed to keep erosion and siltation at the lowest practical level, pipeline and road construction activities would result in erosion and stream siltation."[49]

Addressing the subject of seismicity along the southern two-thirds of the route, it concluded that "the probability that one or more large-magnitude earthquakes would occur in the vicinity of this portion of the proposed route during the lifetime of the pipeline is extremely high, in fact, almost a certainty."[50] The statement also expressed serious fears about the danger of oil spills at Port Valdez and in Prince William Sound. Referring to the local salmon fishery, the authors stated that "it is likely this resource would suffer some damage from pollution associated with the proposed project," and warned that "even with the best available technology and controls, some oil would be spilled. . . ."[51]

Though they did not give them priority, the statement paid appropriate homage to wilderness values and recognized Alaska as their supreme American embodiment: "Nowhere in the United States, except Alaska, are there such large areas where land is essentially untouched by modern man and his technology."[52] It tried to define wilderness and its values, which, it acknowledged, the project would "irreversibly and irretrievably" affect.[53] The statement quoted the Wilderness Act of 1964, which defined a "wilderness area" as "an area of undeveloped Federal land retaining its principal charcter and influence, without permanent improvements or permanent habitation, which is protected and managed so as to preserve its natural condition . . . and which . . . has outstanding opportunities for solitude or a primitive and unconfined type of recreation."[54] It emphasized the ecological value of wilderness as baseline data against which to measure environmental change elsewhere, and its value as a sanctuary for the fraught members of an industrialized and urbanized civilization in search of respite. But the statement stressed that wilderness retained its value regardless of actual use and utility. The authors described its "psychological value" to "those who find reward in knowing that the wilderness is there, whether or not they are able to visit the area. They enjoy the simple and comforting awareness of the existence of such areas."[55] This particular value of wilderness, which some conservationists considered its greatest, was also its least tangible. Despite—or perhaps because of—an abundance of *de facto* wilderness in Alaska, there was to date no Alaskan contribution to the national system of wilderness areas that was being created in accordance with the Wilderness Act of 1964. The statement's authors echoed Bob Marshall's view, expressed thirty years earlier, that nowhere in the United States "is the opportunity to identify and preserve 'wilderness areas' as grand as it is in Alaska."[56]

Alaskan advocates felt that the document did not make a strong enough case for the project. According to Senator Stevens, "it was not written by a proponent."[57] Despite what the Alaskan senator believed, this revised version endorsed the general approval that the Department of the Interior had given to TAPS in the draft impact statement. It recognized the need for an overland gas route through Canada to the American Midwest; it was estimated that 26 trillion cubic feet of recoverable gas existed in association with oil reserves at Prudhoe Bay.[58] However, the statement reemphasized that the national interest in reducing dependency on foreign oil demanded an oil pipeline across Alaska as soon as possible and wholly under American control. It also argued that an Alaskan pipeline would improve the U.S. international balance of payments, provide a boost for the American shipbuilding industry, create badly needed jobs for Alaskans, and essential income for the State of Alaska.[59]

Morton announced a forty-five day "period of grace" to allow for comment on the impact statement before taking further action. Six hundred copies of the statement were printed and could be bought at $42.50 per set. Only seven were available for public inspection, during office hours, at certain government agencies. Lawyers working for the Alaska Public Interest Coalition (APIC) claimed that it took twenty-eight days to obtain copies of the document.[60] APIC and various other groups, including the National Rifle Association and the Boy Scouts of America, supported by a bipartisan bloc of twenty-three senators and eighty congressmen, called for public hearings. Interior refused. According to Undersecretary Pecora, further hearings would be "a circus" and "interfere with a more thoughtful and rational analysis of this complex document."[61] On 4 May, the deputy undersecretary of the interior, Jared Carter, who had had minimal involvement with the project to date, was appointed to review public comments.

The plaintiffs in the conservationist legal action drew up and submitted a 1,300-page rebuttal to the final environmental impact statement.[62] It contained five basic criticisms. The first, and most general, was that Interior had been subservient to the oil industry ever since it began processing the permit application in June 1969. Its authors alleged that Interior's limited conception of its public responsibilities had resulted in "a passive document that blandly accepts at face value the fundamental premises of the oil companies."[63] Secondly, the report claimed that the much vaunted federal construction and operations stipulations were vague and unenforceable. Thirdly, the counterstatement charged that the discussion of meeting energy needs from other sources in the final statement was no more sustained or systematic than the draft statement's had been. Fourthly, the plaintiffs argued that superior transportation alternatives, notably that of a common oil and gas transportation corridor through Canada, had still not been properly investigated. Fifthly, the counterstatement asserted that Interior's economic and national security data, contained in the final three volumes, were specious and distorted, based on a dubious assessment of the unreliability of oil supplies from the Middle East.

Some of the opinions expressed in the final impact statement did not support the official conclusion concerning national security. According to data in the statement, the exploitation of North Slope oil would reduce dependency on Middle Eastern oil from 27 to 20 percent of the total American demand by 1990. Critics denied that this 7 percent reduction was sufficient to justify a pipeline delivering oil to the West Coast, particularly when the area of greatest dependency lay east of the Rockies. This echoed the assertion of the secretary of defense in the final impact statement that North Slope oil "will not solve the U.S. future crude oil deficit, no matter how it is distributed." According to the secretary of defense, no one delivery route was preferable from a national security point of view.[64]

The authors of the counterstatement also queried Interior's claims that Canada would demand majority ownership of a pipeline through Canada, a minimum capacity of 50 percent for transporting its own arctic oil, and preference in jobs and supplying materials during construction. The secretary of the interior considered the possibility that Canada, in a crisis, might cut off the supply of oil through a Canadian pipeline to be one of the most significant drawbacks of a Canadian route. Instead, EDF, FOE, and the Wilderness Society emphasized Canada's reliability as an ally and considered this factor irrelevant to decision- making. Again, the views of the Department of Defense, the secretary of the state, and the president's Office of Emergency Preparedness, contained in the body of the environmental impact statement, provided support for this position. These federal officials also had no doubts about Canada's reliability.[65] Finally, critics questioned whether a pipeline through Canada was genuinely less physically secure than an exposed tanker route down the West Coast. In 1941 the vulnerability of American sea lanes to attack had been the primary justification for the route chosen for a highway to Alaska. Thirty years later, the Department of the Interior felt that national security considerations dictated a tanker route down the West Coast.. Curiously, this official conclusion did not harmonize with the views of the Department of Defense, the secretary of state, and General G. A. Lincoln, the director of the Office of Emergency Preparedness. These officials pointed out that a pipeline through Canada was superior in the sense that it would reduce the amount of American oil that had to travel along vulnerable sea lanes.[66]

On 6 May Deputy Undersecretary Carter examined the protest statement. Two days later, he distributed it to departmental reviewers. These reviews were returned to him on 10 May, when Morton looked over them.[67] On the same day, Richard Nehring, an employee at Interior's Office of Economic Analysis, completed a highly favorable report on the feasibility of a joint oil and gas corridor through Canada. However, his report was not distributed until 16 May.[68] Morton announced his decision to grant the construction permits on 11 May. He explained: "The nucleus of my decision is based on the urgent need to bring North Slope oil and gas into the American market place as rapidly as possible. . . . the trans-Alaska route presents the only feasible means of transporting Arctic oil within an acceptable time frame."[69] Congressman Les Aspin (Democrat, Wisconsin), one of Alyeska's severest congressional critics, denounced this decision. He called it "a blatant example of the interests of the oil industry superseding the public interest. . . . Apparently, contributions from the oil companies to the Nixon campaign are simply more important than the goodwill of environmentalists, and Midwest and East Coast consumers."[70] David Brower recommended that the nation "should do with the Prudhoe Bay oil what the Navy has done with its petroleum reserves—leave the oil in the ground."[71]

Nonetheless, some conservationists were satisfied. In the private sphere, the contented ones were typified by William E. Towell, the executive vice president of the American Forestry Association (1875). While critical of the original proposal and draft impact statement, Towell had always remained confident that the line could and should be built eventually.[72] By the summer of 1972, critics of this type had been won over by the design modifications imposed upon Alyeska and the federal stipulations drawn up to govern construction. Those whose criticism was conditional praised the provisions for elevating a good deal of the line, for the use of thick gravel insulation under the construction work pad, for revegetation, and for "snaking" the line to allow for horizontal and vertical movement. They were satisfied by plans to install an automatic leak-detection and valve shut-off system, underpasses and ramps to allow big game to cross elevated sections of pipe, and by contingency measures to contain oil spills in the Port of Valdez and for treating ballast water at the terminal. They believed that these measures would take care of all problems associated with extreme temperature variations, permafrost thawing, erosion, seismicity, fish spawning, siltation of streams, caribou migration, and land and marine pollution. However, many conservationists continued to oppose the pipeline proposal for some mixture of economic, political, ecological, and other reasons.

In June Congress's Joint Economic Committee held public hearings at which Morton's recent decision to approve the pipeline was discussed. The committee's chairman, Senator William Proxmire (Democrat, Wisconsin), who led congressional opposition to the SST in 1970, had been one of the project's most consistent and vociferous critics. Though he emphasized that the purpose of the hearings was not to discuss the merits of the Alaska Pipeline, these sessions did provide critics with a further platform.

The case against TAPS was strengthened by the appearance of an internal memorandum (undated), which had been prepared at Morton's request by the then deputy undersecretary of the interior, Jack O. Horton, at some stage soon after the final impact statement had been released. Horton had been coordinator of Interior's environmental studies for TAPS. "An Alternative to the Trans-Alaska Pipeline," which compared TAPS with a pipeline system for carrying oil and gas from the American and the Canadian Arctic down the Mackenzie Valley, made a convincing case for the latter. Avoiding pollution in Prince William Sound was just one of the environmental advantages it listed. Another ecological asset that Horton mentioned was the avoidance of spills from tanker traffic down the West Coast. Horton argued that a Canadian line also meant that the pipeline's designers would not have to tackle sensitive permafrost regions in southcentral Alaska and north of the Brooks Range. Furthermore, a Canadian line would avoid the active

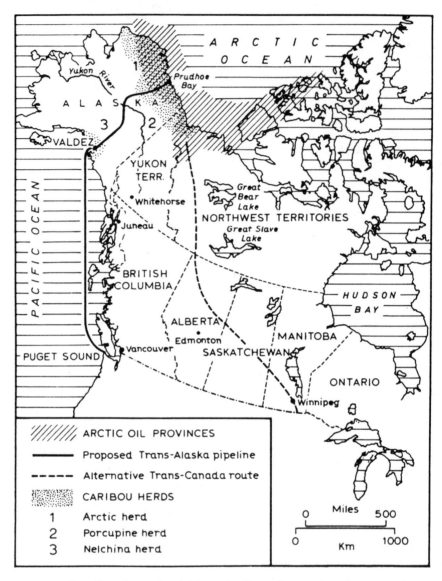

Legend:

///// ARCTIC OIL PROVINCES

—— Proposed Trans-Alaska pipeline

- - - - Alternative Trans-Canada route

CARIBOU HERDS

1 Arctic herd
2 Porcupine herd
3 Nelchina herd

Miles 500
0
0   1000
Km

The trans-Canadian alternative (with arctic oil provinces and caribou ranges) seismic areas of southcentral Alaska. Not least, it would remove the eventual need for two separate pipeline corridors, one for oil across Alaska, and another for oil and gas through Canada.[73]

Economically, Horton argued that a trans-Canadian pipeline would not be prejudicial to the financial interests of the State of Alaska in terms of the revenue it would gain from oil production and taxes. (The section of the final impact statement devoted to analysis of the economic and security aspects of the project had stressed that there would be no difference in terms of consumer price and oil industry profits between the two routes.) Addressing its political advantages, Horton judged that a joint energy corridor through Canada would be far more popular with the American public, the media, the environmentalist community, and the midwestern and eastern electorates. He pointed out that Interior had received over 52,000 letters on the subject of North Slope oil during the past year alone, the "vast majority" of which opposed TAPS.[74]

Charles J. Cicchetti and Richard Nehring were two leading critics who sought to exploit the inconsistency between the contents of Horton's memorandum and the conclusions in the final impact statement. Cicchetti, an economist, was currently a research associate with Resources for the Future (RfF), a conservation think tank based in Washington, D.C. Cicchetti, whose book, *Alaskan Oil: Alternative Routes and Markets*, was published a few months later (August 1972), firmly believed that a route through Canada was economically and environmentally superior.[75] Nehring had resigned from his position at the Department of the Interior's Office of Economic Analysis in protest over Morton's decision to approve the trans-Alaska project and what he considered the censorship of his own favorable report on a joint oil and gas line through Canada. He accused high officials at Interior of a restricted conception of the department's duties regarding investigation of alternatives. He illustrated this by quoting a statement that he attributed to William Pecora, the Task Force chief: "we have imposed alternatives already. We have required more valves, we have required the companies to construct more of the pipeline above ground."[76] Cicchetti and Nehring stressed the dichotomy between comparative environmental data, such as those presented in Volume 5 of the final impact statement, which indicated unequivocally the superiority of a Canadian route (preferably from Prudhoe Bay to Edmonton), and the official conclusion of the impact statement. Nehring accused Pecora of deleting a staff conclusion to the first volume of the statement, which declared that, on balance, a route down the Mackenzie was less damaging environmentally. The implication, he contended, was that economic and national security considerations—themselves of dubious validity—had received priority.[77]

The pipeline was probably the major issue for Alaskans in the presidential election of 1972. Many Democrats in other states were cool, if not hostile toward TAPS. There was considerable discussion among Alaskan Democrats about Senator George McGovern's stance on the pipeline. McGovern was known to be sceptical; he was one of twenty-three senators who had called for public hearings after the final environmental impact statement was issued. Alaska's Congressman, Nick Begich (Democrat), was satisfied that the project had McGovern's approval. But Governor Egan refused to endorse McGovern's presidential candidacy because of his unsupportive attitude toward the pipeline. Instead, Egan backed Senator Jackson, one of the project's leading congressional supporters.

In contrast, the Republican attitude was enthusiastic. In July, Vice President Spiro T. Agnew traveled to Fairbanks to campaign for the reelection of an administration that had adopted the immediate authorization of the Trans-Alaska Pipeline as one of its major second-term legislative priorities. Addressing local party faithful, Agnew attacked the spirit of "anti-growth and pessimism" that prevailed in the lower forty-eight and which he blamed for the delay in authorizing TAPS. Employing time-honored booster rhetoric, he denied that the act of pioneering was an anachronism and defended Alaska and its pipeline advocates as an extension and reaffirmation of the nineteenth-century West and its frontier ethos. As Frederick Jackson Turner had done, Agnew emphasized the significance of the West and the frontier as a process as well as a specific place, and as a phenomenon synonymous with individual opportunity. Geographically, Turner had explained, "The oldest West was the Atlantic coast."[78] However, at bottom, he saw the West as "a form of society, rather than an area" and "another name for opportunity."[79] Frederic Logan Paxson agreed: "The American frontier was a line, a region, or a process, according to the context. . . . As a *process*, its most significant meaning is found."[80]

Agnew, the product of an East Coast, urban environment, identified the West and the frontier, above all, as an attitude. He quoted the poet, Archibald MacLeish, who was fascinated by westward expansion and its mythology, to the effect that "the West is a country in the mind, and so eternal." Fearlessness, as Edward Teller had maintained when he was promoting Project Chariot in Alaska and as Rampart "pumpers" had emphasized, was a prime characteristic of a frontier people. Like such others before him, Agnew played on the Alaskan predisposition to accept new challenges and stressed that to refuse to build the pipeline for fear of the environmental damage it might cause would be "inconsistent" with the adventurous spirit of the frontier.[81]

Organized pipeline opposition also gained new recruits that summer.

In July the Canadian Wildlife Federation and its British Columbian affiliate filed their own suit against Alyeska and the U.S. Department of the Interior. Their major fear was oil spillage from tanker traffic along the British Columbian coast. These concerns applied in particular to the Strait of Georgia and the Strait of Juan de Fuca, which were notorious for fog, high winds, and rip tides. In preparation for the flow of arctic oil, ARCO proposed to build a refinery costing $100 million at Cherry Point on Puget Sound, some ten miles south of the Canadian border.[82] Oil destined for Puget Sound refineries would be shipped through the Strait of Juan de Fuca which separated Vancouver Island from Washington State's Olympic Peninsula. David Anderson, the leader of the Liberal Party in British Columbia and the chairman of the House of Commons Committee on Environmental Pollution, represented the coastal riding (constituency) of Esquimault-Saanich, near Victoria. Anderson entered the lawsuit against the pipeline in his own right.[83] The British Columbian Wildlife Federation achieved legal standing through its members' proprietory interests. The basis for Anderson's participation was Section 101(C) of NEPA, which specified that "Congress recognizes that each person should enjoy a healthful environment and that each person has a responsibility to contribute to the preservation and enhancement of the environment." This provided the individual, regardless of nationality, with the right to sue.

8.3 Denouement: The Shift from the Courts to Congress, August 1972–November 1973

On 15 August 1972 Judge Hart dissolved the injunction he had imposed in January 1972. He ruled that Interior's final environmental impact statement had complied with all NEPA's requirements and dismissed the plaintiff's argument based on the Mineral Leasing Act. On 6 October, however, the U.S. District Court of Appeals in Washington, D.C., decided that the project could not be approved until Congress had amended the Mineral Leasing Act to accommodate a broader right-of-way. On 9 February 1973, in a partial reversal of Hart's ruling, the court ruled unanimously that the rights-of-way and Special Land Use Permits that Morton was ready to issue violated the Mineral Leasing Act: "Congress intended to maintain control over pipeline rights-of-way, and to force the industry to come back to Congress if the amount of land granted was insufficient for its purposes. . . ." "Whether this restriction made sense then, or now," it continued, "is not the business of the courts. And whether the width limitation should be discarded, enlarged or placed in the discretion of an administrative agency is a matter for Congress, not for the Courts."[84] The court declined to rule on the adequacy of the impact statement.

The additional delay naturally aggravated discontent among pipeline advocates in Alaska and even generated some secessionist feelings. The court of appeals' decision provided the impetus for Joe Vogler's formation of the Alaska Independence Party.[85] In March, Senator Gravel formed the Alaska Pipeline Education Committee (ALPECO). Like the Rampart lobby, Yukon Power for America, ALPECO was modeled on the statehood campaign and claimed to represent "the vast majority of Alaskans."[86] ALPECO's members included Tom Kelly, Alaska's commissioner of natural resources under Governor Hickel; George Sullivan, the former mayor of Anchorage; and Frank Murkowski, the president of the National Bank of Alaska, who had served as Alaska's Commissioner of Economic Development from 1967 to 1970. ALPECO's chairman was Les Spickler, the chairman of the Juneau Chamber of Commerce. ALPECO members' efforts to discount the extent of pipeline criticism in Alaska also recalled the Rampart Dam controversy.[87] Like Yukon Power for America, ALPECO received state funds to support its activities ($100,000). Opponents claimed that most of its money came from Alyeska.

Following the Court of Appeals' decision in February 1973, the Nixon administration gave even higher priority to the pipeline's authorization as part of its campaign to decrease American dependency on imported oil. The president's State of the Union address on 15 February 1973 was divided into five subject areas, one of which was "Natural Resources and the Environment." Its tone and content contrasted sharply with Nixon's first address on the environment in 1970, which had also been the first in American history. "The great question of the seventies," he had declared on that occasion, "Is shall we surrender to our surroundings, or shall we make our peace with nature and begin to make reparations for the damage we have done to our air, to our land and to our water?"[88] In 1973 he asserted: "There is encouraging evidence that the U.S. has moved away from the environmental crisis that could have been. . . . I can report to the Congress that we are well on the way to winning the war against environmental degradation—well on the way to making our peace with nature."[89] Nixon's message to Congress on the energy situation on 18 April stressed the importance of TAPS as a solution and opposed any further delay to restudy a Canadian alternative.[90]

The progress that the TAPS proposal was making at this political level was not reflected in the courts or accompanied by any significant shift in public opinion. On 7 April, notwithstanding the appeals of Interior, Alyeska, and the Department of Justice, the Supreme Court refused to review the lower court's February decision. Senator Gravel had admitted that "the American people do not seem, from public opinion, inclined to go for the pipeline."[91] Not only for preservationists in the national conservation community, but also for most of the interested American public, the Arctic was not the exclusive property of the oil industry and development-minded Alaskans. Few Americans were moved by the emphasis Alaskan proponents placed on the economic value of oil development to their state. Accordingly, there was little national public support for the pipeline before the summer of 1973, when the oncoming "energy crisis" enhanced the project's appeal considerably. Gravel conceded as much, both off the record and in public statements: "I have said before that if a national referendum were to be held on the Trans-Alaska Pipeline it would probably be defeated."[92]

On 9 March 1973 the Senate Interior Committee began hearings on the first of a series of bills introduce by the project's congressional supporters. Senator Jackson's bill sought to amend the Mineral Leasing Act to allow the Secretary of the Interior to grant rights-of-way across federal lands above the legal maximum when he deemed them to be in the public interest.[93] Opponents of the Alaska pipeline emphasized that the Mineral Leasing Act had been part of a congressional reaction against the profligate disposal of the public domain during the Gilded Age. Lobbyists for environmental organizations on Capitol Hill often quoted its original purpose: "To prevent monopoly and waste and other lax methods that have grown up in the administration of our public land laws."[94] Conservationists also warned that, if authorized, the project would involve the biggest transfer of land from public to private hands since the railroad era. However, Senator Jackson, commenting on the right-of-way restrictions, declared: "I have not heard and I cannot imagine, an intellectually respectable argument against correcting this situation." (Existing restrictions clearly did not allow enough room for modern machinery to maneuver.) Yet Jackson stressed that the second obstacle to authorization would still remain: "Nothing in this Act shall be construed to amend, repeal, modify or change in any way the requirements of section 102(2)(C) or any other provision of the NEPA."[95]

The type of criticism based on the oil industry's technological unpreparedness and ecological ignorance between 1969 and 1972 was the kind whose beneficial influence Alyeska and its Alaskan allies were prepared to admit at the rights-of-way hearings. Thornton F. Bradshaw, the president of ARCO, confessed: "Early in the game environmentalists blocked us for very good reasons indeed. . . . We did not know how to make an environmentally safe line. They helped us. We learned a great deal from them."[96] Senator Gravel had also apparently undergone a considerable change of viewpoint in this respect. In the summer of 1969 he had told the president of the Valdez Chamber of Commerce that the pipeline "should be built and built quickly."[97] But then, as early as 1971, he had explained to a student audience in his home state of Massachusetts: "In some respects, delay has been a beneficial one. We are in a far better position to weigh the consequences environmental, economic, social. . . ."[98] He was also capable of considerable backtracking. In 1972, in reply to a Canadian who had asked for his opinion of Morton's decision to approve the trans-Alaska pipeline, Gravel had asserted: "My own position has been that we were not in a position to go ahead with the pipeline permit a few years ago. I think that the engineering and regulatory aspects can now be handled safely."[99] At the right-of-way hearings in May 1973, Gravel also employed conciliatory rhetoric. He admitted: "While the four-year delay in construction of the Alaska Pipeline has been costly to the U.S. in balance of payments and a worsening of the energy shortage, it has . . . and I think most of us agree, including the oil industry—served a very useful purpose. A safer line will be constructed today than could have been constructed four years ago."[100] The desire to defuse any remaining environmentalist opposition and strip it of legitimacy can be detected in these peace overtures.

Others were more blunt. In May 1973 the Pioneers of Alaska complained: "We in Alaska are presently facing a deluge of interference in our industry, commerce, land development, public works and business affairs . . . generated by misinformed parties, however well-intentioned, in the name of the Sierra Club, Friends of the Earth, and other like non-resident groups."[101] Resolution Number 44 of the Alaska State Senate proposed a lawsuit (which never materialized) against "the Sierra Club and all the other groups seeking to thwart the orderly development of Alaska." The suit demanded compensation of not less than a billion dollars to cover alleged losses to date.[102] The reference to the Sierra Club, the Wilderness Society, and Friends of the Earth as "non-resident" was slightly misleading. It was true that they were based in the lower forty-eight, where the bulk of their membership resided. But they all maintained active Alaskan chapters. The Sierra Club, for example, enjoyed a higher per capita membership there than in any other state. The club's 381 Alaskan members (at the beginning of 1971) represented one Alaskan per thousand.[103]

The Sierra Club was involved in legal actions against the proposed extension of the Copper River Highway to the Richardson Highway, to provide road access to Cordova. It also opposed a plan of U.S. Plywood Champion Papers, Inc., to build a saw and pulp mill complex in the Tongass National Forest. However, though invariably singled out as "ringleader," the club was not a party to the legal action against the pipeline. Two different explanations of the Sierra Club's nonparticipation have been put forward. David Brower explains that he invited Michael McCloskey, his successor as Sierra Club chief, and Stewart Brandborg of the Wilderness Society, to participate in a suit against TAPS. Brower recalls that only Brandborg responded. McCloskey rejects this reconstruction of events, explaining that the Sierra Club was planning its own suit and felt upstaged when the others acted independently. Nonparticipation indicated an aversion to passive participation, not an unwillingness to tackle TAPS. It was not club policy to commit itself to "tag along" legal actions.[104]

The Sierra Club's Alaska branch certainly advised national headquarters against taking legal action: "Sierra Club intervention at this time could be disastrous as far as the future effectiveness of the Alaska chapter in the state is concerned," warned its local leader, Gerald Ganapole. The other litigants did not carry the same stigma in Alaska and Ganapole believed that "their suit will carry just as much weight whether the Sierra Club joins or not." Ganapole considered it imprudent for the club to hamper its operating capacity in Alaska unnecessarily and to jeopardize whatever progress it had made toward improving its reputation there. Referring to the recent passage of bills in the Alaska Legislature, which the local branch had supported, such as the creation of Chugach State Park and a proabortion measure (as well as many pending bills), Ganapole concluded that "it is a matter of conjecture as to their outcome had the Sierra Club joined the suit."[105]

Many Alaskans whose public involvement stretched back to the 1950s were inclined to perceive the continuing struggle to secure the pipeline as a reenactment of the statehood battle. They argued that statehood would not be invested with substance and meaning until the pipeline was built. In the summer of 1975, Terris Moore, a former president of the University of Alaska (1949–53), denounced the Sierra Club and its cohorts as the latest in a succession of colonialist forces that had plagued Alaska. He drew a parallel between the status of Alaskans since 1958 and that of Black Americans following the landmark Supreme Court decision in *Brown versus Board of Education* (1954). Moore felt that the Civil Rights Act of 1964 bore the same relationship to this earlier breakthrough as the authorization of the pipeline would to statehood: "If statehood is to have real meaning, Alaska must have the opportunity to control its own economic destiny."[106]

In June 1973 Stewart Udall incensed Alaskan pipeline advocates when he came to Alaska on a speaking tour. Addressing an audience in Anchorage at Alaska Methodist University, on behalf of the Wilderness Society, the former interior secretary warned Alaskans against pursuing an excessive rate of economic development. He expressed the hope that Alaska would remain thinly populated and sparsely developed. Going too fast in the direction of growth, he claimed, would "sacrifice this exceptional quality that to many people is the essence of Alaska." His contention that Alaskans were "sharply divided" over the pipeline particularly upset local pipeline proponents.[107]

By the spring of 1973, many of the organizations opposed to TAPS since 1969, private convictions aside, had adopted the so-called trans-Canadian alternative as the best way to fight the Alyeska proposal in Congress.[108] So had some Canadian conservationists. Richard Passmore, the executive director of the Canadian Wildlife Federation, declared, "I cannot personally foresee any grounds on which a reasonable proposal for transport across Canada is likely to be unduly delayed."[109] This represented a considerable change in policy over the past two years. At the confirmation hearings held in January 1971 to consider Morton's nomination as Interior Secretary, the Sierra Club, while agreeing that a pipeline through Canada was preferable, had announced that it continued to favor nonexploitation of arctic oil.[110] The club's president, Phillip Berry, had denied the existence of a genuine energy crisis. He had argued that eliminating wasteful practices, such as the use of large automobiles, would obviate the need for more oil. In 1969 oil provided 43.2 percent of the nation's total energy needs, of which 60 percent was consumed by transportation.[111] Resources for the Future (RfF), a conservation research agency, had also concluded that a 10 percent reduction in gasoline consumption would make Arctic oil development unnecessary.[112] This emphasis on frugality and efficient use was classic utilitarian conservation.

The Sierra Club's stance did not change significantly for more than a year, as the Joint Economic Committee hearings in June 1972 indicated. In principle, the club continued to advocate a total freeze on arctic resource exploitation and a five-year moratorium on oil development. While admitting that a Canadian line was worthy of study, Berry stressed that "neither route would be as desirable as halting North Slope development." He questioned whether Alaskan oil would significantly reduce national dependence on Middle Eastern supplies. He argued that even when flowing at the maximum pipeline capacity at 2 million barrels per day, it would provide for only 9 percent of the nation's total requirements. Imports would still account for 33 percent of consumption. The Sierra Club suggested reduced consumption, federally imposed fuel conservation, and other measures such as improved building design, insulation, and research into non-polluting, renewable forms of energy (wind, solar, and tidal).[113]

Although most accounts date the "energy crisis" from the fall of 1973, there was much discussion of an emerging predicament in the press and among federal officials in the early summer. The oil industry certainly used the specter of a crisis to generate support for the Alaska pipeline. Against the background of this impending "energy crisis"—genuine, contrived, or perceived—environmental organizations were increasingly on the defensive by mid-1973. In contrast to the boom period from 1969 to 1971, these organizations were experiencing zero growth by the fall of 1972.[114] A Canadian route for the pipeline also received considerable public and congressional support from easterners and midwesterners, for economic as well as environmental reasons.[115] A leading proponent of oil delivery to the Midwest was Senator Walter F. Mondale (Democrat, Minnesota). With reference to NEPA and the Mineral Leasing Act, Mondal attributed the three-year delay in authorizing the Alaska pipeline to "the willingness of high government officials to attempt to disobey the law."[116] In July Mondale and Birch Bayh (Democrat, Indiana) introduced the North Slope Energy Resources Act (S. 1565). This gave the secretary of the interior temporary (two-year) authority to grant rights-of-way wider than the legal maximum for all projects other than North Slope oil and gas. In an effort to remove the final decision from Interior, the bill advocated a nine-month study by the National Academy of Sciences of all aspects of alternative routes, to be followed by a final congressional choice. Mondale had often contended that Alyeska wanted a pipeline that would deliver to the West Coast because it was planning to export up to 25 percent of North Slope oil to Japan. "The trans-Alaska pipeline," he declared on 13 July, the day of the Senate vote on S. 1565, "should more properly be called the trans-Alaska-Japan pipeline."[117]

The Environmental Defense Fund had taken the lead among members of the Alaska Public Interest Coalition (APIC) by officially endorsing a pipeline through Canada as early as March 1972. In November 1972, to prepare for the shift in the contest from the legal arena to Congress, the EDF and FOE had begun to urge the other members of APIC to reject the "leave-it-in-the-ground" attitude toward arctic oil as a practical strategy in the political arena.[118] The rest of the coalition fell in line, though the degree of enthusiasm varied. APIC's members gave their collective backing to a Canadian alternative in May 1973. APIC pinned its greatest hopes for further delay upon the Mondale-Bayh amendment, which was supported by the *New York Times*, the *Washington Post*, and the *Christian Science Monitor*. Before the vote on the senators' amendment, George Alderson, APIC's coordinator, sent this advice to all regional representatives of FOE: "Please refrain from saying 'leave the oil in the ground.' It is a fine policy and we all believe it, but if we push it now, its going to be used against us in our attempt in Congress to get a trans-Canadian pipeline or railroad. The pro-TAPS people are trying to find evidence that we'll fight any oil delivery system, and these statements are merely adding fuel to their fire."[119]

Brock Evans, the Sierra Club's representative in Washington, D.C., and the chairman of APIC, justified the new policy to critics—in this instance the Sierra Club of British Columbia—in the following way: "We could not last thirty seconds in any Senator's office by talking about just leaving the oil in the ground—not when the oil companies are shutting down gas stations everywhere."[120] Senator Frank E. Moss (Democrat, Utah) encapsulated the increasingly impatient public mood: "I cannot get overly upset about observing the ritual of the mating season for Alaskan caribou when in the city of Denver last weekend it was almost impossible to find gas."[121] The Mondale-Bayh amendment was defeated in the Senate by 61 to 29.

A number of Canadian conservationists were angry that some of their Canadian colleagues and American counterparts were now actively lobbying for a pipeline through Canada. Lille D'Easum, chairperson of the Energy Committee of the Sierra Club of Western Canada (British Columbia), dismissed a line down the Mackenzie as "a worse alternative."[122] Equally outraged was Jim Bohlen, the chairman of this chapter. Bohlen was an American expatriate who had been a founder of Greenpeace as well as the Sierra Club of Western Canada. He claimed that the "real" voice of Canadian conservation was represented by groups like the Sierra Club of Western Canada, the Society for Pollution and Environmental Control (SPEC), Pollution Probe, and Voice of Women. He also took exception to APIC's assertion in a fact sheet (25 May 1973) that one of the main advantages of a Canadian pipeline was that it would help tap oil in the Canadian Arctic as well.[123] Disturbed by what he saw as double standards, Bohlen wanted the Sierra Club's national leadership to "disassociate itself from any discussion of alternatives on oil shipments from the North on the grounds that it cannot be done safely, environmentally or socially." "The North," he warned, "will be ruined forever."[123]

Replying to these attacks in July, McCloskey denied that it had ever been official club policy to support a pipeline through Canada: "What we have supported is thorough inquiry into all of the various alternatives. The exigencies of blocking the immediate pro-develoment bills in Congress are such that the Club has cooperated with APIC in trying to block the worst bills." He argued that a "leave-it-in-the-ground" stance, however desirable, was now futile and counterproductive. McCloskey was embarrassed by this evidence of internal dissent and realized how damaging it could be. "I hope that we can keep a somewhat low profile on this question until we get through the votes this summer in the American Congress," he confided to Bohlen.[125] Some Canadians were also outraged by FOE's contention in July that "ecosystem tragedies are beginning to occur in the Canadian North that Prudhoe oil and gas would not compound. Canadians will almost certainly push a transportation system along the Mackenzie River to bring their own gas and oil to market."[126] Canada, it is true, was already forging a road link with the Arctic Ocean. Construction of the Dempster Highway between Dawson and Inuvik at the mouth of the Mackenzie, begun in 1957, but effectively dormant until 1968, was currently in progress.[127] However, some Canadian conservationists argued that, if anything, the Canadian Arctic was more pristine than its American counterpart.

Certain elements in the APIC coalition advocated a railroad through Canada to the Midwest as "the wisest of all ways" to move oil from the North Slope. FOE was the most enthusiastic exponent of this method.[128] A railroad, FOE explained, also had the advantage of being a "self-builder." In other words, it did not require a road.[129] FOE cited studies by the Canadian Institute of Guided Ground Transport (CIGGT) and Premier David Barrett of British Columbia (sponsored by the Canadian Transport Commission, Canadian National Railways, the Canadian Northern, and the Canadian Pacific). These studies estimated that twenty-four trains, each consisting of 150 cars and traveling at 40 miles per hour, would be required each day to transport the same amount of oil a pipeline would move.[130] Though caribou experts had noted the number of reindeer killed by trains in Scandinavia,[131] the supporters of a Canadian railroad did not address the threat to wildlife.

While preserving the integrity of the central Brooks Range, both railroad proposals intruded upon the Arctic National Wildlife Range. The CIGGT route ran along its northern coast through caribou calving grounds. Barrett's line cut through the eastern Brooks Range and then continued east along ANWR's southern boundary. A line across to the Mackenzie delta, explained FOE, "would involve only a few hundred miles of virgin terrain, and it would stay north of the Brooks Range instead of knifing through it." But, it did admit, "these are precious miles."[132] Curiously, FOE also preferred a railroad because a pipeline would be temporary. Though there were provisions for removing a pipeline once it became obsolete, FOE claimed that when the oil ran dry, the pipeline would "just sit there and spook caribou." A railroad, however, could be adapted for "perpetual service."[133] The argument that a railroad would provide a better boost to northern development was inconsistent with conservationist fears over the long term impact of the haul road. One might have expected FOE to applaud the ephemerality of a pipeline's intrusion.

These adjustments in policy required by the energy crisis weakened APIC's credibility and strengthened the hand of its opponents. Ecology, the relatively young science of "planetary housekeeping," teaches supranationalism. There is no ecological frontier, using "frontier" in the European sense of political or national boundary, between the American and Canadian Arctic. As FOE emphasized, "tundra and caribou have no nationality."[134] Yet FOE provided no evidence to support its disclaimer that "no American environmentalist entertains the idea of saving the Alaskan Arctic at the expense of the Canadian Arctic."[135] The Alaska Public Interest Coalition's policy on the Canadian pipeline subordinated ecological principles and internationalism not only to political needs but also to nationalistic and cultural requirements. This policy showed less concern with Alaska as part of a last great international arctic wilderness and ecosystem than with wilderness that was specifically American. APIC's yardstick was not the amount of northern wilderness affected by an oil pipeline; the quantity disrupted by a line through Canada would be more, if only because of its greater length (1,738 miles to Edmonton). A pipeline down the Mackenzie would disturb 800 miles of wilderness before reaching established transportation corridors. In contrast, the amount of roadless country traversed by the Trans-Alaska Pipeline would be 360 miles. Of course, the amount of Alaskan wilderness affected by a Canadian line was considerably less—approximately 150 miles.[136]

Advocates of the Alaska pipeline took full advantage of the weaknesses of APIC's position on the Canadian alternative. Criticizing Morton's decision to approve the Alaskan route on national security grounds, FOE emphasized Canada's reliability as an ally, citing cooperation in construction of mutual defense projects such as the Alaska Highway and the DEWline. Moreover, Harvey Manning asserted, "Canadians are sick and tired of being treated as colonial subjects, exploited for American profit."[137] In contrast, Senator Gravel criticized the Canadian alternative as an example of "export pollution." He denounced environmentalists who wanted "other nations to do the drilling, pumping, refining and transportation of oil so that their own water and land will be untouched by man. This is no answer. In fact, it is a new force of imperialism under the guise of the holy cause of the environment. This is the worst kind of conservation. Ecology tells us we are one world with one system."[138]

The extensive studies that the Department of the Interior and the oil industry had been obliged to carry out for the trans-Alaska route put them in a position of considerable advantage in the debate over the Canadian alternative. The most detailed study of the Canadian route to date was *Arctic Oil Feasibility Study* (1972) by Mackenzie Valley Pipeline Review Ltd. (MVPR). Max Brewer, Commissioner of the Alaska Department of Environmental Conservation, had played a prominent part in preparing Interior's pipeline impact studies. In May 1973, Brewer explained that this MVPR report reminded him of some of the early reports from the TAPS consortium to the Department of the Interior.[139] Interior Secretary Morton was able to turn Canada's weaker environmental protection laws to his advantage: "I can give no assurance [that] with respect to a Canadian route similar to the exacting environmental stipulations and strict environmental surveillance that we have developed for Alaska."[140] "The fact that Canada has no NEPA counterpart," Alaska's Democrat in the House (1970–72), Nick Begich, exclaimed, "does not reassure me; it concerns me."[141]

At the rights-of-way hearings in May 1973, Senator Gravel told proponents of a Canadian line that it was "extremely naive" to assume that Canadian environmentalists and nationalists would not "vigorously" oppose it. He considered APIC's claim that there would be no opposition in Canada tantamount to saying that Canadians "are not really interested in environmental denigration as we are in this country. It is part of our new maturity and Canadians are undergoing the same maturity."[142] Ironically, in view of the recent Alaskan situation, American conservationists overlooked Canadian Native land claims as a complicating factor. A political and cultural awareness was also emerging in the Yukon and Northwest Territories. Inspired by the success of their Alaskan compatriots, Native groups such as the Brotherhood and the Committee for Original Peoples Entitlement (COPE) were now pressing their claims under treaties signed in 1899 and 1921.[143]

On 17 July supporters of the Canadian alternative made their last stand during the debate on Senator Gravel's amendment to foreclose further court-ordered delays on the Trans-Alaska Pipeline project. Senator Mondale produced documents that indicated to him that the State Department had deliberately misrepresented the position of the Canadian government to influence the current debate. On 7 July the State Department had reported that the Canadian government would require majority ownership of a Canadian pipeline. However, the Canadians subsequently denied this. The State Department received this corrected information before the vote on the Mondale-Bayh amendment (13 July) but did not release it until afterwards. Mondale charged that the administration had withheld this information to assist the passage of the amendment. (On 23 July State Department officials denied this charge in testimony before the Senate Foreign Relations Committee. Rufus Z. Smith, deputy assistant secretary for Canadian affairs, spoke of the "inadvertant slip-up" in relaying information that Canada would not insist on 51 percent ownership of a Canadian line. This, he explained, had meant that the new information was not received by the State Department until July 16, three days after the vote on the Mondale-Bayh amendment, though it had arrived in Washington, D.C. on 13 July.)[144]

The crucial and most dramatic victory for the pro-Alyeska forces came on 17 July when the Senate narrowly passed the Gravel Amendment. By declaring that the Department of the Interior had fulfilled all the requirements of NEPA, the amendment released the Alyeska project from any further legal delay. At the first vote, it was passed by 49-48. The vote to reconsider, however, was tied at 49-49 because Alan Cranston (Democrat, California), absent for the initial vote, was now present in the chamber, Vice President Agnew exercised his tie-breaking power to cast the deciding vote in favor of the amendment. This was only the second time he had voted as president of the Senate.[145]

Every New England Democrat opposed Gravel's amendment. So did all midwestern Democrats with one exception. Eastern Republicans were evenly split. The most powerful support in both parties came from the South and the West. The closeness of the vote owed something to the wording and intention of the Gravel amendment. Though the amendment was supported by the Nixon administration (probably the only issue on which Gravel and the administration ever saw eye-to-eye), it was opposed by Senator Jackson, sponsor of the official bill. Jackson wanted to authorize the Alaska pipeline without damaging NEPA and argued that Gravel's amendment "invites delay by creating numerous new opportunities for litigation."[146] Jackson's bill was supported by Alaska's other senator, Ted Stevens, Congressman Don Young, many of Alaska's state leaders and its press. As late as the day of the vote, all of them considered the Gravel amendment a potentially counter productive gamble.[147]

The *Los Angeles Times*, which was pro–Alaska Pipeline, disapproved of the Gravel amendment because of its consequences for NEPA. It described the amendment (19 July) as "an unfortunate case of doing the right thing in the wrong way." Liberals were dismayed. Since entering the Senate in 1968, Mike Gravel had won an enthusiastic national following among Americans of this persuasion. Support was especially strong among dissaffected and disillusioned young Americans who had been won to his side initially by his theatrical exposure of the Pentagon Papers (29 July 1971). Gravel had earned further respect from liberals for his picket of the White House to protest the nuclear weapons test on the Aleutian island of Amchitka in 1971, his sponsorship of a nuclear power moratorium, his promotion of alternative energy (particularly solar), and his criticism of the Vietnam War. His pipeline boosting distressed and confused these admirers as much as Gruening's support for Rampart had upset liberals and conservationists in the 1960s.

Many conservationists, needless to say, were outraged. The Wilderness Society commented of the Gravel amendment, "When political and economic pressure is applied, NEPA is not worth a damn."[148] The majority of eastern and midwestern papers also roundly condemned the amendment. The *Baltimore Sun* ("NEPA Emasculated," 19 July), described it as a "notorious act of panic and hysteria" encouraged by the oil industry through a "putative energy crisis which Alaskan oil would not remedy anyway." The editorial reprimanded one of Maryland's senators, Glen Beall, for supporting the amendment. The *Minneapolis Tribune* (19 July), which dismissed arctic oil as "a quick fix for America's profligate tastes in fuel," believed that the breach of NEPA was a greater calamity than the authorization of the pipeline itself. The *New York Times* "Debacle in the Senate," 18 July) called it "a shocking show of irresponsibility." The *Washington Post* (17 July) deplored Gravel's measure as an "open assault" on NEPA. The *Trenton Times-Advertiser* (19 July), which praised New Jersey's senators for standing firm against the amendment, hoped that the House of Representatives would not disgrace itself by following the Senate.

On 2 August the House debated H.R. 9130. Section 20(1)(d) declared that "the actions of the Secretary of the Interior heretofore taken with respect to the proposed trans-Alaska oil pipeline shall be regarded as satisfactory compliance with the provisions of the National Environmental Policy Act of 1969."[149] John Dellenback (Republican, Oregon) and Wayne Owens (Democrat, Utah) introduced an amendment to delete this part of the bill, which, Morris Udall (Democrat, Arizona) commented, "would have been laughed out of the chamber a year ago. . . . Tonight a lot of those who helped to write the National Environmental Policy Act into law are preparing to gut it."[150] The Dellenback-Owens amendment was rejected by 221-198.

John Saylor, the House Interior Committee's ranking Republican member, and a long-standing TAPS critic, claimed that the energy crisis had been fabricated to justify "pressure" legislation.[151] Congressman Robert Kastenmeier (Democrat, Wisconsin) denounced the proposed legislation as "a private bill for the benefit of Atlantic-Richfield, British Petroleum, and Exxon." Ron Dellums (Democrat, California) condemned capitulation to "scare tactics" and Bella Abzug (Democrat, New York) described it as "a blatant sell-out" that "short-circulated NEPA." An amendment sponsored by Morris Udall, John Anderson (Republican, Illinois) and John A. Blatnik (Democrat, Minnesota), designed, in Udall's words, "to give the environmentalists the one thing they want . . . objective study of the Canadian alternative," was rejected by voice vote. Supporters of the Canadian alternative, such as John Anderson, were still claiming that "Canadian environmentalists are not expected to fight the [Canadian] pipeline."[152] The House passed H.R. 9130 by 356-60.

In spite of the backing the pipeline project had received from the Department of the Interior, the Nixon administration, organized labor (the AFL-CIO endorsed the pipeline on 9 May 1973), the oil industry, and the State of Alaska, congressional authorization had not been inevitable. In 1941 it had taken a national security crisis to build a highway to Alaska. In 1973, another crisis, the energy panic, was required to secure widespread congressional and popular support for the pipeline project.[153] At the hearings on the draft impact statement for the pipeline, which were held in February 1971, Evangeline Atwood had commented that a national crisis (like the one that forced the construction of the Alaska Highway) would put paid to the debate over the pipeline. Events had vindicated her assessment.

The passage of the Gravel amendment was the first victory for Alaskan boosters since the Kenai Moose Range was opened to oil development in the late 1950s. This was also the first major conservationist defeat since then, despite the fact that this constituency's interest in Alaska, its political power, and national policy on conservation of natural resources and environmental protection were much stronger than ever before. Indeed, this antipipeline coalition may have been the most powerful and broad-based conservationist force in world history. On the other hand, Alaskan boosters had never enjoyed such massive national support.

In a postmortem on the conservationist performance during the struggle against TAPS, Brock Evans emphasized how powerful the advocates had been. In the context of the energy panic, the intense pressure exerted by the oil industry with its powerful allies had proved irresistible.[154] Edgar Wayburn, the club's vice president, agreed: "We were up against forces too big, too powerful and too well-financed for us to beat."[155] For Evans, the outstanding feature of the congressional struggle was not that the Sierra Club and its allies were defeated but that Alyeska's victory was so narrow.[156] In short, the club had put up a brave fight and had suffered no disgrace. Evans and Wayburn also denied that the defeat was a colossal tragedy. In contrast to the apocalyptical way preservationists had treated the pipeline proposal for four years, these officials now tended to downplay their dismay and the project's significance as a destructive agent. In fact, Evans denied that the recent struggle was the crucial engagement in the campaign to protect the nation's last frontier wilderness: "We lost a big battle but not the whole war. We are ready to go on to the other battles ahead."[157] According to Wayburn, the fight against TAPS was "only one aspect of the formidable conservation war we are fighting in all of Alaska."[158]

On 10 September President Nixon's message on legislative priorities established authorization of the pipeline as the administration's priority for the remainder of the current session.[159] In Congress that fall, pipeline advocates blamed the energy crisis on the conservationists who had stalled the project's progress. Congressman Craig Hosmer (Republican, California) claimed that if construction had been allowed to proceed as planned in 1969, there would now be no closed schools or gas stations:

"To preserve the 7,680 acres that would be occupied by the pipeline seems an inordinate price to pay for fuel rationing, cold homes, cold schools, and blackmail by the Arab world."[160] On 6 October, Egypt and Syria had invaded Israel. To retaliate for American military aid to Israel, the Arab members of OPEC imposed an embargo on oil exports to the United States. On 8 November, in the context of this embargo, Nixon delivered a further message on the energy crisis, which reaffirmed the urgent need for Alaskan oil.[161] Only fourteen congressmen voted against the final authorization bill on 12 November, while 361 voted for it. The next day the Senate passed the bill by 80-5. Shortly after signing the Trans-Alaska Pipeline Authorization Act, the president introduced "Project Independence" to the nation in a televised address: "Throughout its history, America has made great sacrifices of blood, and also of treasure, to achieve and maintain its independence. In the last third of this century, our independence will depend on maintaining and achieving self-sufficiency in energy."[162] As far as the pipeline's Alaskan supporters were concerned, a greater measure of autonomy, long-thwarted, had at last been achieved for Alaska. Alaska's value as a national storehouse of resources had been finally recognized. Congressman Don Young triumphantly hailed authorization as "a Second Statehood Act."[163]

The authorization act barred further judicial review on the basis of NEPA and restricted further legal action to the act's constitutionality. A sixty-day period was granted for those who might wish to challenge this. David Brower was prepared to continue litigation.[164] The Environmental Defense Fund recommended against it. John Dienelt, one of EDF's leading attorneys in the case felt: "The possible extent and effect of negative reaction to continuing the TAPS fight on the plaintiffs specifically and the environmental movement generally is admittedly speculative. . . . We believe the risk of harm from a strong and anti-environmentalist reaction is a very serious one which is not worth taking." Dienelt advised fellow-litigants that "research and analysis has produced no theory on which we believe even a colorable argument that Congress's action violated the Constitution can be based."[165] EDF dropped its legal fight after Nixon signed the authorization act. On 16 January 1974, after the Wilderness Society and FOE announced that they would not challenge the act's constitutionality, Judge Hart dissolved the injunction against the pipeline project.

The Arctic Institute of North America (AINA) wanted to effect a reconciliation between the adversaries in the recent struggle. Toward the end of 1973, the institute had begun to organize a group called the Arctic Environmental Council "to maintain continuous dialogue and to promote mutual understanding among the energy companies and the organizations [conservationist] initially involved with the TAPS construction and operation."[166] Friends of the Earth declined an invitation to join the council. It explained that "it would be irresponsible for us to lend our support to an institution that purports to unite representatives of the public in amicable alliance with representatives of the oil interests."[167] The Wilderness Society also kept its distance, believing that its participation would "play into the hands of Alyeska" by giving the Arctic Environmental Council the seal of approval as "the official civilian environmental surveillance group."[168]

The Sierra Club, however, did join the council. The club believed that the general concept of citizen surveillance of major construction projects was a valuable one worthy of support. It later defended the Arctic Environmental Council and its decision to join as "an experiment in something new."[169] Other representatives of conservation organizations who agreed to serve as members of the Arctic Environmental Council were Hamilton Pyles, the executive secretary of the Natural Resources Council; P. W. Schneider of the National Wildlife Federation; Raymond A. Haik of the Izaak Walton League, and, briefly, Robert B. Weeden of the Alaska Conservation Society.[170] Weeden was replaced by David Klein as the sole Alaskan member.

On 29 April 1974, using the Hickel Highway to transport building equipment to the North Slope, Alyeska began work on the haul road and construction work pad north of the Yukon. Unlike the infamous Hickel Highway, the haul road was built upon a gravel base at least five feet thick to insulate the permfrost. The $125 million, 360-mile road was finished on 29 September.

9
The Pipeline Project

9.1 CONSTRUCTION

In the fall of 1969, a grave marker bearing the inscription *Alaska: 1867–1969* was planted in the ground next to a section of the newly arrived, Japanese-made pipe displayed on the campus of the University of Alaska-Fairbanks.[1] For many, the pipeline was an event with funereal significance. For, if built, they believed the pipeline would mark the death, or at least the beginning of the end, of Alaska's wild frontier innocence. Three-and-a-half years later, when authorization of the project was imminent, this view had not changed. One journalist described the pipeline as "the first giant step toward destroying the last authentic wilderness available to those living under the U.S. flag."[2]

It was not until work began on the haul road that a number of critics began to perceive this part of the project—an adjunct that had never been subject to the same environmental impact statement requirements and had largely escaped public and congressional debate—as the fatal act. The Sierra Club's representative on the Arctic Environmental Council, Brock Evans, traveled along the recently completed gravel road in a bus with the other members of the council on its first surveillance trip in the fall of 1974. Evans scribbled in his notebook, in words reminiscent of George Marshall's comments on the Hickel ice road five years earlier: "The wilderness is broken forever. True, there is lots more—'the road only takes up 12 square miles of Alaska' they are fond of saying. But that is not the point. It is the psychology of the North that is gone."[3] Soon afterwards, Evans tried to explain to Frank Fisher, the manager of Alyeska's Environmental Protection Unit, what he meant by saying the "spell of the north" had been broken. With the notorious Hickel Highway apparently forgotten, Evans described the "feeling" that had existed before "that there was nothing; No roads within tens or dozens of miles, not just a little bit away . . . The feeling of a whole mountain range, a whole region where there was essentially nothing. That's what has been changed."[4] The Alaska Sportsmen's Council maintained that

251

the road would be a prime recreational asset by improving public access to the Brooks Range. But Evans felt that "it's a road to see what no longer exists."[5] The damage was psychological, intellectual, sociological, and emotional as much as environmental. "The old empty North that Jack London and others wrote about," he mourned, "is simply gone."[6] He pronounced the last American frontier dead long before the first section of pipe was laid.

Harvey Manning, the author of Friends of the Earth's Cry Crisis! Rehearsal in Alaska (1974), echoed this conviction that the pipeline project marked the crucial break with the past in Alaska. Manning agreed, too, that the main damage would be symbolic and psychological: "The injury would be mostly to an idea. Those who care about wilderness would have lost the big one; their efforts for Alaska would become less spirited, their minor victories would taste of defeat. It is the difference between defending a virgin and a whore."[7] From this perspective, the project's ecological impact was not necessarily the crucial factor in the controversy.

For the residents of Wiseman, the hamlet made famous by Bob Marshall, the noted outdoorsman and conservationist, the impact was more tangible. According to Rick Reakoff, a big game guide based in the village, all ten permanent residents opposed the project when the road builders arrived on the other side of the Middle Fork of the Koyukuk River in July 1974. Harry Leonard, a seventy-six-year-old prospector who had spent forty winters in what was now a quasi-ghost town, parked his tractor on the road that summer to protest what he considered the infringement of his mining claims. Construction work was halted for six hours. The following day, Leonard walked into a construction camp, produced a gun and demanded that the crew leave. Wisemanites stressed that they were not as worried by the pipeline itself as by the road's potential for opening up the area. As Reakoff's wife explained, "If they would build a pipeline and just go away, that would be O.K."[8] Boosters, on the other hand, wanted unrestricted public access and believed that an "Arctic Wilderness Parkway" would be a tremendous tourist attraction.[9]

The first section of pipe was laid beneath the Tonsina River, 70 miles north of Valdez, in late March 1975. More Alaskans now began to take note of the less desirable changes that oil was bringing to Alaska along with the financial benefits. Newspapers that for six years had been impatient for construction to begin, started reporting on the declining quality of life in Alaska due to an acute housing shortage and rising crime, prices, smog levels, divorce rates, and prostitution. Valdez, the site of the terminal and operations center, and Fairbanks, the project headquarters and the community at the center of the construction boom, bore the brunt of the socioeconomic impact.[10] Tom Snapp's *All-Alaska Weekly*, based in Fairbanks, carried a regular *Pipeline Watch* in 1975 and 1976 compiled by Richard Fineberg, a political scientist at the University of Alaska. In addition to the human cost, Fineberg's column examined issues like project mismanagement, alleged violation of state and federal stipulations, weld X ray certification forgeries,[11] embezzlement, thievery, and bribery. (Allegations of corruption, it should be noted, were also common in the wake of the construction of the Alaska Highway.)[12]

In the media outside Alaska, comparisons were rife between Fairbanks and Dodge City, the buffalo hunters' mecca and cattle town of legendary wickedness. Both "lusty" and "brawling" towns had large numbers of transient males and plenty of saloons and whores. Fairbanks served as a place for pipeline workers to "whoop it up" just as Dodge City had catered to soldiers, buffalo hunters, and, later, cowboys. Alaska had certainly become a frontier of opportunity once again for the oldest profession. A pimp from Las Vegas, whose business was flourishing in Fairbanks, related, "I gathered my little flock together, rented us a Winnebago and we headed up the Alcan to the promised land."[13] However, the press focused on the sordidness and the sadness of the spectacle rather than the romance and the tangible evidence of frontier vitality in the reemergence of these classic western ingredients. The theme of vanishing innocence on "the lost frontier" loomed large. "America's last frontier is being conquered," reported the *Los Angeles Times*, "A lifestyle is slowly—and for some sadly—surrendering its purity to the nation's thirst for oil and the individual's appetite for wealth."[14] The Alaskan tourist industry sought to capitalize on this sombre and nostalgic national mood. "See the real Alaska. Now," urged a Wein Air Alaska advertisement. "While there's still time. America's last great frontier won't stay that way forever."[15]

The belief that the "real" Alaska was succumbing, rapidly losing its exceptionalism, and being replaced by a blander region with no special characteristics, brings to mind Frederic Logan Paxson's elegy entitled *When the West Is Gone* (1930). Paxson had been concerned with the end of the West, and, with it, the possible loss of American distinctiveness: "Shall we become merely a part of western civilization?"[16] In the same way, those who regretted recent events in Alaska feared that Alaska was becoming merely a part of American civilization. These commentators felt that Alaska was lapsing into conformity with the rest of the nation due to the processes of advancing homogenization. For them, Alaska had gone in the sense that the pipeline project and associated boom were erasing the state's special identity.

Some had no qualms or reservations about what was happening. City fathers and community champions like William R. Wood, the former president of the University of Alaska, were unquenchable optimists and men with inextiguishable visions who usually dreamed of future glory. For these boosters, Fairbanks days of fame and plenty had finally arrived. While serving as mayor of Fairbanks, Wood affectionately recalled the pipeline construction years as "a lovely time, packed with excitement I wouldn't have missed for the world."[17]

Just as Americans were divided in their attitude toward what was happening in Alaska during the pipeline construction years, there was no consensus about the role of the four-year delay that had preceded construction. Having secured its pipeline at last, there were times (as seen toward the end of the previous chapter), when Alyeska had magnanimously acknowledged the benefits of the delay. In 1975, Peter De May, Alyeska's Director of Project Management, admitted to students at the Tanana Valley Community College (Fairbanks) that, as late as 1972, there had been "not one mile of pipeline design in existence."[18]

However, Alyeska's attitude toward the four-year delay and the conservationists who had opposed the project remained ambivalent. In the introduction to the Arctic Environmental Council's first report, Edward L. Patton, the company's president, asserted: "if Alyeska had been free to set its own environmental goals in the project the resulting package probably would not have differed radically from the goals represented by the Department of Interior stipulations."[19] The final draft of the Arctic Environmental Council's report following its second inspection of the project (1975) contained a statement by Alyeska concerning its environmental policies to which the Sierra Club's Brock Evans strongly objected. The offending statement read: "In spite of these efforts [to minimize impact], a long and intense national debate over the benefits of the pipeline . . . preceded its Congressional authorization." This implied to Evans that four years of criticism had been superfluous. He protested that "it was only because of the intense national debate that Alyeska has made whatever efforts it has to protect the environment here."[20] The Arctic Institute of North America heeded this complaint and amended the final report to read: "these efforts followed a long and intense national debate over the benefits of the pipeline as opposed to the environmental and social costs which would result from it."[21]

Alyeska nevertheless continued to offer another version of recent history. In 1980, in a special Alyeska report ("The First Three Years"), E. F. Rice, professor of civil engineering at the University of Alaska, Fairbanks, recalled: "Early on, it was realized that a buried pipeline containing naturally warm oil . . . would heat the surrounding ground. . . . the decision had to be made to elevate the pipeline. . . ." Rice omitted the detail that it was federal geologists who reached this conclusion early on and then waged a long campaign to convince the oil industry.[22]

The first crude oil began to flow south through the pipeline on 20 June 1977. It did not flow smoothly. An earth-filling machine bumped into a vent fitting and started a leak that interrupted the first flow. At Pump Station 8, 500 miles down the line, the pipe cracked as cold liquid nitrogen was injected in front of the moving oil. Once again, the system was shut down. On 8 July, at the same pump station, oil accidentally sprayed inside the building started a fire that killed one man and injured a number of others. Oil finally reached Valdez on 31 July. The first crude-laden tanker, the ARCO *Juneau*, left on 1 August and docked at ARCO's Cherry Point refinery on Puget Sound on 5 August.

At this point, those who are not engineers and scientists may find helpful a rudimentary description—by a self-confessed layman—of the finished project's various components and how they are intended to work.

From Pump Station 1 at Prudhoe Bay, where crude oil enters the pipeline, oil moves south on its first leg across the hundred-mile-wide North Slope. Much of the pipe is raised above this treeless plain. The pipe gradually gains altitude as it climbs into the Brooks Range, reaching the highest point along its 800-mile route at the summit of Atigun Pass (almost 4,800 feet above sea level). From this continental divide, drainage flows either northward or southward. The pipe descends sharply from the Brooks Range and then crosses the rolling, boreal country of the interior north of the Yukon River. The pipeline skirts Fairbanks in the heart of the interior and climbs into the Alaska Range, which it crosses at Isabel Pass (3,500 feet above sea level). After descending, it rises again, traversing the coastal Chugach Range at Thompson Pass. Twenty-five miles later, after following the route of the former WAMCATS line, the pipeline reaches its destination at Valdez.

Alyeska buried 380 miles of pipe and elevated 420 miles. The pipe could only be buried in three types of material: permanently thawed soils, bedrock (frozen or thawed), and varieties of permafrost which, when thawed, were considered unlikely to damage the pipe. Most of this thaw-stable permafrost was a mixture of sand and gravel. The techniques employed in this so-called conventional burial departed nonetheless in several ways from designs for places like Texas. Burial depth varied considerably, though it was usually between 3 and 16 feet. In some places, to avoid unstable soils, the ditch for the pipe was dug to a depth of 30 feet. Throughout the buried sections the pipe was supported on a bed of material composed of sand and small gravel and was cocooned in a compact jacket of the same material. All buried pipe was wrapped with a tape specially designed to prevent corrosion.

To provide additional protection throughout the life of the pipeline (and to compensate for any tape accidentally damaged during pipe installation), "sacrificial" ribbons of zinc wire known as anodes were placed in the ditch and connected to the pipe at intervals to prevent electrochemical corrosion through ionization. Zinc, a more active metal than steel, prevents the flow of electric current created by soil chemicals away from the steel pipe.[23]

In all areas of unstable ("marginal") permafrost the pipe was elevated. Elevated sections range from thirty miles in length to stretches of a few hundred feet; most are between 700 and 1,800 feet long. Elevated pipe is supported on crossbeams installed between vertical supports that are found at sixty foot intervals. The pipe is anchored at 800–1800-foot intervals by groups of four closely spaced vertical support members. Thermal devices known as heat pipes were installed to maintain the permafrost around the pipe's vertical support members in a stable state. These heat pipes contain a refrigerant that vaporizes heat from the ground and condenses it in the finned radiators that form the external extension of the heat pipes and are located above the vertical supports. The heat pipes act to remove ground heat whenever the air temperature is lower than the soil temperature.[24] By keeping the permafrost frozen, the heat pipes prevent the heaving and lifting motion caused by the process of freezing and thawing known as jacking. On the North Slope, heat pipes were not required due to the stability of the soil.

The insulation system around the elevated pipe itself is intended to prevent the temperature of the pipe's half-inch wall (0.462 and 0.562 inches to be precise) from falling below minus 20 degrees Fahrenheit when the pumping rate is as low as 600,000 barrels a day. This insulation system is also designed to guarantee that the level of heat generated in the pipe by pumping and friction does not exceed 145 degrees Fahrenheit when the pumping rate attains its maximum level of 2 million barrels a day. Insulation also protects all valve and support points. The aim of this system is to ensure that, if a shutdown occurs, the oil in the pipe will be kept in a pumpable condition for up to three weeks.[25].

For those places where neither buried nor elevated construction techniques employed along the rest of the line could be used, Alyeska developed two special designs. These were needed for three permafrost areas along the southern portion of the route that were considered thaw-unstable and where the pipe would normally have been elevated. Where the pipeline intersected with the Glenn Highway at Glennallen and in two places where it crosses caribou migration routes, a total of 4 miles of pipe were buried under refrigerated conditions. This involved insulating the pipe with 3 inches of polyurethane foam covered with a resin-reinforced fiberglass jacket and then burying it between coolant lines. At each of these places, a refrigerated brine coolant (5 degrees Fahrenheit) is pumped from a mechanical refrigeration unit through 6-inch diameter coolant lines laid in the bed of the ditch on both sides of the pipe. According to Alyeska, this system keeps the surrounding permafrost frozen "much as the ice in indoor skating rinks is kept frozen."

Alyeska used a second special burial method at twenty-three animal migration points within elevated sections. Here, the pipe angles sharply down and disappears underground for 100 feet before reemerging. North of the Brooks Range, where ground temperatures are colder than in southern Alaska, Alyeska insulated the pipe in these buried game crossings (known as "sag bends") with polystyrene foam "planks." South of the Brooks Range, where the temperature of permanently frozen soil stays close to 32 degrees Fahrenheit, better protection against thawing was demanded. At buried game crossings in this section, free-standing thermal devices were installed next to the insulated ditch to draw off the heat from the soil around the pipe.[26]

The 420 miles of elevated pipe were designed to accommodate expansion and contraction of the pipe caused by temperature changes. The pipe can withstand a temperature range of 215 degrees. The lowest temperature it can experience is circa minus 70 degrees Fahrenheit when the pipe is empty in the coldest parts of winter. The highest is likely to be around 145 degrees Fahrenheit when the pipe is full at the maximum daily pumping rate of 2 million barrels. Between these two extremes, the pipe may expand lengthwise by 18 inches in a typical elevated section. To permit this expansion and contraction, the elevated pipe is constructed in zigzags that transform changes in its length into lateral move-

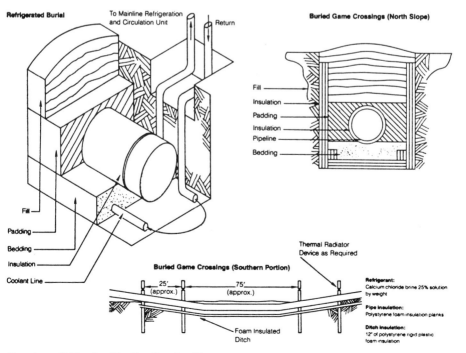

Courtesy of Alyeska Pipeline Service Company.

ment. These zigzags explain why the pipe became 800 rather than 789 miles long. This flexible, trapezoidal design allows a maximum of 96 inches of lateral motion due to heat and a maximum of 50 inches of lateral motion in the opposite direction due to cold. It is also supposed to permit 24 inches of lateral motion in the event of seismic activity up to 8.5 on the Richter scale. Where the pipeline crosses the Denali Fault in the Alaska Range, the design is intended to permit up to 20 feet of horizontal motion and 5 feet of vertical motion. To permit sideways motion, the pipe is mounted off-center on the crossbeams between the vertical supports.[27] Movement of the pipe (up to 12 inches in either direction longitudinally) at these anchors is intended to be possible only during an earthquake.[28]

Forty-two percent of the 100,000 girth welds needed to connect the pipe joints were made under factory conditions at Valdez and Fairbanks. During this operation, two 40-foot lengths of pipe were joined. All remaining girth wields were carried out in the field. Industry and government regulations stipulate that welds are subjected to destructive laboratory tests and radiographed (X-rayed) to check for possible defects. According to the regulations governing the project, no section of pipe could be covered with insulation or buried until each weld had been certified. In 1975 it was revealed that a number of weld certification records had been falsified. Remedial action demanded the excavation of some sections of buried pipe.[29]

Pipe valves of two types were installed to reduce the extent of damage from spills, "Gate" valves, designed to close in four minutes and function in air temperatures of minus 70 Fahrenheit, allow the pipe to be opened and closed. The vast majority of the seventy-one gate valves are located at significant stream crossings, other sensitive environmental spots, and near ten populations centers. Gate valves (operated by remote control) were usually installed on pipe crossing flat terrain and downhill slopes. In addition to these blocking valves, there are eighty "check" valves, located mostly on uphill sections of line. Check valves are held open by flowing oil and are designed to close automatically when the flow stops or reverses. Their function is to prevent the flow of oil backwards in the event of a break upstream (north). These two types of valves are intended to restrict any spillage from a break in the line to an average of 10,500 and a maximum of 64,000 barrels (in one steep section on the south slope of the Brooks Range).[30]

The pipeline crosses more than 800 substantial streams. In some cases, the pipe was buried under the scour level. But in most instances, elevated pipe crosses water on supports. Fourteen bridges were built. The longest, at 2,290 feet, is the Yukon River Bridge, which was incorporated into the highway bridge. Standard plate girder structures were built across ten rivers. Suspension bridges were built across the Tanana and the

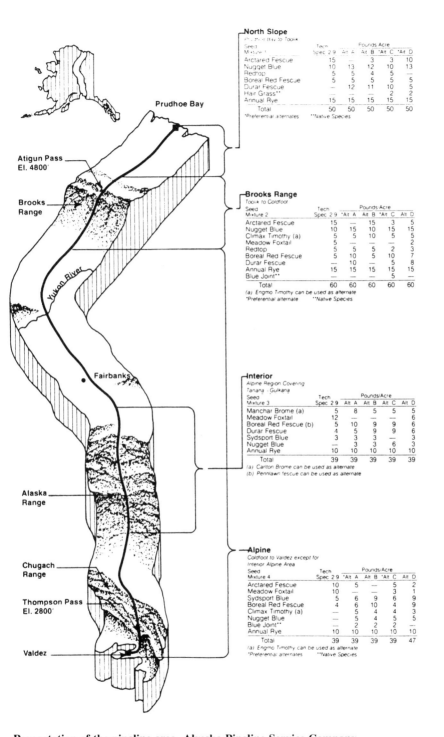

North Slope
Prudhoe Bay to Toolik

Seed Mixture 1	Tech Spec 2.9	Pounds/Acre			
		Alt A	Alt B	*Alt C	*Alt D
Arctared Fescue	15	—	3	3	10
Nugget Blue	10	13	12	10	13
Redtop	5	5	4	5	—
Boreal Red Fescue	5	5	5	5	5
Durar Fescue	—	12	11	10	5
Hair Grass**	—	—	—	2	2
Annual Rye	15	15	15	15	15
Total	50	50	50	50	50

*Preferential alternates **Native Species

Brooks Range
Toolik to Coldfoot

Seed Mixture 2	Tech Spec 2.9	Pounds/Acre			
		*Alt A	Alt B	*Alt C	Alt D
Arctared Fescue	15	—	15	3	5
Nugget Blue	10	15	10	15	15
Climax Timothy (a)	5	5	10	5	5
Meadow Foxtail	5	—	—	—	2
Redtop	5	5	5	2	3
Boreal Red Fescue	5	10	5	10	7
Durar Fescue	—	10	—	5	8
Annual Rye	15	15	15	15	15
Blue Joint**	—	—	—	5	—
Total	60	60	60	60	60

(a) Engmo Timothy can be used as alternate
*Preferential alternate **Native Species

Interior
Alpine Region Covering Tanana - Gulkana

Seed Mixture 3	Tech Spec 2.9	Pounds/Acre			
		Alt A	Alt B	Alt C	Alt D
Manchar Brome (a)	5	8	5	5	5
Meadow Foxtail	12	—	—	—	6
Boreal Red Fescue (b)	5	10	9	9	6
Durar Fescue	4	5	9	9	6
Sydsport Blue	3	3	3	—	3
Nugget Blue	—	3	3	6	3
Annual Rye	10	10	10	10	10
Total	39	39	39	39	39

(a) Carlton Brome can be used as alternate
(b) Pennlawn fescue can be used as alternate

Alpine
Coldfoot to Valdez except for Interior Alpine Area

Seed Mixture 4	Tech Spec 2.9	Pounds/Acre			
		*Alt A	Alt B	*Alt C	Alt D
Arctared Fescue	10	5	—	5	2
Meadow Foxtail	10	—	—	3	1
Sydsport Blue	5	6	9	6	9
Boreal Red Fescue	4	6	10	4	9
Climax Timothy (a)	—	5	4	4	3
Nugget Blue	—	5	4	5	5
Blue Joint**	—	2	2	2	—
Annual Rye	10	10	10	10	10
Total	39	39	39	39	47

(a) Engmo Timothy can be used as alternate
*Preferential alternates **Native Species

Map labels: Prudhoe Bay; Atigun Pass El. 4800'; Brooks Range; Yukon River; Fairbanks; Alaska Range; Chugach Range; Thompson Pass El. 2800'; Valdez

Revegetation of the pipeline area. Alyeska Pipeline Service Company.

Tazlina and a 400-foot-long, tied-arch steel bridge was raised above the Gulkana.

Before and during construction, conservationists and wildlife biologists were concerned about the integrity of valuable anadromous runs of salmon, trout, grayling, Arctic char, and whitefish, not only in major drainages such as the Copper River, the Tanana, the Yukon, and the Koyukuk, but also in the numerous small streams intersected by the pipeline project. Fish migrate into small creeks in the spring and must vacate before freeze-up. Alyeska made do with "low-water" fords whenever it considered stream flow or drainage insufficient to interfere with the passage of vehicles across the construction work pad—a 35-to-65-foot-wide gravel surface used for vehicular traffic and the operation of machinery during construction. Where stream flow was stronger, Alyeska resorted to bridges and culverts. Culverts often become blocked with ice and thaw waters have difficulty flowing through them in springtime. They can also prevent fish migrating upstream to spawn if too small and improperly installed. According to Alyeska, however, since many streams were dry or frozen in winter, it timed much of its construction activity for this season. And when siltation did result from excavation work carried out during stream flow periods, the company stressed that it built settling ponds that trapped sediment and prevented its discharge into drainages.[31]

Alyeska insulated the permafrost under the construction work pad north of the Brooks Range with layers of polystyrene "planks" buried in gravel. South of the Brooks Range, where the permafrost is warmer, Alyeska considered it unlikely that thawing could be prevented from taking place eventually even if large quantities of polystyrene were used. However, the use of polystyrene meant that less gravel was required for the project. Alyeska estimated that if gravel alone had been used for insulation, the work pad would have been between 5 and 9 feet thick north of the Brooks Range. Polystyrene reduced its total thickness there to an average of 24 inches—2 to 4 inches of foam and 20 to 22 inches of gravel.[32]

Federal stipulations required Alyeska to revegetate areas disturbed during construction to protect the pipeline, control erosion, and restore wildlife habitat. Because seeds of grasses native to the areas crossed by the line were unavailable in commercial quantities, Alyeska's agronomists tested varieties of cold-resistant grasses from other parts of the world. They selected four basic seed mixtures (each containing five to seven grass types) to correspond to each of the four major vegetation areas into which they divided the pipeline corridor. The first exotic grass mixture was sown between Prudhoe Bay and the northern foothills of the Brooks Range. The second mixture was used in the Brooks Range and its southern foothills as far as Coldfoot. A third mixture was devel-

oped for the alpine region of the Alaska Range between Tanana and
Gulkana. A fourth was applied to remaining areas between Coldfoot
and Valdez. Seedings, at the rate of 40–60 pounds per acre, were made
by hand, air blower, hydroseeder, airplane, and helicopter. Alyeska anticipated that native grasses would reclaim these revegetated areas within
the next decade.[33]

In addition to the pipeline itself, which Alyeska calls the "passive"
element, there are two major "active" components of the pipeline system:
twelve pump stations and the Marine Terminal at Valdez. At eight
of the stations, pumps driven by 13,500 horsepower jet-aircraft-type en-
gines/power turbine drivers (now upgraded to 17,500 horsepower) move
the oil south. The main source of fuel for the four pump stations north
of the Brooks Range is natural gas from the Prudhoe Bay oilfields,
supplied by a 147-mile, small diameter (8–10 inch) auxiliary pipeline
parallel to TAPS. Small refineries (known as topping plants) able to
produce turbine fuel for jet engines from oil taken from the pipeline
were built at Stations 6, 8, and 10 to supply fuel for themselves and
the other pump stations south of the Brooks Range. Stations 8, 9, 11,
and 12 buy commercial electricity. The rest are supplied from private
generators.[34] Each station is equipped with fuel, power, water, communications, fire detection and fighting, housing, sewage, heating, trash disposal, catering, recreation, shower, and laundry facilities to support
operating crews. All facilities are insulated and enclosed hallways connect withdrawn buildings that are designed to withstand minus 60 degrees
Fahrenheit temperatures and 100 mile per hour winds.

Three pump stations have microwave facilities. A network of 40 microwave stations serves as a communications system for control and operation of the pipeline. There are master control stations at Prudhoe Bay,
Fairbanks, and Valdez. The rest are built on high ground, accessible
by helicopter, away from but parallel to the pipeline corridor. Five pump
stations, three north of the Brooks Range (1, 2, and 3) and two south
of the range (5 and 6) were built on unstable permafrost. Under pump
station buildings, six inches of foam insulation, covered with polyurethane sheeting, were buried in the sand. Refrigeration plants circulate
chilled brine through a network of coils buried in gravel underneath
the foam.

The 1,000-acre shore site at the Valdez Marine Terminal (across the
bay from the town of Valdez) encompasses the Operations Control Center
and storage and loading facilities. Valdez lies in a humid maritime zone
with mild temperatures throughout the year and high precipitation (up
to 300 inches per annum). Accordingly, the eighteen storage tanks have
cone-shaped roofs to withstand and shed the weight of snow. These
tanks have a capacity of about six days of pipeline throughput at a

pumping rate of 1.52 million barrels per day. Tanker loading facilities consist of three fixed and one floating berth. Shore facilities also include a vapor recovery plant, oil spill contingency equipment, fire-fighting systems, water treatment and sewage systems, and a ballast treatment plant.

Federal and state stipulations require all oily ballast water from the holds of Alyeska's tankers arriving at the Marine Terminal to be pumped ashore for treatment before being discharged into the Port of Valdez. Alyeska's original system treated ballast in four stages. In the primary separation stage, ballast water is pumped into gravity storage tanks for a six-hour period that allows oil to float to the surface where it is removed by floating-boom skimmers. The water is then discharged into an airflotation basin where a coagulant and an a polyelectrolyte are mixed in, causing material in suspension to form particles of oil and chemical known as "floc." During the third stage, pressurized water containing dissolved air is mixed in. Air bubbles cling to the floc and bring it to the surface where it can be skimmed off. Meanwhile, solids such as grit settle to the bottom of the basin. Finally, the ballast water flows into a "pH adjustment system" where a proper acid/base balance is restored by introducing sodium hydroxide. According to regulations, if the standard meets the 8 parts of oil per million permitted by the Environmental Protection Agency (EPA), the treated ballast water is discharged into the sea at depths between 200 and 375 feet, and between 700 and 1,050 feet offshore.[35]

The scale of the undertaking eclipsed that of any construction project in Alaskan history. Whereas DEWline construction in the 1950s had required 7.34 million cubic meters of gravel, Alyeska needed 73 million–32 million for the construction work pad alone. Three-thousand men had worked on the DEWline at any given time. The peak pipeline workforce was 28,072.[36] Still, the hyperbolic media coverage should be placed in historical perspective. All who have commented approvingly on any twentieth-century Alaskan construction project have been susceptible to immoderate claims and the use of inflated language.

Like those who built the Alaska Highway, the pipeline's builders sought to blend the old images of pioneering with new ones. One advertisement for Frontier Companies of Alaska showed a six-mule team wagon hauling 48 inch steel pipe while Wesway Steel Company described its off loading ramps as "Dog Sleds for the Alaska Pipeline trail."[37] Nevertheless, the construction of the pipeline did contribute something fresh to juxtapose against the traditional litany of herculean exploits in a ruthless frontier setting, and what Leo Marx has called "the rhetoric of the technological sublime." For the pipeline project was heralded by advocates as a groundbreaking example of painstakingly planned and carefully executed pioneering largely stripped of conquest and destruction. Alaska's Senator

Gravel had described the project as the most exhaustively studied scheme of the twentieth century, which the exception of the space program.[38]

The public utterances of Alyeska and its compatriots were characterized by a tactful tone that spoke of making "the last frontier" also "the best."[39] This produced a curious mixture of pride and humility toward the pipeline as an engineering structure. Two years into construction, one of its designers reflected: "While in other places and other times aesthetics means a quality of the structure itself, how good it appeared in contrast to the things around it, how it stood out, like the grandeur of a great railway terminal or the marvel of the Brooklyn Bridge as the first suspension bridge, the aesthetics of a structure in Alaska at this time is inevitably judged by how it complements its surroundings; not how it stands out from them."[40]

According to one journalist, $7.7 billion and 132 million man-hours had "bought a tube . . . as stiff, strong and unyielding as a column, yet, in its entirety, as supple and resilient as a rubber hose and, as leakproof as man can make it despite bends, curves, snow loads, earthquakes and heat-caused movement."[41] As well as an engineering *tour de force*, Alyeska portrayed the project as a showpiece of corporate ecological enlightenment. "No other project has had to live under such a relentless and unwavering public spotlight," commented one journalist writing for Alyeska. "And no other project has ever been so prepared to respond."[42] Pictures of blooming fields of golden poppies with drilling rigs rising innocuously in the background, and of caribou grazing undisturbed in the vicinity of pumping stations, accompanied the text of a panegyrical history of the Trans-Alaska Pipeline.[43] The authors emphasized that Alyeska accommodated environmental considerations at enormous cost. The example they chose was Alyeska's rerouting of an 18-mile section of haul road (at an extra cost of $3 million). This was done to avoid the nesting areas of raptors such as peregrine falcons, gyrfalcons, and rough-legged hawks on the Franklin and Sagwon bluffs on the Sagavanirktok River, which rise above the otherwise flat terrain of the North Slope.[44] What the account did not mention was that Alyeska did not reroute the road voluntarily. The Endangered Species Act of 1973 required this action.

Cited as further evidence of Alyeska's good faith and exemplary behavior was the deferral of haul road construction in one area until the Dall sheep lambing season was over.[45] Other examples of sensitivity to which the authors gave prominence included the relocation of a beaver pond and the planting of faster-growing, lusher, nonnative grasses in areas disturbed during construction. They also referred to the introduction of 1.4 million willow trees to disguise the pipeline visually, "so that moose and caribou will not be confused by the addition to their environment."[46]

James P. Roscow, a business affairs journalist writing in 1977, also emphasized the extreme lengths to which Alyeska went to comply with government stipulations and to placate conservationists: "Alaska's wild creatures got priority again and again. Early in the road work, a crew was sent elsewhere until a nearby hibernating bear woke up."[47] He also drew attention to Alyeska's obedience to strict federal antilitter regulations: "A visitor from the litter-addicted lower forty-eight will be startled to see the crew of a pickup truck stop abruptly on the oil field's spine road, in the half-light of January with the temperature dropping toward minus 50°, and scramble frantically over the snowy tundra to retrieve empty packing cartons that had blown out of the truck bed. Workers on the North Slope have been ordered to pick up and leave for not being as careful as this."[48]

Others were highly critical of Alyeska's performance and the effectiveness of surveillance by the Joint State/Federal Fish and Wildlife Advisory Team (JSFWAT). JSFWAT had been established in May 1974 to monitor the protection of fish and wildlife resources during construction. It exercised its advisory authority through two offices set up at state and federal levels; the State Pipeline Coordinator's Office and the Alaska Pipeline Office. Its purpose was to oversee construction and ensure compliance with state laws, federal stipulations, and the terms and conditions of the right-of-way agreement.[49] Gil M. Zemansky, a sanitary engineer, was employed by the Alaska Department of Environmental Conservation (ADEC) to monitor pipeline construction for compliance with state environmental laws during the second half of 1974 and the beginning of 1975. Zemansky, the ADEC's only full-time pipeline monitor, left this position in February 1975 after a disagreement with his employer. He subsequently strove to publicize what he saw as Alyeska's widespread noncompliance with state water pollution control laws.[50]

Though many conservationists considered construction a great defeat from the standpoint of preserving arctic wilderness values, some were relatively happy with what had been achieved in terms of mitigating the project's enormous potential for ecological damage. During his second project inspection as a member of the Arctic Environmental Council in 1975, Brock Evans did not dwell on defeat, as he had done the previous year. Instead, he reflected on what had been achieved by the struggle conservationists had waged to prevent construction: "I became impressed with the magnitude of what we really won."[51]

In contrast, David Brower of Friends of the Earth never relaxed his stance. Shortly before the pipeline was finished, he denounced it as "the greatest environmental disaster of our time." Whereas Alaskan proponents had approached the authorization of the pipeline as a second statehood act, Brower spoke of the need for a second Alaska purchase;

to buy the state back from Texas oil interests.[52] In general, the national press also felt that Alyeska's deeds had fallen short of its promises.[53]

Even before the pipeline was finished, those who had advocated a pipeline through Canada to the Midwest were claiming a moral victory on economic grounds. Most West Coast refineries, the destination for Alaskan arctic oil, were unable to handle its high-sulphur content. On the other hand, plenty of low-sulphur oil from Indonesia was available to them. Because of this oversupply on the West Coast, Prudhoe Bay oil would have to be shipped through the Panama Canal to the East Coast.[54] Accordingly, Alaska's congressional delegation and successive governors began their efforts, unsuccessful to date, to persuade Congress to lift the export ban. This would enable Alyeska to supply the lucrative Japanese market. The export ban was imposed as part of the authorization act on national security grounds to ensure that Alaskan oil would be available for domestic use in the event of another embargo.

9.2 Assessing the Pipeline's Impact

In a collection of poetry (1983), William R. Wood described the pipeline as modest, benign, romantic, and certainly beautiful:

> A silken thread, half hidden
> across the palace carpet.[55]

Moreover, by 1979, Alyeska and its allies and supporters were already drawing triumphant conclusions about the project's impact on wildlife. Roscow reported that "caribou graze placidly beside the humming air-strips," and that the pipeline had in fact enhanced the environment for these creatures: "As the oil companies raised the base camp pilings above the tender tundra, caribou used the air space beneath to shelter from Prudhoe Bay's long hours of summer sun and notorious swarms of flies."[56]

It would be presumptuous to judge the pipeline as an aesthetic structure and it is still much too soon to make a significant assessment of the ecological consequences of a pipeline that has been in operation only ten years. Eventually, there will be a comprehensive investigation of the pipeline's effects on the physical environment and its biotic commu-nity, either by an environmental historian conversant with scientific ecol-ogy or by an ecologist or wildlife biologist with a taste and aptitude for history. This task lies far in the future. It will demand scrutiny over many generations to gauge, for instance, the long-term repercussions on the caribou that provided the focus of much public concern.

A scientist's research is not inherently more detached, objective, and exact than an historian's. Perspective can also be distorted by closeness to events. Some industry-sponsored projects are unrealistically designed to deliver definitive data in too short a period of time. Granted, some short-term conclusions are worth considering. As far as consequences for caribou are concerned, studies suggest that the impact of the pipeline is most noticeable on the Central Arctic herd, one of Alaska's smallest, which lives year-round on the North Slope. Cows with calves tend to avoid the corridor because of traffic on the haul road. Cows with calves also avoid the pipeline corridor just as they would avoid tall shrubs along rivers; the pipeline and the roadbed provide concealment for predators. Adult male caribou seems to have adapted more easily. They are often attracted to these new features to feed on revegetated areas. Some have also been observed frequenting the road surface and the abandoned construction work pad in an effort to gain relief from insects. It appears that traffic and other forms of human activity influence caribou behavior as much as the physical presence of the pipeline and road.[57] (A similar pattern of behavior has been noted among Dall sheep in the Brooks Range.)

In the oilfield area, certain environmental consequences have already been recorded. By 1977 there were 200 kilometers of gravel road in the region between the Kuparuk and Sagavanirktok Rivers, not to mention a complex of feeder pipelines. Gravel roads, acting like dikes, can inhibit drainage. Water backs up and forms ponds that drown plant life. Some experts believe that oilfield development at Prudhoe Bay, which covers an area of 650 square miles, has significantly reduced the coastal acreage to which the Central Arctic herd can retreat for calving and the area available for grazing and relief from mosquitoes; caribou seek sanctuary from these insects in breezy coastal areas and often wade into the sea in an effort to shake off the insect menace.[58] Concerning the elevated sections of the pipe, it has been argued that the clearance provided for animals such as caribou, bison (in the Delta region), and bear was often not sufficient (a minimum of 10 feet), especially in winter, in areas where snow is deep.[59]

The oil industry believes that these observations are contradicted by the fact that the Central Arctic herd has doubled in size over the past ten years. However, since this increase also applies to the other two arctic herds in Alaska, some wildlife biologists believe that long-term factors (possibly climate) are responsible. Animal populations fluctuate widely over time and a number of biologists believe that the Arctic is currently in the midst of a phase of faunal abundance as far as some species are concerned. At least one expert has argued that the Central

Arctic herd's growth also reflects the dramatic decrease in the size of the wolf and grizzly populations in the area affected by the project, largely as a result of illegal trophy hunting facilitated by the road. Apart from man, these are the caribou's main predators, which, in the past, helped to keep the herd's size down. Now the Central Arctic herd has virtually no natural predators.[60] To date, the Nelchina herd in the Copper River basin, which declined steadily in size between 1967 and 1973, probably due to heavy sport hunting, does not appear to have been affected by the pipeline project. Living in an open spruce forest, unlike their arctic, "barren ground" cousins accustomed to open terrain, these woodland caribou are more tolerant of obstructions.

Certain wildlife biologists also point out the drawbacks of revegetation with exotic grasses. Because of the heat from the pipe, introduced grasses grow unnaturally lush. This attracts many herbivores, particularly in winter, for heat from the pipe also means that this new forage supply remains free of snow. Over time, these animals could become concentrated along the pipeline corridor, where they will be easy pickings for predators. Moreover, since the animals tend to graze the tasty new grasses to the ground, the erosion controlling function of the new grasses will probably be less effective than was hoped. A recent report by Fish and Wildlife Service officials contends that revegetation efforts have met with minimal success and that "reinvasion" by native plants has been "extremely slow in all but the most favorable sites."[61]

Critics also contend that riparian habitats have suffered considerable damage. They claim that construction activities and gravel excavation have had a detrimental effect upon vegetation along many streams. In particular, they point to the loss of willow trees that grow in sheltered spots along the major drainages of the Arctic. These willows provide critical food and shelter for the small moose population that lives in the Brooks Range and on the North Slope. According to a 1987 report by federal fish and wildlife officials, 21 percent of the riparian moose habitat along the Sagavanirktok River was disturbed during pipeline construction, of which over half cannot be rehabilitated.[62]

Another major complaint concerns the incidence of air pollution, particularly around pump stations where topping plants emit hydrocarbons such as sulphur dioxide (SO_2). Lichens, as indicated in connection with Project Chariot and radioactive fallout, have a peculiar physiology. As a result, they can build up large concentrations of SO_2, despite relatively low atmospheric levels of the pollutant, for the same reasons that they are vulnerable to radionuclides.[63]

There have been spills of diesel and crude oil, jet fuel, and petrol. Alyeska does not deny this. What the oil industry and its critics argue about is how many spills have been serious. Between 1970 and 1986, the BLM's Pipeline Monitoring Branch recorded over 300 spills of more than 100 gallons. A total in excess of 10,000 metric tons of crude has been lost since 1977, though some was recovered. One of the biggest (June 1979) was a leak of 5,267 barrels from pipe buried on the north side of Atigun Pass, the highest point along the pipeline route (4,739 feet), where the pipe cannot be elevated due to avalanche danger. At this point the pipe cracked after sagging due to thawing in what Alyeska described as an "unrecorded" ice-rich zone. From the headwaters of the Atigun River, the spilled oil flowed north toward the Beaufort Sea.[64] Refrigeration tubes and supports fifty feet below the pipe were subsequently installed. Another spill, at mile 734, also resulted from stress and cracks from melting of the permafrost.[65]

In 1985 the Cordova District Fisheries United (formerly the Cordova District Fisheries Union) threatened to file suit against Alyeska for mismanagement of ballast water treatment at Port Valdez. The fishermen's specific charge was that toxic wastewater, unaffected by the treatment process because these malign substances are dissolved, have been discharged into the bay.[66] In July 1985 the federal Environmental Protection Agency, from which Alyeska holds the permit to operate its ballast treatment plant, accused Alyeska of violating its permit and degrading water standards in Valdez Bay.[67]. Studies funded by the National Marine Fisheries Service and the EPA indicate that the macoma clam population in the mudflats of Valdez Bay declined by 85 percent between 1978 and November 1984. These tiny filter feeders, which are the dominant invertebrate in the littoral zone, are known to be especially sensitive to sublethal quantities of hydrocarbons. Unfortunately, most toxic compounds in oil are water-soluble. According to some marine scientists, the incremental buildup of low-level oil pollution is a greater long-term environmental threat than the occasional, conspicuous, well-publicized large spill. Oil toxins interfere with the chemoreceptors needed for gathering food and for reproduction. Though they may not kill outright, these hydrocarbons reduce resistance to disease and injury. Again, however, the investigating specialists caution that it is too soon to draw any firm conclusions from evidence such as the clam studies.[68]

Events have overtaken this book. The oil spilled into Prince William Sound in March 1989 that has affected a large coastal area of southcentral Alaska has overshadowed in severity all the environmental impacts to date off Valdez, at Prudhoe Bay, and along the Trans-Alaska Pipeline. The oil industry's inadequate response to this catastrophe, well documented in a host of publications, has exposed the fraudulence of the promises and reassurances of Alyeska and its members and the impotence of their technology.

A section of the Trans-Alaska Pipeline is prepared for burial in the tundra about 40 miles south of Prudhoe Bay. Alyeska Pipeline Service Company.

Below-ground sections of the pipeline are buried in trenches 8 to 16 feet deep and compacted with select material before burying. Alyeska Pipeline Service Company.

Welders begin first "passes" around 48-inch pipe. Alyeska Pipeline Service Company.

Construction of the Thompson Pass section of the pipeline system nears completion. Alyeska Pipeline Service Company.

A section of 48-inch-diameter pipe is lowered onto the pipeway of the Yukon River bridge, where the pipe will be welded and installed in supports. Alyeska Pipeline Service Company.

A ditching machine clears soil away prior to emplacement of 8- and 10-inch-diameter pipe for a 148-mile natural gas pipeline from Prudhoe Bay to Pump Station 4. The small diameter line will provide fuel to power the Trans-Alaska Pipeline's pump stations north of the Brooks Mountain Range. Alyeska Pipeline Service Company.

This ditching operation just south of Prudhoe Bay is for a buried section of the pipeline. Alyeska Pipeline Service Company.

Workers prepare a joint of 48-inch-diameter pipe for a tie-in of a buried section of the pipeline. Alyeska Pipeline Service Company.

Large rollers are used to support concrete-coated 48-inch mainline pipe with floats attached as it is pulled into a water-filled ditch across the Sagavanirktok River near Happy Valley camp on the Trans-Alaska Pipeline. The coated pipe weighs one ton per foot, and is pulled by a hydraulic "pulling machine" from the opposite end of the ditch. Alyeska Pipeline Service Company.

Work crews bolt a clamp to an elevated section of the pipeline south of Prudhoe Bay. Alyeska Pipeline Service Company.

Filling in the trench, location unknown. Alyeska Pipeline Service Company.

Joining two sections of pipe, location unknown. Alyeska Pipeline Service Company.

The Gulkana River bridge for the pipeline was constructed of materials already available to the project. These included surplus 48-inch-diameter mainline pipe and 18-inch-diameter vertical support members (VSMs). The large diameter pipe and the VSMs form the backbone of the bridge's superstructure—its foundation piers on either side of the river. Alyeska Pipeline Service Company.

The superstructure of the Gulkana River bridge for the Trans-Alaska Pipeline consists of nearly 50 separate pieces of steel. The steel was specially fabricated into tie girders, struts, floor beams, arch ribs, and smaller joints for construction of a tied arch bridge to carry the pipeline across the river, north of Glennallen. Alyeska Pipeline Service Company.

The Trans-Alaska Pipeline crosses the South Fork of the Koyukuk River, about 260 miles south of Prudhoe Bay, on a plate and girder bridge. Alyeska Pipeline Service Company.

Work continues during a snowfall on a bridge crossing for the pipeline across the Hammond River, about 130 miles north of the Yukon River. Alyeska Pipeline Service Company.

Sections of 48-inch-diameter pipe are lowered onto a pipeway along the side of a bridge across the Yukon River. The pipe was later welded and installed in the shoe assemblies, foreground. Alyeska Pipeline Service Company.

A suspension bridge carries the pipeline across the Tazlina River, one of nearly 800 streams and rivers along the route from Prudhoe Bay to Valdez. Alyeska Pipeline Service Company.

Road building equipment climbs a slope north of the Yukon River (visible in background), near Five Mile camp on the first day of construction of the pipeline project, 29 April 1974. Alyeska Pipeline Service Company.

Long arctic summer days assisted construction of the 360-mile pipeline construction road from Prudhoe Bay to the Yukon River. This photo, taken shortly before midnight in mid-June, 1974, is of earth-moving equipment near Galbraith Lake camp, about 140 miles north of the Arctic Circle. Alyeska Pipeline Service Company.

A pickup truck heads south past rolling tundra towards the Brooks Mountain Range on a portion of the all-weather construction highway. Alyeska Pipeline Service Company.

One of the first link-ups of the 360-mile Prudhoe Bay to Yukon River highway occurred when construction crews working from the north and south met at this site on the tundra between Prudhoe Bay and Franklin Bluffs construction camp. Alyeska Pipeline Service Company.

The final link-up of the haul road took place on 27 September 1974 at the South Fork of the Koyukuk River, about 100 miles north of the Yukon. Alyeska Pipeline Service Company.

Grader operators work 24 hours a day maintaining the haul road, using their CB radios to keep in touch with one another. Alyeska Pipeline Service Company.

Arctic ice floats in the background as a tug pulls two barges through the Beaufort Sea, delivering cement, modular camp units, insulations, and other materials for construction to Prudhoe Bay. Alyeska Pipeline Service Company.

Forty-eight-inch-diameter pipe for the pipeline was originally stored at pipe yards at Prudhoe Bay (pictured), Fairbanks, and Valdez. Beginning in 1971, nearly 165 miles of 40-foot and 60-foot joints of pipe filled the Prudhoe Bay yard, awaiting transfer to field staging sites. Alyeska Pipeline Service Company.

Units unloaded at the Prudhoe Bay dock will become housing and office quarters at construction camps north of the Yukon River. Alyeska Pipeline Service Company.

At this site about 100 miles south of Prudhoe Bay, a short section of the pipeline is buried, allowing animals reluctant to cross under the elevated pipeline an opportunity to cross over it. Alyeska Pipeline Service Company.

The short section of buried pipe provides a crossing for animals, such as caribou. Heat pipes will be placed in the small vertical pipes alongside the buried section of pipe to help keep the permafrost soil frozen. Alyeska Pipeline Service Company.

A buried game crossing, known as a sag bend, within an elevated section of pipe. Location unknown. Alyeska Pipeline Service Company.

Caribou grazing on revegetated area under the pipeline. Location unknown. Alyeska Pipeline Service Company.

Lone bull caribou approaching elevated pipeline. Location unknown. Alyeska Pipeline Service Company.

Reflecting a noonday winter's sun, insulation panels glisten on a stretch of aboveground pipe. The insulation system consists of the steel-jacketed fiberglass panels, weatherproof expansion joints, and special foam insulated modules at support and anchor assemblies. Alyeska Pipeline Service Company.

Pipeline with icicles. Location unknown. Alyeska Pipeline Service Company.

Atigun construction camp, three miles north of the continental divide in the Brooks Mountain Range. Alyeska Pipeline Service Company.

The only access to remote communications sites on the pipeline route is by helicopter. This site, located about 40 miles north of Valdez, will be part of a permanent microwave communications system that will monitor the operation of the pipeline. Alyeska Pipeline Service Company.

A plate and girder bridge designed to support an aboveground section of the pipeline traverses the South Fork of the Koyukuk River. Alyeska Pipeline Service Company.

The elevated pipe, supported by H-shaped steel assemblies. Alyeska Pipeline Service Company.

An aboveground section of the pipeline winds across a wooded area about 35 miles north of Fairbanks. Alyeska Pipeline Service Company.

A helicopter equipped with an infrared video camera examines heat pipes installed in aboveground supports. The system is used to verify that the heat pipes are transferring heat from the permafrost to the air. Alyeska Pipeline Service Company.

The elevated pipe in an area of ice-rich permafrost, just north of the Brooks
Mountain Range. Alyeska Pipeline Service Company.

The first tanker to call at the Valdez Terminal prior to the pipeline's startup was the 120,000-deadweight-ton oil tanker, ARCO *Fairbanks*. Shown here docked at Berth 4, the ship spent the month of April in Port Valdez and Prince William Sound involved in training exercises. Alyeska Pipeline Service Company.

One of four berths being constructed at the Valdez Terminal. Alyeska Pipeline Service Company.

The ARCO *Juneau*, a 165,000-deadweight-ton tanker, sails from the Valdez Terminal with the first load of crude oil from Prudhoe Bay. Alyeska Pipeline Service Company.

A containment boom encircles the 900-foot-long ARCO *Fairbanks* docked at Berth 4 during an oil spill equipment drill at the Valdez Terminal. The boom would prevent the spread of oil in the event of a spill. Alyeska Pipeline Service Company.

10
After the Pipeline

The caribou love it [the pipeline]. They rub up against it,
and they have babies. There are more caribou in Alaska than
you can shake a stick at.
— George Bush, campaigning for the presidency, 1988

At the time that I was here last more than twenty years ago,
the North Slope was not yet there . . . or to be more accurate,
nobody was there, except the North Slope. What I have seen
is simply incredible. A couple of thousand people working
hard and continuously and producing half the value the whole
of Alaska is producing. What was really new to me . . . while
the North Slope was growing, so was the caribou, from 3,500
to 18,000. Whatever the Congress thinks of the North Slope,
the caribou quite obviously approve.
— Edward Teller, 1987

10.1 RECENT ISSUES

In the summer of 1976 it was revealed that one of Alyeska's contracting
companies had falsified X rays for some faulty pipeline welds. In response, David Brower advocated a fresh suit against the Department
of the Interior for failure to properly oversee construction. James
Kowalsky, FOE's Arctic representative, claimed that further legal action
would be too costly and would damage conservationist efforts on other
fronts.[1] For many pragmatic conservationists, especially in Alaska, the
pipeline was becoming a dead issue. According to William A. Rice,
the executive director of Trustees for Alaska, an umbrella organization
of Alaskan conservationists: "The pipeline is built. It was not built the
way we wanted it built, but it was better than Alyeska wanted to build
it."[2]

There had been no respite for conservationists, and other issues had
arisen in Alaska that were more pressing; offshore oil leasing in the
Gulf of Alaska, Bristol Bay, and the Bering Sea; renewed exploration
in Pet 4; and the proposal to build a dam on the Susitna River.[3] Most
urgently, there were a number of gas pipeline proposals and the future

of the pipeline haul road to consider. Rice saw no value in conservation-ists concerning themselves with "the incremental benefits that we might be able to gain from devoting such a major portion of our energies to strict enforcement of the [pipeline] stipulations." For him, it was "far more important to the future of this state" that they commit their energies and resources to fighting the aforementioned "potential atroci-ties."[4]

Soon after TAPS was authorized, a number of proposals were advanced for transporting natural gas from Prudhoe Bay to the lower forty-eight. The Alaska Arctic Gas Company favored a pipe along the coast of the Arctic National Wildlife Range, through its middle, or around its southern boundary, proceeding down the Mackenzie. Other concerns preferred a route along the existing TAPS corridor as far as Fairbanks and then parallel to the Alaska Highway. Since the latter proposal fol-lowed established rights-of-way, the Alaska Arctic Gas Company plan was more upsetting to North American conservationists and Canadian Natives.

These groups worried about the integrity of critical caribou calving grounds should the last ecologically undisturbed stretch of American Arctic coastal plain be invaded.[5] The debate over the Alaska Arctic Gas Company proposal's impact on wilderness, and the fragile, subtle, and intangible character of this commodity, echoed what had been a prominent aspect of the recent TAPS controversy. To underline the small physical area involved, one pro–gas pipeline newspaper on the East Coast commented that "you could put four Yellowstones and a Yosemite within the ANWR and still have room for L.A." In contrast, it argued, the 4-inch gas line would make "a thin, almost imperceptible line." Averil Thayer, who was ANWR's manager from 1969 to 1981, agreed that the area directly consumed would be insignificant. But he compared the line to "a razor blade slit across the face of the Mona Lisa."[6]

A Canadian government commission was appointed to investigate the socioeconomic and environmental effects of a gas line on northern Can-ada ("the last great wilderness area of Canada"). The commission's chair-man was Judge Thomas R. Berger of the Supreme Court of British Columbia. Its report (May 1977) recommended against the construction of the Alaska Arctic Gas Company pipeline in view of the unacceptable impact on ecology, wildlife (especially the Porcupine caribou herd), and Native peoples. At one of the Berger Commission's hearings, a historian drew an analogy between the proposed gas line ("iron snake") and the impact of the Canadian Pacific Railroad in the 1880s, which for him marked "the symbolic triumph of white industrial society over the tradi-tional lifestyle."[7] Although it expressed preference for a gas line paralleling TAPS and the Alaska Highway, the report advocated a ten-year moratorium on oil and gas development in the Canadian Arctic.[8]

The report's opening sentence placed the Alaska Arctic Gas Company's proposal in Turnerian historical perspective: "The history of North America is the history of the frontier: of pushing back the wilderness, cultivating the soil, populating the land. . . ." The question to be answered was: "Should we open up the North as we opened up the West?"[9] The report's concept of wilderness was strongly influenced by American attitudes and definitions. However, Berger believed that any assault upon its arctic stronghold was also an attack on the Native way of life. His definition introduced a distinction not contained in the American Wilderness Act; that between Euro-American "industrial" practices and "non-consuming," traditional subsistence cultures as types of human activity. This enabled him to apply the term wilderness to the "homeland" of the indigenous inhabitants of the Northwest Territories.

This illustrates the confusion surrounding the meaning of wilderness and the often ambiguous use of the term. Some anthropologists and ethnohistorians are bothered by these conceptual inconsistencies and the semantic muddle. They are troubled by the apparent conflict between *wilderness* as a term denoting a remote, unpopulated area, devoid of any sign of human activity, and its application to environments in Alaska where there is incontrovertible evidence of long-term (and continuing) intensive non-recreational use of the land by hunting and gathering indigenes. They consider *wilderness* a shibboleth and prefer the term *wildland*.[10] Richard K. Nelson argues that what white Americans perceive as "virgin lands" "are no more wilderness than are farmlands to a farmer or streets to a city dweller." The illusion of wilderness is preserved over much of Alaska, he contends, because Native use has not been accompanied by significant environmental change. The land, argues Nelson, "functions much as it would if no humans inhabited it. . . . the transition from utilized to unutilized environment is not perceivable."[11]

The Berger Commission nominated 9 million acres adjoining the Arctic National Wildlife Range for a Northern Yukon Wilderness Park, thus reviving the idea put forward by some American conservationists in the early 1950s for an international park in the northeast North American Arctic. The commission's report also showed how seriously the American conservationists who had advocated a pipeline through Canada in 1973 as an alternative to TAPS had misread the Canadian situation. Ironically, the report's opposition to oil and gas development in the Canadian Arctic was greeted enthusiastically by the same American groups that had favored an oil pipeline across this region in 1973 instead of one through Alaska.

The major obstacles to a gas route parallel to TAPS were financing and regulatory arrangements, not conservationists' objections. In 1978 the Northwest Alaska Pipeline Company proposed the Alaska Highway Pipeline System, a buried, chilled pipe across Canada to the lower forty-eight. The company secured presidential and congressional approval in 1981. The line's estimated cost in 1979 was $12 billion and the proposal entered financial limbo due to the unwillingness of the State of Alaska and the federal government to provide assistance. In 1982 Alaska's Governor Hammond appointed an Economic Committee on North Slope Natural Gas, co-chaired by former governors Hickel and Egan. In 1983 they recommended the 820-mile Trans-Alaska Gas System (TAGS), a buried, chilled gas pipeline with its terminal at Nikiski on Cook Inlet. Here the gas would be liquified and exported to Japan, which had been receiving gas from Cook Inlet since 1969 and from Korea.[12] Later that year, Hickel, whose enthusiasm for grand designs has never flagged, formed the Yukon-Pacific Corporation to lobby for TAGS (estimated cost $25 billion). By 1985 the anticipated cost of the proposed Alaska Highway Pipeline System had risen to $43 billion and cheap supplies of gas from Canada and Mexico continue to obviate the need for arctic gas.[13]

Of greater concern to Alaskan conservationists than gas pipelines was the future of the gravel haul road. In the fall of 1974 the Alaska Transportation Commission had conducted hearings in Fairbanks to consider applications by private bus companies that wanted to operate a tourist service to Prudhoe Bay. The Alaska Conservation Society opposed any nonoil industry use of the road at least until the project was finished.[14] The Alaska Conservation Society continued to argue that the value of the Brooks Range and the Arctic would "always become greater if the area is not connected to the vehicle population of the United States."[15] The society believed that it should continue to be operated as a private road until aircraft could perform all maintenance and supply functions, removing the need for any road at all.

In the fight to keep the road closed to the public, conservationists and the oil industry have found themselves on the same side. Alyeska had minimal future requirements for the road since it planned to carry out much of its maintenance and surveillance by air. Moreover, more people meant more security headaches: "The greatest fear is man's simple vandalism—potshots from high-powered rifles in the hands of thoughtless hunters, which could damage pipe insulation and auxiliary facilities. A lesser concern is the pipe itself, whose half-inch steel walls should resist even high-powered rifles."[16]

On 22 July 1977, Larry D. Wertz, a young, sometime trapper and miner, tried to blow up an elevated section of pipe just north of Fairbanks.

The blast blew apart the casing and insulation but did not damage the pipe itself. Alyeska's surveillance forces discovered the damage five days later. The motives of Wertz, who underwent psychiatric examination, remained unclear. However, there was nothing in his testimony to suggest that this was a conscious act of eco-terrorism to match the literary fantasies of popular writers. On 15 February 1978, following another act of sabotage (a few miles south of Fairbanks), 8,000 barrels of crude oil spilled from a small hole in the pipe before the leak was detected by a private pilot. Luckily, the oil did not reach the Chena River. Again, many questioned the efficacy of Alyeska's pipeline security system and computer for automatic leak detection and alarm, which is programmed to pick up changes in the pressure and flow of the oil. The culprit(s) was never found.

Early in 1978, Governor Hammond, who believed in "cracking the road open conservatively," announced his support for opening the road in summer to mass transportation by private companies.[17] In October 1979 the haul road came under the control of the State of Alaska, which opened it to public traffic as far as the Yukon River Bridge (Mile 56). Prodevelopment interests in Fairbanks lobbied to open the road year-round to Deadhorse, seven miles short of the Arctic Ocean, where the road finishes.

The Sierra Club, FOE, the Audubon Society, and Trustees for Alaska feared increased hunting pressure, more direct road kills, and the proliferation of lodges, motels, campsites, and service stations north of the Yukon River.[18] Together with the Arctic Slope Regional Native Association, the North Slope Borough (the regional local government of northern Alaska since 1972), the Tanana Chiefs Conference, and Secretary of the Interior Cecil Andrus, they advocated "industrial use" only. By industrial uses are meant all those relating to the development of natural resources. (i.e., oil and gas, coal, and placer mining). This also covers access to villages for residents.[19] In 1982 the State of Alaska opened the road—which the state legislature had renamed the James Dalton Highway in honor of the pioneering arctic engineer—in summer (June, July, August) to Dietrich Camp (Mile 206). Here, in the southern Brooks Range, the state set up a checkpoint at an annual cost of $350,000. If only for financial reasons, there now seems little likelihood that the road will be opened all the way to the Prudhoe Bay oilfields. Maintenance costs are prohibitive, especially for state governments on an austerity drive in response to the decline in world oil prices that has drastically affected state revenues.

During the late 1970s national interest in preserving wilderness all over Alaska reached unparalleled heights. The most impassioned national conservation issue as far as preservationists were concerned was the campaign to secure federal protection for those areas set aside for study as part of the Alaska Native Claims Settlement Act (ANCSA) (1971). Section 17, clause d(2) of ANCSA authorized the Secretary of the Interior to set aside "national interest" lands to be considered for national park, wilderness, wildlife refuge, or national forest status. The act established a deadline of 18 December 1978 for action by the Secretary of the Interior on these proposals. This crusade increasingly absorbed the attention of conservation organizations, who embraced "d(2)" as an attractive traditional cause and one which was more "winnable" than the fight against TAPS had been.[20] The d(2) campaign ended in 1980, when the Alaska National Interest Lands Conservation Act (ANILCA) placed 104 million acres of public domain under some sort of federal protection.[21]

The national interest lands campaign was the climax of seventy-five years of expanding conservationist involvement in Alaska, and ANILCA was the final fruit of twenty years of increasingly intensive efforts to protect Alaskan wildlife, wilderness, ecological systems, and, not least, "frontier" values. Oil development in the Arctic, the Hickel Highway, the Trans-Alaska Pipeline, and the haul road had destroyed the hope, first expressed by Bob Marshall in the late 1930s, that the frontier could be embalmed throughout northern Alaska. However, for many conservationists, ANILCA provided ample consolation and compensation. The act arrested the traditional American process of development and environmental transformation at a comparatively early stage in many parts of Alaska, north and south of the Yukon. For other conservationists, and many Alaskans, the conversion of de facto "old" wilderness into designated and managed conservation units pronounced the death of the frontier's spirit and the loss of its essence as surely as any oil pipeline or gigantic dam. Ironically, though many frontier environments have been protected, they feel that the land has been stripped of the quintessence of wilderness and frontier; absence of human control and management.

But ANILCA is another story. Suffice to say that the national interest lands debate ran along well-worn and predictable grooves. The lines were well-rehearsed and the roles were wearyingly familiar. Despite the solemn announcements of the death of the last frontier as a result of the profane construction of the oil pipeline, the conceit of Alaska as a New World reborn was successfully and spectacularly resurrected for the d(2) campaign. In 1977, John Seiberling, the congressman from Ohio who had fought for the Canadian pipeline route in 1973, believed it was important to know that "somewhere in this world there are pristine areas comparable to what this whole North American continent was like when the first Europeans landed on its shores."[22] It was said ad nauseam that d(2) presented the nation with its last chance to learn from the past, to ensure that Alaska's caribou did not suffer the historic fate of the buffalo. John Kauffmann, a National Park Service planner, described Alaska as "the last remnant of what the New World used to be."[23] In the hands of conservationists, the last frontier image appeared to have lost little of its power and persuasiveness. The sullied whore had become a precious virgin again.

Many Alaskans were outraged at the prospect of the d(2) settlement, which they denounced as the biggest "lock up" in Alaskan history ("the Great Terrain Robbery"). In 1979, Secretary of the Interior Cecil Andrus used the Antiquities Act (1906) to withdraw 56 million acres of Alaskan public domain. As a result, he and President Carter were given the same treatment in Anchorage and Fairbanks that was accorded to Gifford Pinchot in Cordova and Katalla in 1911 during the coal lands controversy; they were burned in effigy. The Alaskan congressional delegation, local politicians, businessmen, and many other Alaskans objected vehemently to d(2). Editorials attacking the proposals in the late 1970s are virtually indistinguishable from those written in the 1920s denouncing the proposal to establish Glacier Bay National Monument. In 1980, when ANILCA became law, it was the turn of Alaskan boosters to lament the death of the last frontier as a place of socioeconomic opportunity and a source of American cultural inspiration and national virility.

One of the units established as part of ANILCA was the 8.63 million-acre Yukon Flats National Wildlife Refuge, which encompassed seven of the Native villages that would have been inundated if the Rampart Dam project had gone ahead. The act also created Gates of the Arctic National Park in the central Brooks Range where Bob Marshall had roamed. ANILCA also enlarged the Arctic National Wildlife Range from 8.9 to 19 million acres, changed its name from range to refuge, and designated approximately 8 million of its acres as wilderness. ANILCA (Title X) Section 1002 instructed the Department of the Interior to study the resources (mineral and wildlife) of the refuge's 1.5-million-acre coastal plain ("1002 area"), which had not been included in the wilderness classification. The House of Representatives had favored wilderness designation for this region—the most critical ecological part of the entire refuge—but the Senate felt this would be unwise before a thorough assessment of the oil and gas potential had been made.

In 1986 Trustees for Alaska and other conservationist groups filed a lawsuit against the Department of the Interior on the basis of the National Environmental Policy Act (NEPA). These groups objected to the department's plans to circulate the environmental impact statement it had prepared on oil leasing in ANWR for public comment at the same time that the statement was submitted to Congress. The conservationist litigants argued that NEPA required public participation in the

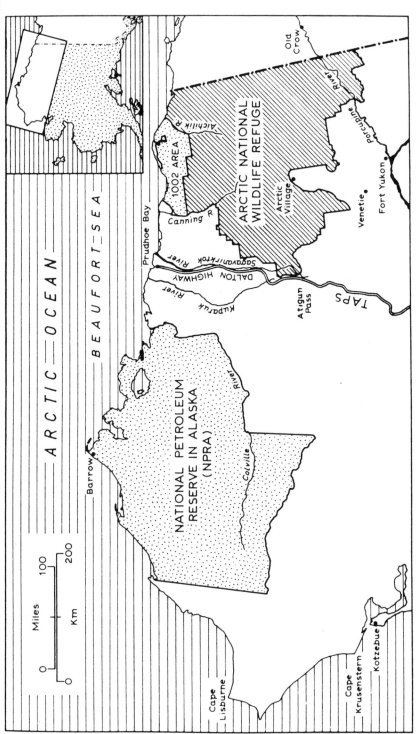

The Arctic National Wildlife Refuge (1987)

preparation of an environmental impact statement. Obliged by the court, the U.S. Fish and Wildlife Service held public hearings in January 1987 to discuss the environmental impact statement it had prepared on oil leasing in the 1002 area.[24]

The Alaska Department of Natural Resources, Division of Oil and Gas, estimates that the reserves of this "frontier" stand at 29 billion barrels—over three times the size of the Prudhoe Bay field.[25] The governor of Alaska and its congressional delegation want to open the refuge to the oil industry in its search for a second Prudhoe Bay. According to a poll carried out by the Alaska Oil and Gas Association, seven out of ten Alaskans feel that the area should be exploited.[26] Of 1,362 letters received for the public record from Alaskans as part of the recent hearings, 1,063 believe the range should be opened.[27] Among other Americans, Texans and Oklahomans are most supportive. In February 1987, Alaska's congressman, Don Young, introduced legislation (H.R. 1082) to open ANWR to oil leasing. He calls this a "critical energy security issue."[28] Young holds the strategic position of vice chairman of the House Interior Committee. He is also the ranking Republican member of the Fish and Wildlife Subcommittee of the Merchant Marine and Fisheries Committee. H.R. 1082's counterpart in the Senate is S.B. 1217, introduced by Alaskan senators, Stevens and Murkowski.

Interior's final report on oil leasing in ANWR (April 1987) was more sober in its appraisal of the size of recoverable proven oil reserves in the 1002 area (9.2 billion barrels) than the State of Alaska had been. Nevertheless, it agreed that the area contains the highest oil and gas potential of any unexplored American onshore "frontier" region and recommended to Congress that the coastal plain be opened for "orderly" oil exploration.[29] To justify opening the range, the Secretary of the Interior, Donald P. Hodel, has used the same basic argument as those federal officials who supported the construction of the Trans-Alaska Pipeline; the nation badly needs more domestic oil to stem increasing dependency on imports from war-torn regions such as the Middle East and the Arabian Gulf.[30]

The State of Alaska is elated, for production from the Prudhoe Bay field recently passed its high point. Despite the world oil glut, the State of Alaska is hoping for a new boom that will halt and reverse the current exodus.[31] Hodel and the Alaska congressional delegation emphasize that a fresh find will prolong the active life of the Trans-Alaska Pipeline, through which the last barrels of oil from the Prudhoe Bay and Kuparuk River fields (the latter brought into production in 1981) is expected to flow early next century.[32] Commenting on Alternative E—wilderness classification for the coastal plain—Hodel stresses that there is already sufficient wilderness in Alaska, both statutory (as a result of ANILCA) and nonstatutory *(de facto)*.[33]

Just as various projects from the past were used as evidence to bolster positions during the TAPS controversy, the Trans-Alaska Pipeline and the Prudhoe Bay industrial complex themselves have now become a source of analogy for the purposes of advocacy. Those who want to open the refuge invariably cite the "minimal impact," especially on caribou, of twenty years of oil activity on the North Slope and ten years of pipeline operation as proof that arctic oil development does not degrade the environment.[34] According to a poll conducted by the Alaska Oil and Gas Association, 86 percent of Alaskans believe that conduct at Prudhoe Bay illustrates that oil and gas development can be carried out in ANWR without ecological harm.[35] Indeed, the Alaska Oil and Gas Association cites the growth of the Central Arctic herd as proof that oil development is not a disadvantage for what they deny is an easily stressed, "wilderness-dependent" animal.[36]

Some Americans, especially in the Northeast and the Midwest, oppose oil development and demand wilderness designation instead. Of the 11,361 letters received during the public comment period for the environmental impact statement, 3,707 take this position.[37] In January 1987, Morris K. Udall (Democrat, Arizona), the chairman of the House Interior Committee, introduced H.R. 39 to grant wilderness status to the 1002 area. These critics question the economic wisdom of oil development in ANWR and deny that the environmental impact at Prudhoe Bay and along the Trans-Alaska Pipeline has been negligible. Accusing the Department of the Interior of having deleted the more critical findings of its own scientific experts, they denounce the impact statement as a rationalization for a decision in favor of leasing already taken and argue that the official conclusion contradicts the data in the impact statement which predicts environmental damage. Critics claim that Interior has failed to obey NEPA, which requires it to consider alternative ways of meeting American energy needs, and to look into ways of reducing energy demands.

Those who want to keep ANWR closed to oil activity are hoping to exploit a 60-page report on the ecological impact of oil development at Prudhoe Bay and along the Trans-Alaska Pipeline north of the Brooks Range by officials at the U.S. Fish and Wildlife Service's office in Fairbanks. This report of December 1987, which has not received official approval, was leaked to the *New York Times* and several environmental organizations in May 1988. A comparison of predicted and actual environmental impacts, it was prepared in response to a request from Congressman George Miller (Democrat, California), the chairman of the Water and Power Resources Subcommittee of the House Interior Committee.

Miller, who does not oppose oil development in ANWR in principle, requested this report so that those who are deliberating over the fate of the 1002 area would have data available to help them reach a decision. The report's basic thrust is that oil development in northern Alaska has caused substantially more environmental damage than was predicted in the environmental impact statements that the Department of the Interior prepared for the Trans-Alaska Pipeline project in 1971 and 1972.[38] The report claims that the amount of land devoted to oilfield development has exceeded predictions; 800 square miles rather than 550—and that total road miles are 30 percent more than was expected.[39] Specifically, its authors insist that there has been considerable freshwater pollution on the North Slope as a result of oil spills and erosion;[40] that the quality of marine waters has deteriorated substantially due to the expansion of oil production offshore—a development not anticipated in the environmental impact statements for TAPS.[41] Turning to wildlife, the report claims that there has been significant destruction of avian habitat at Prudhoe Bay and that most bird species have declined in population.[42] In addition, they believe that the loss of vegetation (and, hence, animal habitat) has been twice as great as predicted in the impact statements. This is partly a reflection of gravel requirements, which, according to the authors, exceeded predictions by 400 percent.[43] Moreover, they contend that revegetation measures have largely been a complete failure.[44] The report confirms that the pipeline generally parallels caribou routes but notes that the normal summer movements of the Central Arctic herd have been interfered with and that the harvest has increased considerably despite hunting restrictions. The report reiterates earlier observations that the growth of the Central Arctic caribou herd is a reflection of the decline in the number of predators in the area.[45] In these ways, the debate over the impact of arctic oil development and the Trans-Alaska Pipeline has been reopened.

Conservationist critics hark back to the pioneering work of Bob Marshall, Olaus and Margaret Murie of the Wilderness Society, and Lowell Sumner and George Collins of the National Park Service in the cause of arctic wilderness preservation. Some of the conservationists who have spoken out against oil leasing in ANWR testified in favor of the establishment of an arctic wildlife range in the late 1950s. They argue that leasing will violate the original purposes for which the range was created in 1960; the protection of wilderness and a complete ecosystem as well as wildlife.[46] Only 40 miles of coastal plain in ANWR, from the Aichilik River to the Canadian border, are not part of the 1002 area. Many conservationists insist that the wilderness character of the last sizeable section of American Arctic coast not already sacrificed to oil development will be destroyed and that its intact ecosystem, the largest remaining in the United States, will be irreversibly affected by the exploration phase alone.[47] Though the authors of Interior's report deny that oil development will cause significant damage in an area whose ecological uniqueness and preciousness they acknowledge, they freely admit that its "unquantifiable" wilderness character will be "irretrievably . . . eliminated."[48]

Like proponents of oil development in the 1002 area, conservationist critics are also evoking the region's frontier status. Naturally, what they stress are not its storehouse attributes but, once again, the scope for "pioneering" in a replica of the original New World. The Northern Alaska Environmental Center of Fairbanks claims that "travel across the area by primitive means is reminiscent of the hardships, challenge, drama and peril faced by early American people, but which is becoming increasingly difficult to experience today."[49] Michael McCloskey of the Sierra Club, restating Frederick Jackson Turner's view of the frontier as the "crucible" of the American character, has emphasized that "wilderness areas epitomize our national heritage. We carved our American society out of the natural world around us."[50]

In 1987 the Sierra Club established the protection of ANWR as its top priority. The club has no interest in an arrangement that permits oil leasing but imposes the strictest controls to ensure that wildlife and ecology are protected. For impact on wilderness character and values cannot be mitigated. According to the club's executive director, "No compromise is possible here."[51] In this view, when wilderness values are at stake, anything less than total victory amounts to total defeat.

The Government of Canada and Canadian Natives are equally upset by the prospect of leasing in ANWR. Representatives of the Yukon Territory and the Northwest Territories stress that the 3-million-acre Northern Yukon National Park is located immediately adjacent to ANWR and argue that the United States has no right to jeopardize the well-being of "international" animals and birds such as caribou, musk oxen, polar bear, and snow geese.[52] Ecological fears revolve around the future of North America's sixth largest caribou herd, the Porcupine (180,000), which ranges between Canada's Yukon Territory and its well defined "core" calving grounds in the 1002 area. Any elevated east-west pipeline feeding oil from ANWR to the Trans-Alaska Pipeline (and accompanying road) will cut across its migration route.

North American Native peoples are divided in their attitudes toward oil development in ANWR. In Alaska, the city council of Kaktovik and the North Slope Borough support limited oil leasing in ANWR. George N. Ahmaogak, Sr., the mayor of the North Slope Borough, believes that "wildlife and oil development can co-exist in the Arctic" and denies that pipelines and roads will restrict caribou movements.

Native politicians like Ahmaogak perceive the interests of their community on this matter as being much the same as those of other American oil consumers. The mayor shares the secretary of the interior's concern about the nation's growing dependence on "insecure" sources of foreign oil. As Warren O. Matumeak, the North Slope Borough's land management administrator, representing the mayor, explained at the hearings: "Residents of the North Slope Borough have only recently become accustomed to having fine schools, modern housing, police and fire protection, and other services and facilities long taken for granted by most Americans. Responsible development of ANWR would help enable us to continue providing these services far into the future."[53]

The residents of Kaktovik are dependent on the Porcupine caribou herd only to a limited extent. The most heavily dependent Native communities are Arctic Village, a settlement in Alaska to the south of ANWR, and Old Crow, just across the border in Yukon Territory. Caribou are the single most important food source for these two villages, which, unlike Kaktovik, do not have access to marine mammals. The Alaskan residents of Venetie, Fort Yukon, and Chalkyitsik, and a number of other Canadian villages also hunt the Porcupine herd. These villages oppose oil leasing in ANWR. The council of Arctic Village has pleaded with the decision-makers, "Don't do to the caribou what your ancestors did to the buffalo."[54]

During 1988, advocates of oil leasing in the Arctic National Wildlife Refuge made considerable progress in Congress, a number of Senate committees voting in favor of exploration. Then, in the spring of 1989, the Energy and Natural Resources Committee sent a bill to the Senate recommending exploration. The oil spill that devastated Prince William Sound and the surrounding coastal area in March 1989 has complicated the situation.

Proponents of oil leasing in the Arctic National Wildlife Refuge see no connection between the disastrous tanker accident off Valdez and proposed drilling in the refuge. However, they accept that in view of the public reaction to the Exxon spill there is little point in proceeding this year with efforts in Congress to open the range. Nevertheless, they do not think that the spill will have any long term negative impact on their cause.

The coming congressional struggle over oil leasing in ANWR promises to deliver a decisive verdict on just how successful conservationists have been in Alaska over the past century. Wildlife protection, the original sphere of conservationist concern in Alaska, has provided some of the biggest successes for conservationists. Profligate practices that shocked Elliott, Dall, and Grinnell in the late nineteenth and early twentieth centuries were reversible and sometimes have been reversed. By

1935 fur seal numbers were already back to 1867 levels. Protected initially by the treaty of 1911, which settled the Bering Sea controversy, and further by the Marine Mammals Act of 1972, the sea otter thrives again in Alaskan waters. (In view of the March 1989 oilspill in Prince William Sound, this conclusion may have to be revised.) It has even been transplanted to repopulate the full extent of its historic range from the Aleutians to southern California.

Other clear evidence of success are the national forests, national monuments, national parks, wildlife refuges, and wilderness areas that have been set aside in abundance.

Moreover, Alaskan boosters have not objected to the extension of formal protection to certain parts of Alaska. According to Walter Hickel, "some areas seem designed by God for development. . . . I have often said, thank God, Alaskan oil is on the North Slope, one of the most barren spots on earth, instead of the beautiful Wrangell Mountains. . . . No-one is ever going to buy a lot and retire at Prudhoe Bay."[55] The Wrangell Mountains—one part of Alaska where Hickel thoroughly approved of park creation—contained spectacular alpine scenery, and the area for which he supported park status was worthless in terms of extractive resource value.

Before the decision on oil leasing in ANWR, however, it is hard to assess the capacity of conservationists to protect ecologically valuable areas which, unlike the Wrangell Mountains, are neither monumental nor devoid of commercially valuable minerals. In the late 1950s many Alaskans supported the proposal to create ANWR because they thought the area was good for little but scenery, and on condition that range status would not interfere with the use of any valuable subsurface resources that might eventually be found there. Conservationists have not managed to secure absolute protection for areas of Alaska containing (or thought to contain) resources like oil.

The current ANWR debate may also be the ultimate test of the durability of "the last frontier" image and the continuing ability of conservationists to employ it to captivate and mobilize Americans and their congressmen in support of their Alaskan causes. If oil development in the refuge does proceed, if further massive oil spills occur off the coast of southcentral Alaska, and should Alaska become filled with even more of humankind's technology and works, it is worth asking whether the symbolic landscape of the last American frontier will recede in step. This uncorrupted image fulfills deep cultural and historical needs. By offering a vision of what Alaska ought to be, it provides a cherished link with an idealized national past. This helps to soothe the century-old fear of being a nation without a frontier. So it is likely that the image will grow stronger and more popular. Perhaps it is invincible.

Thirteen feet high and mounted on a seven-foot base, the five-figure Pipeline Monument stands at the Pipeline Terminal, Valdez. Commissioned by Robert O. Anderson, chairman of Atlantic Richfield Company, in 1977, it was dedicated in 1980. The figures *(left to right)* are: "The Surveyor," "The Teamster" (a woman), and "The Workman" (a Native American).

10.2 PERSPECTIVES ON THE FRONTIER

Alyeska Pipeline Service Company is encouraging the belief that the Alaskan frontier remains unconquered. On 20 September 1980 a Pipeline Monument was unveiled and dedicated at the Valdez Terminal. Thirteen feet high and mounted on a seven-foot base, it commemorates "the larger-than-life qualities demanded of the pipeline planners and builders—bravery, strength, tenacity, confidence, and imagination." Dedicated to these "Twentieth Century pioneers," the monument is an assemblage of five symbolic figures. Its sculptor, Malcolm Alexander, specializes in depicting "heroic" Americans. He describes one of them, "The Surveyor," who faces the coastal range: "Looking toward those mountains . . . he has a feeling of awe, watching the ingenuity of man complete an almost impossible task. . . ." "The Teamster" is called "a unique woman . . . a woman who wanted to be a pioneer, to build something in possibly the last frontier in the world." A third figure, "The Workman," a Native Alaskan, is introduced as "a man from another culture, who came together with many other people and did his part."[56] This celebration of the supreme act of latter-day pioneering is devoid of all traditional references to conquest. The pipeline has become part of the fabric of the frontier.

The Pipeline Monument. *Left to right:* **"The Welder," "The Workman," "The Teamster."**

The pipeline monument is also innovative in another significant respect. The five sculpted figures, as indicated, include a woman and a Native Alaskan. Historians have largely neglected the role of women on the nineteenth-century frontier (though during the last ten years, historians have begun to rectify the situation). Native Americans were usually victims of the pioneering process, not participants who shared the benefits and glories. According to Alyeska, however, between 5 and 10 percent of the work force which built the Trans-Alaska Pipeline was female and between 14 and 19 percent was minority.[57] The official image of corporate pioneering in late-twentieth-century Alaska, where women and Natives operate bulldozers, fly airplanes, and program computers, is neither exclusively male nor white.

This study has been littered with observations on the mythology, imagery, and symbolism of the frontier in Alaskan history. There have also been some scattered reflections on the meaning of the frontier as an historical concept and its application and relevance to Alaska. It is now time to confront more squarely the problem of the Alaskan frontier and the significance of Alaskan history to frontier scholarship.

It does not help that there is still no conceptual agreement among scholars or broad public consensus as to what the term *frontier* means. Within the framework of Turner's understanding of the sort of place the frontier was, Morgan B. Sherwood has characterized contemporary Alaska as "postfrontier." Sherwood accepts that many historic, widely accepted frontier ingredients persist in Alaska. Those he cites are large numbers of wild animals, considerable subsistence hunting, extensive tracts of land which are physically substantially unaltered from what they were centuries ago, and a low overall population density. By 1970 Alaska's population was 300,000. Over half was urban (concentrated in or near Anchorage, Fairbanks, and Juneau), and the Alaskan occupational structure was essentially no different from that of any other state. Yet the overall population density in 1971 was only 0.4 per square mile. This figure, the authors of the draft environmental impact statement prepared for TAPS had commented, was "roughly equivalent to the density in the American West of a century ago."[58]

Despite these features, Sherwood argues that innovations in communications and technology have rendered the frontier—as Turner, his predecessors, and his legatees understood and applied it—an obsolete description of environmental conditions in contemporary Alaska. Because the Second World War brought so many of the crucial changes in communications and technology to Alaska, Sherwood identifies it as "the beginning of the end of frontier Alaska." "Subsequent changes" he claims, "were a matter of degree, not kind."[59]

At this point, it is worth recalling that Turner used the terms *frontier* and *wilderness* virtually interchangeably when he discussed the frontier as a geographical area. One of his doctoral students, Avery Craven, explained in 1941 that Turner "used the term 'frontier' as men of his day used it to refer to any border country where men are few and nature dominated and where the first crude steps were being taken to reverse that situation."[60] In 1914, in a restatement of his central idea, Turner said of Americans: "Their most fundamental traits, their institutions, even their ideals, were shaped by this interaction between the wilderness and themselves."[61] It would not have been strange if Turner had called his thesis "The Significance of the Wilderness in American History."

In view of the synonymity of the terms *frontier* and *wilderness* within the framework of Turnerian frontier historiography, the assessment of Alaska as postfrontier has serious implications for the axiomatic characterization of many parts of Alaska as wilderness. Sherwood argues that a less anachronistic criterion than population density is required for a satisfactory measurement of contemporary Alaska's frontier and wilderness status and qualities. Because advances in technology and communications have destroyed the concept of a remote, inaccessible place, and have enhanced environmental vulnerability, he proposes the ratio of machines to land area as a more meaningful analytical tool than Turner's classic yardstick of people per square mile. He also stresses that though they cannot be seen, traces of intrusion are nonetheless present in apparently undisturbed and ecologically intact areas. He refers in particular to atmospherically transported pollutants, such as radioactive fallout. He concludes that between the Second World War and the 1960s, wilderness in Alaska "became only a word, not a place."[62]

Sherwood does not mention the Trans-Alaska Pipeline as one of the subsequent changes of degree. However, if one accepts that technology is the critical variable, then it appears that the construction of the Trans-Alaska Pipeline has been assigned a false prominence, despite the enormity of its scale. The pipeline seemed to stand out from the past like a lone drilling rig on the tundra and impressed many contemporaries, especially critics, as the most imposing aspect of the process of environmental change and dilution of frontier qualities in Alaska. If the historian assumed the role of pathologist and performed an autopsy on the corpse of "the last American frontier," the postmortem might indicate that the causes of death were multiple and incremental. Historical perspective reveals that the pipeline was not the fatal part of the Alaskan frontier's demise. What it delivered was the *coup de grace.*

At least one historian sees the relationship between technology and the frontier differently. Patricia Nelson Limerick's book *The Legacy of Conquest: The Unbroken Past of the American West* (1987) is a criticism of the exclusion of the twentieth-century West from American frontier history. Limerick is concerned with the essential continuity, appearances to the contrary, between the history of the West in the nineteenth and the twentieth centuries. In her view, some of the most misleading evidence of discontinuity is provided by technological advance. She looks at the attitudes behind the far more sophisticated technology at human disposal nowadays and the processes of development which it advances and finds them much the same as they were in the nineteenth century: "Are Geiger counters and airplanes less frontier-like than picks and shovels?" she asks, and answers emphatically in the negative. She dismisses as sentimental and pedantic the view that "frontiers involve mules, horses, and oxen but not jeeps; pickaxes and pans but not air drills and draglines . . . bows and arrows but certainly not nuclear tests."[63]

In *The Legacy of Conquest,* Limerick acknowledges the existence of Alaska and seeks to include Alaskan history within the orbit of her arguments; "Many patterns explored here apply also to Alaska, but limits of space and time have prohibited its full inclusion."[64] In truth, limits of space and time prohibited any inclusion of Alaska whatsoever. She refers to Alaska once, in passing, in the context of oil development throughout the West in the 1970s.[65] This is a pity, for Alaska provides some of the most compelling, if neglected evidence to support many of her observations about the persistence late into the twentieth century of typical nineteenth-century issues and processes; especially conflicts over land, which she calls "the emotional center of Western history." (In truth, there is nothing uniquely Western or nineteenth-century about land conflict in the United States.)

Limerick has been looking for the essence of the frontier. Her search assigned a high profile to the phenomenon of "rushes" to exploit natural resources, especially minerals, and highlights the role of the boom-bust economic cycle. She gives particular attention to mining because: "No industry had a greater impact on Western history than did mining. . . . mining set a mood that has never disappeared from the West: the attitude of extractive industry—get in, get rich, get out."[66]

This focusing of attention on natural resources began with the work of critics of the frontier thesis such as James Malin, Henry Nash Smith, Richard Hofstadter, and David Potter, who all condemned Turner's agrarian bias. Turner, they argued, mistakenly conceived of wealth and opportunity largely in terms of fertile, unsettled land.[67] Potter argued that this limited conception of the sources of abundance blinded Turner to the possibilities of wealth based on the nonagrarian resources of a frontier region.[68] Potter also stressed the importance of variables such as access to and demand for resources, pointing out that resources do not possess intrinsic worth and often become valuable only with the advent of technology capable of exploiting them and moving them to market.[69] As Malin remarked (without reference to Alaska), "everything that was claimed for the hither edge of free land, was almost equally applicable to the exploitation of free subsoil resources."[70] Interestingly, the catholic definitions of the frontier which Ray Allen Billington favored in the 1960s gave as much prominence to general resource wealth as they did to free land.[71]

Malin, a fierce critic of the Turner thesis, had already taken another approach to the study of the frontier. His approach concentrated on the conflict over natural resources between people who were already there and those who came from elsewhere. He focused on how "cultural technology" determined access to and ability to use resources. Cultural technology, he explained, referred not simply to mechanical inventions but to "all manner of ideas and social inventions." "Cultural technology," he asserted, "gave to the people of the invading European culture an overwhelming power to overrun, and to displace, or to annihilate the more primitive peoples. . . . To possess any measure of validity, the so-called frontier thesis would have to be treated within the anthropologist's formula of competition of cultures [i.e., the "sum total of a way of life"], but historians do not deal with it in that manner."[72]

Howard Lamar and Leonard Thompson have taken up Malin's challenge. They define the frontier as a territorial zone of conflict where a struggle for control occurs between the indigenes and an invading (European) culture.[73] In such environments, which may vary considerably in physical character and economic value, they identify a process "that takes place whenever one people intrudes into terrain occupied by another."[74] "The frontier 'opens' in a given zone when the first representatives of the intrusive society arrive; it 'closes' when a single political authority has established hegemony over the zone." As Malin had been, Lamar and Thompson are critics of Turner's "closed space" doctrine, which treats the American frontier as a physical commodity that expired with the West in the 1890s. They cite twentieth-century Alaska as an example of an "open frontier zone."[75] The criteria to which they refer are an abundance of comparatively unexploited natural resources and a low overall population, which includes a relatively intact Native component. From this perspective, the frontier in Alaska opened when the first Russian fur traders arrived.

Since 1867 there has been conflict in Alaska on three levels, two of them cultural. Its first and most obvious manifestation is the non-cultural physical struggle between the intrusive culture and the natural environment. The first form of cultural conflict, which Lamar and Thompson stress, is the contest between the invaders and the indigenes.

The second form of cultural conflict—and the one with which this study has been principally concerned—is the antagonism between the ideologies and policies of boosters and conservationists. The outcome of interhuman conflict in Alaska has not been the establishment of hegemony by what Lamar and Thompson refer to as a "single political authority." Control is divided among a number of contending, and sometimes cooperating, interests—the State of Alaska, the oil industry, the federal government, Native Alaska, and the conservationist community. The process of apportionment began in 1958 with statehood, which gave the state the right to select 27 percent of the unappropriated public domain. Native Alaska received a 12 percent slice in 1970. The oil industry won its pipeline in 1973 and its conservationist critics gained their share of Alaska (28 percent) with the passage of ANILCA in 1980.

The degree of resolution, however, is deceptive. Conflict persists between all interested parties that seek to enlarge or defend their hard-won portions of Alaska. As indicated, the latest battleground is also an old one; the northeast Arctic. By implication, the frontier in Alaska remains open. Frontier historians as well as environmental historians ignore Alaska at their peril.

Notes

INTRODUCTION

1. For the sake of brevity, the acronym TAPS will be used throughout this study to denote the project and its proposer. Strictly speaking, the TAPS proposal became the ALPS proposal when the Alyeska Pipeline Service Company (ALPS) superseded the Trans-Alaska Pipeline System in August 1970. However, the original acronym stuck in most contemporary references and is employed here in deference to popular use.

2. The "lower forty-eight" is a term Alaskans use to denote the forty-eight contiguous American states. It will be employed in this sense (without quotation marks) throughout the study.

3. *Seattle Post-Intelligencer,* 6 September 1973.

4. Growe to Rogers C. B. Morton, 18 May 1972, Oil and Gas File, box 42 (Pipeline Correspondence), Fairbanks Environmental Center Collection, University of Alaska Archives, Fairbanks (UAF).

5. Frank Moolin, Alyeska Course, Tanana Valley Community College, Fairbanks, Fall 1975, Tape U75-10, Oral History Collection, Alaska and Polar Regions Department, University of Alaska, Fairbanks (UAF).

6. This particular observer interpreted such "inflated" American comments about these "improvements" as evidence of a patriotism verging on bigotry. See Isaac Candler, *A Summary View of America* (London: T. Cadwell, 1824), 121–22.

7. The North American Water and Power Alliance (NAWAPA) project, advanced in the mid-1960s, but which has not materialized, might have eclipsed the Trans-Alaska Pipeline in size, complexity, and cost. The Ralph M. Parsons engineering company of Pasadena, California, put forward a proposal to divert water from the rivers of Alaska and northern Canada to the American Southwest for irrigation and the generation of electricity via an elaborate series of dams, reservoirs, tunnels, canals, and aqueducts. The entire system was expected to take between twenty and thirty years to build. One of the main features would have been a 500-mile long reservoir filling the Rocky Mountain trench in British Columbia. In addition to serving the Southwest, NAWAPA was intended to divert clean water to the Great Lakes.

8. There are a number of parallels between the two controversies. In both cases, preservationists in the conservationist community objected to decisions by the Department of the Interior to approve engineering projects. Critics suggested alternatives, but proponents of these projects denied their practicality. The final decisions were made by Congress, and both sides lobbied heavily. For details of the Hetch Hetchy controversy, see Roderick Nash, *Wilderness and the American Mind,* 3d ed. rev. (New Haven: Yale University Press, 1982), 161–81.

9. Brown's thirty-two part series began in the *Anchorage Daily News* on 22 July 1969. It won the Scripps-Howard Foundation's Edward J. Meeman Award for 1969.

10. Derek Lambert, *Great Land* (London: Bantam, 1978).

11. Milt Machlin, *Pipeline* (New York: Pyramid Books, 1976). Valdez Narrows is a channel two miles long and a mile wide. It connects the southwestern end of Port Valdez to Prince William Sound. The channel for deep draft vessels is reduced to a width of a half mile by a small island and pinnacle of rock located at the north end of the narrows.

Life, according to the old saying, can be stranger than fiction. On the night of 23 March, 1989, the Exxon *Valdez* left the pipeline terminal bound for Long Beach, California. The tanker's captain, a man with a history of alcohol abuse that was known to his employers, retired to his cabin after the pilot who had guided the ship through Valdez Narrows had quit the ship. The third mate, who lacked the special U.S. Coast Guard certificate required for controling vessels in these waters, was left alone on the bridge. In the early hours of 24 March, the ship hit rocks. Two miles further on, and far outside the ten-mile wide shipping lanes, the tanker struck a well charted reef, Bligh Island, and ran aground. 240,000 barrels of crude (over ten million gallons) were discharged into Prince William Sound, making this the worst oil spill in American history.

12. Peter M. Hoffman, "Evolving Judicial Standards under the National Environmental Policy Act and the Challenge of the Alaska Pipeline," *Yale Law Journal* 81 (1972): 1592–1639. For a contemporary summary from an economist's point of view, see Richard Norgaard, "Petroleum Development in Alaska: Prospects and Conflicts," *Natural Resources Journal* 12 (1972): 83–107.

13. Henry R. Myers, "Federal Decision-Making and the Trans-Alaska Pipeline," *Ecology Law Quarterly* 4 (1975): 915–61.

14. For a semi-scholarly, general history of the pipeline, the controversy, and the related boom, covering geology, oil economics, Native claims, union politics, and weld scandals, see Robert Douglas Mead, *Journeys Down the Line: Building the Trans-Alaska Pipeline* (Garden City, N.Y.: Doubleday and Co., 1978). John Hanrahan and Peter Gruenstein, *Lost Frontier: The Marketing of Alaska* (New York: W. W. Norton, 1977), also belongs in this category.

15. Alfred W. Anderson, "Alaska's Riches: Promise of the American Arctic," Library of Congress, 4 November 1969, prepared for Howard W. Pollock, Publicity File 1969–70, box 13, Howard W. Pollock Papers, UAF. Pollock was Alaska's representative in Congress (Republican) from 1966–70.

16. 20th Alaska Science Conference, Fairbanks, 24–27 August 1969, as reported in the *New York Times,* 31 August 1969.

CHAPTER 1. THE FRONTIER IMAGE AND ENVIRONMENTAL REALITY, 1867–1940

1. Some of the first scholars to devote their attention to the processes of man-induced change in the natural environment were historical and cultural geographers like Carl Ortwin Sauer and pioneering ecological historians such as James C. Malin. In *The Grassland of North America,* Malin approached human history as part of a larger natural history. Sauer wanted to reconstruct the original character of the environment and to trace how, through successive patterns of land use, settlement, and economy, man had "filled the natural world

with his works." See Carl O. Sauer, "Foreword to Historical Geography," *Annals of the Association of American Geographers* 31 (1941): 1–24.

2. I do not mean to overlook the role of the Native American as a modifying agent (through the deliberate use of fire, for example). Nor am I suggesting that the process of environmental change did not begin until the arrival of Europeans. However, enduring changes were post-Columbian. Wilbur R. Jacobs has suggested that historians pay more attention to the environmental consequences of westward expansion: "The Great Despoliation: Environmental Themes in American Frontier History," *Pacific Historical Review* 67 (February 1978): 1–26. Also useful as a starting point is John Opie, "The Environment and the Frontier," in *American Frontier and Western Issues: A Historiographical Review*, edited by Roger L. Nichols (Westport, Conn.: Greenwood Press, 1986), 7–25. Opie calls environmental history "a revisionist critique of traditional frontier expansionism" (7). In its broadest sense, the environmental change under discussion has been characterized as bio-ecological imperialism. See Alfred W. Crosby, *The Columbian Exchange: Biological and Cultural Consequences of 1492* (Westport, Conn.: Greenwood Press, 1972). Crosby's *Ecological Imperialism: The Biological Expansion of Europe, 900–1900* (Cambridge: Cambridge University Press, 1986) deals with this process in a global context.

3. This focus on these two particular groups is one aspect of a larger field of inquiry: Ralph H. Brown referred to the relationship between "fancy" and "fact" in his *Historical Geography of the United States* (New York: Harcourt, Brace and Co., 1948), 1. (His treatment of the subject ends in 1870.) John K. Wright, another historical geographer, has argued that the study of geographical knowledge, past and present, for which he coined the term "geosophy," must extend beyond professional and scientific geographical knowledge to include "the geographical ideas, both true and false, of all manner of people—not only geographers, but farmers and fishermen, business executives and poets, novelists and painters, Bedouins and Hottentots—and for this reason it necessarily has to do in large degree with subjective conceptions." See John K. Wright, "Terra Incognitae: The Place of the Imagination in Geography," *Annals of the Association of American Geographers* 37 (March 1947): 12. The study of geographical and environmental perception is now often called "image geography." See J. Wreford Watson, "Image Geography: The Myth of America in the American Scene," *Advancement of Science* 27 (September 1970): 71–80; J. Wreford Watson and Timothy O'Riordan, eds. *The American Environment: Perceptions and Policies* (Chichester, England: John Wiley and Sons, 1976), touches on images and Alaska in places. See Watson, chap. 1, "Image Regions," 26–28; chap. 5, "The Image of Nature in America," 73–75. Roger M. Downs and David Stea, *Maps in Minds: Reflections on Cognitive Mapping* (New York: Penguin Books, 1977), discuss how we collect, organize, store, recall, and manipulate information about the spatial environment to produce a cognitive map that is a reflection of the world as we believe it to be. For a discussion of the Arctic in general as an "image region," see Barry Lopez, *Arctic Dreams: Imagination and Desire in a Northern Landscape* (New York: Bantam, 1987), chap. 7, "The Country of the Mind," 226–70. There has been one historical geographical study of Alaska: John Coyne, "Alaska: Image of a Resource Frontier Region, 1867–1900" (M. A. Birkbeck College, University of London, 1974). Coyne deals with impressions contained in official reports of federal government agents sent to investigate the latest American territorial acquisition.

4. *New York Review of Books* 19 (5 October 1972): 16. Barzini's comment

was made with reference to Denis Mack Smith, the historian, and his exposure of the myths of the Italian *risorgimento*. It is equally pertinent to the present subject.

5. Richard A. Bartlett proclaims that the conquest of the West was the most exciting period of American history. Indeed, he goes even further, asserting that "maybe it was the happiest time for a whole people in all history." Richard A. Bartlett, *The New Country: A Social History of the American Frontier, 1776–1890* (New York: Oxford University Press, 1974), 448.

6. Richard Slotkin, *The Fatal Environment: The Myth of the Frontier in the Age of Industrialization, 1800–1890* (New York: Atheneum Books, 1985), 20.

7. For a general discussion of the relationship between the "antihistory" upon which the United States was founded and the development of nostalgia and reverence for tradition, see David Lowenthal, "The Place of the Past in the American Landscape," from *Geographies of the Mind: Essays in Historical Geosophy (In Honor of John Kirtland Wright)*, edited by David Lowenthal and Martyn J. Bowden (New York: American Geographical Society and Oxford University Press, 1976), 89–117.

8. Frederick Jackson Turner, "The Significance of the Frontier" (1893), from *The Frontier in American History* (1920) (Huntington, N.Y.: Robert E. Krieger Publishing Co., reprint, 1976), 38.

9. U.S. Congress, Senate, Committee on Interior and Insular Affairs, testimony of Stewart Brandborg, hearings, *Rights-of-Way Across Federal Lands: Transportation of Alaska's North Slope Oil*, 93d Congress, 1st session, part 1, 9 March 1973 (Washington, D.C.: GPO, 1973), 262. Brandborg was executive director of the Wilderness Society.

10. Turner, *The Frontier in American History*, 2.

11. Harold P. Simonson, *The Closed Frontier: Studies in American Literary Tragedy* (New York: Holt, Rinehart, and Winston, 1970), 5.

12. During the TAPS controversy, environmentalists attributed the "second chance" phrase to Bob Waldrop of the Sierra Club. See Harvey Manning, *Cry Crisis! Rehearsal in Alaska* (San Francisco: Friends of the Earth, 1974), 185. The popularization of the phrase in the postwar era can be attributed to Paul Brooks. See Paul Brooks, "Alaska: The Last Frontier," *Atlantic Monthly* 210 (September 1962): 78. Brooks was editor-in-chief at Houghton Mifflin of Boston and a member of the Sierra Club Board of Directors. However, the origins of the phrase were much earlier, in the prewar era, as this chapter will show.

13. As noted by Henry Nash Smith in *Virgin Land: The American West as Symbol and Myth*, 1950, 2d ed. (Cambridge: Harvard University Press, 1971), 4.

14. Smith, *Virgin Land*, preface, 1st ed. (1950) from the 20th Anniversary ed. (1971), xi.

15. Ray Allen Billington, "The Image of the Southwest in Early European 'Westerns,'" in Billington and Albert Camarillo, *The American Southwest: Image and Reality* (Los Angeles: William Andrews Clark Library and the University of California, 16 April 1979), 19–20. See also Billington, "Cowboys, Indians, and the Land of Promise: The World Image of the American Frontier," proceedings, 14th International Congress of the Historical Sciences, New York, 1976, 60–79.

16. My intention here is not to reinforce the enduring fallacy that the vast majority of American opinion opposed the purchase of Alaska. This misconcep-

tion, still repeated in many standard histories of the United States, was exposed by Thomas A. Bailey a long time ago. "Why the United States Purchased Alaska," *Pacific Historical Review* 3 (1934): 39–49. See also Virginia Hancock Reid, *The Purchase of Alaska: Contemporary Opinion* (Long Beach, Calif.: author, 1939). Reid argues that the "legend" that most Americans opposed purchase had hindered Alaskan development and the aim of her study is to reveal the true economic value of Alaska. See also Richard E. Welch, "American Public Opinion and the Purchase of Russian America," *American Slavic and East European Review* 17 (1958): 481–94; Ronald J. Jensen, *The Alaska Purchase and Russian-American Relations* (Seattle: University of Washington Press, 1975). True, the House of Representatives did not approve the appropriation bill ($7.2 million) until nine months after the United States took possession of Russian America. Yet other factors, such as timing, constitutional procedure, and, not least, President Johnson's unpopularity, accounted for this as well as a low opinion of Alaska's value. Of 43 Congressmen who voted against appropriation (113 voted in favor), 41 had voted to impeach Johnson. In general, opposition was stronger on the East Coast than in the West. However, even Greeley's opposition was less than absolute. His paper did not oppose appropriation and on 7 May 1867, the New York *Daily Tribune* acknowledging (grudgingly) that fisheries, timber, and minerals might be of some value, referred to Alaska as an "American Norway." Bailey, Welch, and Jensen demonstrate convincingly that many Americans appreciated the value of Alaska commercially as well as strategically.

17. Hubert Howe Bancroft, *History of Alaska, 1730–1885,* 1886 (New York: Antiquarian Press, reprint, 1960), 587–88.

18. See, for example, the New York *World,* 9 April 1867. The American-Russian Commercial Company maintained ice ponds and ice houses on Kodiak Island and operated a lucrative trade for almost thirty years, supplying California, Mexico, and even South America. The trade collapsed in the early 1880s when the "Ice Company" was unable to compete with ice brought from the Sierra Nevada by the Southern Pacific Railroad and the artificially made substance. See Edward L. Keithahn, "Alaska Ice, Inc.," *Pacific Northwest Quarterly* 36 (April 1945): 121–31.

19. *Harper's Weekly* 11 (13 April 1867): 226; as quoted by Reid, *The Purchase of Alaska: Contemporary Opinion,* 31.

20. *Philadelphia Inquirer,* 17 April 1867.

21. Oliver Risley Seward, *William H. Seward's Travels Around the World* (New York: Appleton, 1873), 35–36.

22. Bancroft, *History of Alaska,* 598.

23. Seward described the Aleutians as "stepping stones" to Asia and saw Alaska's significance largely within the context of trade with Asia and the establishment of American commercial supremacy in the Pacific. See Ernest N. Paolino, *The Foundations of the American Empire: William H. Seward and United States Foreign Policy* (Ithaca: Cornell University Press, 1973), 106–18; Jensen, *The Alaska Purchase,* 91–92, 140. Howard I. Kushner has placed great stress on American interest in the value of Alaska's natural resources as a motive for purchase, which he interprets as an integral part of a pattern of American economic expansionism in the Pacific Northwest. See Howard I. Kushner, *Conflict on the Northwest Coast: American-Russian rivalry in the Pacific Northwest* (Westport, Conn.: Greenwood Press, 1975), xi–xii, 158; "The Significance of the Alaska Purchase to American Expansion," in *Russia's American Colony,*

S. Frederick Starr, ed., Special Study of the Kennan Institute for Advanced Russian Studies of the Woodrow Wilson International Center for Scholars (Durham, N.C.: Duke University Press, 1987), 295–315.

24. For the text of a subsequently published version, greatly refined and extended, of Sumner's original three-hour speech on 8 April 1867, see Archie W. Shiels, *The Purchase of Alaska* (College: University of Alaska Press, 1967), 22–127. Twenty of the forty-eight pages of Sumner's essay were devoted to appraising Alaska's climate and resources. See *Speech of the Honorable Charles Sumner, of Massachusetts, on the Cession of Russian America to the United States* (Washington, D.C.: *Congressional Globe* Office, 1867), as quoted in Reid, *The Purchase of Alaska: Contemporary Opinion*, 118. Sumner was chairman of the Senate Committee on Foreign Relations so his support was crucial to ratification. Jensen argues that after considerable initial personal scepticism, Sumner came to appreciate the national geopolitical and ideological importance of annexation in terms of creating a North American republic (Jensen, *The Alaska Purchase*, 83–86, 91–92, 140–41). In the House of Representatives, the treaty's supporters also stressed these factors. (Ibid., 115–17; Shiels, *The Purchase of Alaska*, 178–81.)

25. Shiels, *The Purchase of Alaska*, 84, 94.

26. Jensen, *The Alaska Purchase*, 113. Virginia Hancock Reid, on the other hand, claims that congressmen placed great emphasis upon Alaska's resource value. See *The Purchase of Alaska: Contemporary Opinion*, 51–54.

27. Bancroft, *History of Alaska*, preface, viii.

28. Charles Hallock, *Our New Alaska: or, The Seward Purchase Vindicated* (New York: Forest and Stream Publishing Co., 1886), 50. Hallock, the author of a number of books on hunting and fishing, was a leading advocate of fish and game laws and a conservationist in many respects. However, he drew a clear distinction between the reformist policies required for natural resource management in the rest of the United States and what Alaska needed, which was economic growth. See Morgan B. Sherwood, *Big Game in Alaska: A History of Wildlife and People* (New Haven: Yale University Press, 1981), 21.

29. John J. Underwood, *Alaska, An Empire in the Making* (New York: Dodd, Mead and Company, 1913), foreword, xv–xvi.

30. *Alaska-Yukon Magazine* 7 (February 1909): 347–54.

31. Hallock, *Our New Alaska*, 40–65; Underwood, *Empire in the Making*, 211–12, 220–21.

32. Thomas Willing Balch, *The Alaska Frontier* (Philadelphia: Allen, Lane and Scott, 1903).

33. As quoted in Fulmer Mood, "Notes on the History of the Word *Frontier*," *Agricultural History* 22 (1948): 81.

34. *Alaska Almanac 1909*, edited and compiled by E. S. Harrison, The Arctic Club (Seattle: The Harrison Publishing Co., 1909), 62. The Arctic Club, formerly the Alaska Club, was described as a home for men "that are developing and civilizing the Northland" (ibid., 145).

35. "The Prospector," from "Ballads of a Cheechako" (1910), *Robert Service, Collected Verse* (London: Ernest Benn Limited, 1930), 200.

36. Grindall Development Company, "Go to Alaska 'Our Frontier Wonderland,'" Seattle, ca. 1910.

37. Alaska Bureau, Seattle Chamber of Commerce, *Alaska 'Our Frontier Wonderland'*, Seattle, 1921, 5.

38. Merle Colby, *A Guide to Alaska: Last American Frontier* (New York:

Federal Writers' Project and Macmillan Publishers, 1939). One of the author's aims was to dispel "Popular Errors About Alaska" (xliii–v.)

39. Frederick Jackson Turner, "The West and American Ideals," 1914, in *The Frontier in American History*, 296. And in 1926, six years before his death, he commented that "the opening of the Alaskan wilds furnished a new frontier and frontier spirit to the Pacific Northwest as well as to the nation"; Turner, "The West—1876 and 1926: Its Progress in a Half Century," *World's Work* 52 (July 1926): 327.

40. Frederic Logan Paxson, *The Last American Frontier* (New York: The Macmillan Company, 1910); *History of the American Frontier, 1763–1893* (Boston: Houghton Mifflin, 1924); *When the West is Gone* (New York: Henry Holt, 1930).

41. Ray Allen Billington, *Westward Expansion: A History of the American Frontier* (New York: Macmillan Company, 1949), 745. In reaction, Morgan B. Sherwood has argued that Alaska offers "a living laboratory for study of the Turner thesis." From Robert A. Frederick, ed., *Proceedings of the Conference on Alaskan History: Frontier Alaska: A Study in Historical Interpretation and Opportunity*, Anchorage, Alaska Methodist University, 1967, *Alaska Review*, Special Issue (1968), 21. Two historians have applied aspects of Turner's ideas to Alaska: Orlando W. Miller, *The Frontier in Alaska and the Matanuska Colony* (New Haven: Yale University Press, 1975); and Melody Webb, *The Last Frontier: A History of the Yukon Basin of Canada and Alaska* (Albuquerque: University of New Mexico Press, 1986).

42. The *New York Times*, 23 January 1972, overlooked the historical origins of the last wilderness image of Alaska when it commented on this third image "taking shape" in the early 1970s alongside the well-established icebox and store-house images. James Lee Cuba, "A Moveable Frontier: Frontier Images in Contemporary Alaska" (Ph.D. diss., Yale University, 1981), also neglects the historical evolution of the frontier image in this sociological study of Anchorage as a community. Three Ph.D dissertations in the field of literary history are also useful on the subject of frontier and images in Alaska: Frank E. Buske, "The Wilderness, The Frontier and the Literature of Alaska to 1914: John Muir, Jack London and Rex Beach" (University of California, Davis, 1976); Hilton John Wolfe, "Alaskan Literature: The Fiction of America's Last Wilderness" (Michigan State University, 1973); and Marilyn Jody, "Alaska in the American Literary Imagination: A Literary History of Frontier Alaska" (University of Indiana, 1969). Parts of Wolfe's dissertation are especially interesting; chap. 1, "Frontiers: History and Romance," 17–97; "Conclusion," 308–17. Wolfe considers it essential that frontier historians study Alaska, despite its atypicality among American frontiers. He is also aware of the value of comparisons between Alaska and non-American frontiers.

43. Richard Pollak, introduction, Tom Brown, *Oil On Ice: Alaskan Wilderness at the Crossroads* (San Francisco: Sierra Club, 1971), 1.

44. See Roderick Nash, *Wilderness and the American Mind*, 3d ed, rev. (New Haven: Yale University Press, 1982), Preface, xiv, prologue, 1.

45. Michael McCloskey, "The Wilderness Act of 1964: Its Background and Meaning," *Oregon Law Review* 45 (June 1966), 288.

46. In 1946 Clifford C. Presnall observed that wilderness "exists only in the mind of civilized man. To the primitive Indian, there was no such thing." See "Our Wilderness Heritage," *Living Wilderness* 11 (September 1946), 6. (Presnall was an employee of the U.S. Fish and Wildlife Service.) See also Nash, *Wilderness and the American Mind*, preface, xiv; "Ideal and Reality in 'Ultimate'

Wilderness: Aviation and Gates of the Arctic National Park," *Orion,* Spring 1983: 6. For further comments by Nash on the ethnocentricity of the wilderness concept, see "Tourism, Parks and the Wilderness Idea in the History of Alaska," *Alaska in Perspective* 4 (1981), 16.

47. Nash, *Wilderness and the American Mind,* 317.

48. Bil Gilbert, "Power and Light on a Lonely Land," *Pipeline,* 1974, clipping, without page numbers, Oil and Gas File, box 43 (105), Fairbanks Environmental Center Collection, University of Alaska, Fairbanks (UAF).

49. Smith, *Virgin Land,* Twentieth Anniversary ed, preface, ix–x. For useful observations on the role of landscape impressions and caricatures, see Marwyn S. Samuels, "The Biography of Landscape," Donald W. Meinig, ed. *The Interpretation of Ordinary Landscapes: Geographical Essays* (New York: Oxford University Press, 1979), 70–72.

50. James C. Malin, *The Grassland of North America: A Prolegomena to its History* (Lawrence, Kans.: Author, 1947), 154; *History and Ecology: Studies of the Grassland,* Robert P. Swierenga, ed. (Lincoln: University of Nebraska Press, 1984), 65–66.

51. Malin, *The Grassland of North America,* 153–55. See also "Grassland, 'Treeless,' and 'Subhumid,' A Discussion of Some Problems of the Terminology of Geography," from *History and Ecology,* 22–23, 65–66. Malin denounced the use of subjective terminology that conveyed deficiency and inadequacy, such as "arid" and "semiarid," preferring quantitative, nonevaluative terms such as "low-rainfall" and "dry" (22–23).

52. James C. Malin, "On the Nature of the History of Geographical Area (With Special Reference to the Western United States), 1960, in *History and Ecology,* 133.

53. Vilhjalmur Stefansson, *The Friendly Arctic; the story of five years in Polar Regions* (New York: Macmillan, 1921). See chap. 2, "The North That Never Was," 7–26. His book *The Northward Course of Empire* (New York: Macmillan, 1924), also contained a chapter entitled "The North That Never Was."

54. For critical comments on Stefansson's attitude to the Arctic and his proposals for its economic development, see Barry Lopez, *Arctic Dreams,* 346–48. For a different view, see William R. Hunt's sympathetic recent biography: *Stef: A Biography of Vilhjalmur Stefansson, Canadian Arctic Explorer* (Vancouver: University of British Columbia Press, 1986), 172–85, 206–21.

55. Malin vehemently denounced the view that human activity (especially that of white colonists) desecrated nature and that the Europeans who arrived there had destroyed a perfect state of nature or dramatically altered the ecology of the New World. He hated the application of terms such as *virgin* to the environment: "Custom of long standing has settled upon the term *virgin* as applied to forests, or grassland, or soil, that has not been disturbed by civilized man. The term is a misnomer in the first place, because it is a sex term that has no proper application to vegetation, or to the soil. In the hands of conservation propagandists, the term *rape* of the virgin continent, . . . has often been introduced, carrying over into the discussion of nature the idea of sex crime. Nature does not offer any such parallel or analogy and all such terms should be eliminated." See Malin, *The Grassland of North America,* 152. For further insight into Malin's attitudes toward the natural environment and toward conservationists (he was usually critical of them), see Robert W. Johannsen, "James C. Malin: An Appreciation," *Kansas Historical Quarterly* 38 (Winter 1972): 463–65.

56. Of relevance here are Donald Worster's comments on Malin's inconsistencies. See *Nature's Economy: A History of Ecological Ideas* (1977) (Cambridge: Cambridge University Press, reprint, 1985), 242–49.

57. A study that touches on this subject (but not explicitly) is Orlando W. Miller, *The Frontier in Alaska*. Miller examines the dichotomy between traditional expectations and cultural patterns and new environmental settings by looking at the agrarian frontier image in relation to efforts to promote the settlement of the Alaskan frontier according to the old agrarian pattern.

58. Samuel P. Hays, *Conservation and the Gospel of Efficiency: The Progressive Conservation Movement* (Cambridge: Harvard University Press, 1959). Though he uses the term utilitarian to describe Pinchot and his policies, Hays avoids the now standard appellation "utilitarian conservation."

59. Ibid., 141–46, 189–98.

60. For details of the conflict between Pinchot and Muir, see Nash, *Wilderness and the American Mind,* 133–38.

61. Two scholars have pursued the analogy between the Siberian and American frontiers, that Turner himself suggested in 1904. See Donald W. Treadgold, *The Great Siberian Migration* (Princeton: Princeton University Press, 1957). Elsewhere, Treadgold indicates that the Siberian frontier in the nineteenth century also bore a popular icebox stigma (6); "Russian Expansion in the Light of Turner's Study of the American Frontier," *Agricultural History* 26 (October 1952): 147–51. See also Andrei Anatolevich Lobanov-Rostovsky, "Russian Expansion in the Far East in the Light of the Turner Hypothesis," in *The Frontier in Perspective,* edited by Walker D. Wyman and Clifton B. Kroeber (Madison: University of Wisconsin Press, 1957), 83, 87–88. Though Russian America represented the easternmost outpost of Tsarist expansion, neither author extended the scope of inquiry to include the Russian share of the New World.

62. In fact, this image was rather contradictory. The New York *World* did not explain how a "sucked orange" could have potential.

63. See Nelson McGeary, *Gifford Pinchot: Forester & Politician* (Princeton: Princeton University Press, 1960), 86–87; Gifford Pinchot, *The Fight for Conservation* (London: Hodder and Stoughton, 1910), 40–44.

64. *Alaskan Searchlight,* 18 February 1895. As quoted in Ted C. Hinckley, "Alaska and the Emergence of America's Conservation Consciousness," *Prairie Scout* 2 (1975): 104. Elliott has been rescued from obscurity by a more environmentally conscious generation in search of prescient forerunners. See James Thomas Gay, "Henry Wood Elliott, Crusading Conservationist," *Alaska Journal* 3 (Autumn 1973): 211–16; Robert L. Shalkop, "Henry W. Elliott, Fighter for the Fur seals," *Alaska Journal* 13 (Winter 1983): 4–12. For further details of the depletion of marine resources in the pre-Pinchot-Ballinger era, see Hinckley, "Alaska and the Emergence of America's Conservation Consciousness," 79–111.

65. Since assuming the leadership of the Western Union Telegraph Expedition's (1865–67) scientific corps upon the death of Robert Kennicott, Dall had amassed a more extensive knowledge of things Alaskan, from invertebrates to ethnology, than any other living person. For details of his career, see Morgan B. Sherwood, *Exploration of Alaska: 1865–1900* (New Haven: Yale University Press, 1965), 36–56. For biographical details, see Michael J. Lacey, "The Mysteries of Earth-Making Dissolve: A Study of Washington's Intellectual Community and the Origins of American Environmentalism in the Late Nineteenth Century" (Ph.D diss., George Washington University, 1979), 60–70.

66. See William H. Dall, "On the Preservation of the Marine Animals of the Northwest Coast," *Smithsonian Institution Annual Report for 1901* (Washing-

ton, D.C.: GPO, 1902). Also useful is Hinckley, "Alaska and the Emergence of America's Conservation Consciousness," 85–89 (otters) and 91–92 (walrus).

67. William H. Dall, *Alaska and its Resources* (Boston: Lee and Shepard, 1870), 242, 444–46.

68. Henry W. Elliott, *A Report Upon The Condition of Affairs In The Territory of Alaska* (Washington, D.C.: GPO, 1875), 5. For similar sentiments see "Ten Years' Acquaintance with Alaska," *Harper's New Monthly Magazine* 55 (November 1877): 801–16.

69. Henry W. Elliott, "The Loot and Ruin of the Fur-Seal Herd of Alaska," *North American Review* 185 (June 1907): 436.

70. For details, see Thomas A. Bailey, "The North Pacific Sealing Convention of 1911," *Pacific Historical Review* 6 (March 1935): 1–14.

71. In this study, I follow the now standard practice that reserves the term "Native Alaskan" exclusively for indigenous, nonwhite Alaskans.

72. John R. Bockstoce, *Whales, Ice and Men: The History of Whaling in the Western Arctic* (Seattle: University of Washington Press, 1986), 9, 98.

73. John Muir, *The Cruise of the Corwin: Journal of the Arctic Expedition of 1881 in Search of De Long and the Jeanette,* edited by William Frederic Badè (Boston: Houghton Mifflin, 1917), 142–43. For details of the walrus slaughter, which occurred mainly between 1868 and 1883, and the plight of the indigenes, see Bockstoce, *Whales, Ice and Men,* 129–142.

74. This was also part of an effort to elevate the nomadic hunter-gatherer to the more "civilized" rank of pastoralist.

75. Morgan B. Sherwood, "The End of American Wilderness," *Environmental Review* 9 (Fall 1985): 203; *Big Game,* 84.

76. A. Starker Leopold and Frank Fraser Darling, *Wildlife in Alaska: An Ecological Reconnaissance* (New York: Ronald Press Co., 1953), 68–82.

77. The term *outside* was (and still is) used by Alaskans to refer to anywhere that is not Alaska. It is also sometimes applied within Alaska by rural Natives and whites to denote the cities of Fairbanks and Anchorage.

78. Miller, *The Frontier in Alaska,* 16.

79. In 1880 Alaska's population was 33,426. By 1890 it had fallen to 32,052. A couple of years after the Klondike-Alaska gold rushes, it had swollen to 63,592, of which 29,000 were Native. Twenty years later, it had declined to 55,036.

80. See Nash, "Tourism, Parks and the Wilderness Idea in the History of Alaska," 4–8. Also useful is Marilyn Jody, "Alaska in the American Literary Imagination," 20–51.

81. Ted C. Hinckley, "The Inside Passage: A Popular Golden Age Tour," *Pacific Northwest Quarterly* 36 (1965): 67–74.

82. William H. Goetzmann and Kay Sloan, *Looking Far North: The Harriman Expedition to Alaska, 1899* (New York: Viking Press, 1982), introduction, xii–xiii.

83. S. Hall Young, *Alaska Days With John Muir* (New York: Fleming H. Revell Company, 1915), 210–11.

84. John Muir, *Our National Parks,* 1902 (Madison: University of Wisconsin Press, reprint, 1981), 12.

85. John Burroughs, *Harriman Alaska Expedition,* vol. 1, *Narrative of the Expedition* (New York: Doubleday, Page and Company, 1902), 78–80.

86. John Muir, *Travels in Alaska* (Boston: Houghton Mifflin Company, 1915).

87. Muir, *Our National Parks,* 7.

88. As quoted in Alfred Runte, *National Parks: The American Experience* (Lincoln: University of Nebraska Press, 1979), 48.

89. Muir, *Our National Parks,* 11.

90. John Muir, *Steep Trails,* edited by William Frederic Badè (Boston: N. S. Berg, 1918), 341.

91. Henry W. Elliott, *An Arctic Province: Alaska and the Seal Islands* (London: Sampson Low, Marston, Searle and Rivington, 1886), 395, 436, 439.

92. Muir, *Our National Parks,* 10.

93. Ernest Thompson Seton, *The Arctic Prairies* (1911) (London: Constable, 1915), 15, 259, 269.

94. One could discuss at length the nature of the relationship between sportsmen and preservationists and the validity of the distinction. Samuel P. Hays classifies the Boone and Crockett Club as preservationist. George Bird Grinnell, one of its founder members, is described likewise by Alfred Runte, with reference to his role in the creation of Glacier National Park in Montana and as the most vocal critic of the inadequate enforcement of game laws pre-1894 that permitted mass poaching in Yellowstone National Park; Hays, *Conservation and the Gospel of Efficiency,* 40; Runte, *National Parks,* 75. Sportsmen, John F. Reiger and James B. Trefethen have argued, were in the vanguard of American efforts to conserve wildlife and timberlands and to set up parks. See Reiger, *American Sportsmen and the Origins of Conservation* (New York: Winchester Press, 1975), 11–24; Trefethen, *Crusade for Wildlife: Highlights in Conservation Progress* (Harrisburg, Pa.: Stackpole Co., 1961), 15–26. Muir's distaste for killing animals and his disapproval of hunting for pleasure was wellknown. He explained sportsmen's interest in conservation matters as being "partly, I suppose, because the pleasure of killing is in danger of being lost from there being little or nothing left to kill." Muir to H. F. Osborn, 16 July 1906. William Frederic Badè, ed., *The Life and Letters of John Muir* (Boston: Houghton Mifflin, 1924), 2:350.

95. George Bird Grinnell, *A Brief History of the Boone and Crockett Club* (New York: Forest and Stream, 1910), 4, 60; Reiger, *American Sportsmen,* 114–32.

96. Sherwood, *Big Game,* 38.

97. Edward W. Nelson, "Notes on the Wild Fowl and Game Animals of Alaska," *National Geographic* 9 (4 April 1898): 132. Nelson's contact with Alaska began in 1877 as a meteorologist with the U.S. Signal Corps at St. Michael, northeast of the Yukon delta. For biographical details, see Anna E. Riggs, "E. W. Nelson, Unpaid Collector," *Alaska Journal* 10 (Winter 1980): 91. Nelson was an expert on the plight of the sea otter.

98. Grinnell, *Brief History,* 37–41; Sherwood, *Big Game,* 26–28.

99. Charles Sheldon, *The Wilderness of Denali: Explorations of a Hunter-Naturalist in Northern Alaska* (New York: Charles Scribner's Sons, 1930). Denali (meaning "the Great One") is the aboriginal name for Mount McKinley. Sheldon traveled extensively in the region between 1906 and 1908.

100. For an account of the park's origins, see Gail E. H. Evans, "From Myth to Reality: Travel Experiences and Landscape Perceptions in the Shadow of Mount McKinley, Alaska, 1876–1938" (master's thesis, University of California, Santa Barbara, February 1987), 131–37, 149–58. Those who lobbied for the park were not so concerned with protecting predators (perceived as competitors by sport hunters) such as wolves and grizzlies.

101. See the testimony of Belmore Browne and others, U.S. Congress, Senate Committee on Interior and Insular Affairs, Subcommittee on Territories and Island Possessions, hearings, *S. 5716, A Bill to create Mount McKinley National*

Park, Alaska, 64th Congress, 1st session, 5 May 1916 (Washington, D.C.: GPO, 1916), 6–7, 9, 11–13.

102. Belmore Browne, *Plea for Mount McKinley National Park,* Committee on Conservation of Forests and Wildlife, Camp Fire Club of America, Boone and Crockett Club, American Game Protective Association, New York, 1916, 4–5.

103. Morgan Sherwood explains that in general there was little market hunting in Alaska due to game laws. The railroad's main impact on wildlife was direct. Trains killed between 100 and 300 moose annually. See *Big Game,* 73.

104. See George Bird Grinnell, "The Salmon Industry," in *Harriman Alaska Expedition,* vol. 2, *History, Geography and Resources,* 349–50. Afognak Island became part of Chugach National Forest in 1907.

105. Grinnell, "The Salmon Industry," 2:350–51, 339.

106. William T. Hornaday, *Our Vanishing Wildlife: Its Extermination and Preservation* (New York: New York Zoological Society, 1913), 179.

107. Hornaday began his career as a specimen collector and taxidermist. For biographical details, see Michael J. Lacey, "The Mysteries of Earth-making Dissolve," 45–46.

108. Dave Bohn, *Glacier Bay: The Land and the Silence* (San Francisco: Sierra Club, 1967), 82–94.

109. Goetzmann and Sloan, *Looking Far North,* xii–xiii.

110. Harold J. Lutz, *Aboriginal Man and White Man as Historical Causes of Fires in the Boreal Forest, with Particular Reference to Alaska,* Yale School of Forestry, Bulletin Number 65 (New Haven: 1959): 23–32. Also useful by Lutz is *Ecological Effects of Forest Fires in the Interior of Alaska.* U.S. Department of Agriculture, Technical Bulletin 1133 (Washington, D.C.: GPO, July 1956). See also Stephen J. Pyne, *Fire in America: A Cultural History of Wildland and Rural Fire* (Princeton: Princeton University Press, 1982); chap. 8, "Fire and Frost: A Fire History of Alaska," 497–98, 501–2 (gold rush) and 502–3 (Alaska Railroad).

111. In 1909, however, Idaho had the largest percentage (39) of its total territory in forest reserves. Paul W. Gates, *History of Public Land Law Development* (Washington, D.C.: Public Land Law Review Commission, GPO, 1968), 580–81.

112. This act had laid down a maximum of 160 acres per individual, 320 acres for each "association of individuals," and up to 640 acres for more than four individuals. Corporations found a way around these restrictions through dummy entries or by entries processed under agricultural land statutes. Gates, *History of Public Land Law,* 724–45.

113. For details see James Penick, Jr., *Progressive Politics and Conservation: The Ballinger-Pinchot Affair* (Chicago: University of Chicago Press, 1968).

114. For details of Alaskan support for Ballinger see Herman Slotnick, "The Ballinger-Pinchot Affair in Alaska," *Journal of the West* 10 (April 1971): 340–41, 345–46. Evangeline Atwood, *Frontier Politics: Alaska's James Wickersham* (Portland: Binfords and Mort, 1979), also reports that Ballinger was a hero to many Alaskans (212). For contemporary reactions to federal conservation policies, expressions of support for Ballinger, and refutations of the accusation that the Alaska Syndicate was "gobbling" up Alaska, see *Alaska-Yukon Magazine* between 1910 and 1912. "Too Much Cry of Wolf Retards Alaskan Development," *Alaska-Yukon Magazine* 9 (February 1910): 174–77; "Conservation that Locks Up," ibid. 9 (April 1910): 290–93; Falcon Joslin, "The Conservation Policy in Alaska," ibid. 9 (May 1910): 341–49; J. Y. Ostrander. "What Conservation Does

to Alaska," ibid. 11 (January 1911): 45–47; George E. Baldwin, "Conservative Faddists Arrest Progress and Seek to Supplant Self-Government with Bureaucracy," ibid. 13 (February 1912): 44–46, 57–58.

115. In 1906 Alaska was granted a delegate, who served as a member of the House of Representatives, where he was eligible to speak, to serve on committees, and introduce bills, but not to vote. James Wickersham served as delegate from 1909 until 1917, in 1918 and 1920, and again from 1931 until 1933.

116. *Skagway Alaskan,* 11 October 1910; as quoted in Atwood, *Frontier Politics,* 223.

117. Casey Morgan, "A Land to Loot," *Collier's,* Special Alaska Issue, 6 August 1910, 21.

118. *Alaska-Yukon Magazine* 9 (May 1910): 425.

119. Apparently, no arrests were made. The local law enforcement officer happened to be out-of-town. For these details, see Jeannette Paddock Nichols, *Alaska: A History of Its Administration, Exploitation, and Industrial Development During Its First Half Century Under the Rule of the United States* (Cleveland, Ohio: Arthur H. Clark Co., 1924), 369–70.

120. Ibid.

121. For Alaskan attitudes to Pinchot, see Slotnick, "The Ballinger-Pinchot Affair," 341–44. A further notorious incident occurred in 1914. In defiance of federal law, 500 gold miners continued to mine local coal at Candle Creek, to the north of Nome. The miners needed coal to thaw frozen auriferous gravels and took their action rather than use expensive Canadian coal. *Fairbanks Daily News-Miner,* 26 December 1914.

122. The office of commissioner in Alaska was set up by the Organic Act of 1884. A commissioner in Alaska had the jurisdiction and powers of commissioners of the U.S. circuit courts and justices of the peace. They also served as public notaries and coroners. They originally served only in Sitka, Wrangell, Unalaska, and Juneau, but as settlements were established throughout the territory, new posts were created. St. Michael was an important base for traders, missionaries, and federal customs, military, and law enforcement agents, as well as the U.S. Signal Corps. It was the major port of entry for steamers traveling along the Yukon and the port of departure for boats to San Francisco. Stephenson spent five years at St. Michael.

123. William B. Stephenson, Jr., *The Land of Tomorrow* (New York: George H. Doran Company, 1919), x.

124. Ibid., 238–39.

125. See Stuart Ramsay Tompkins, *Alaska: Promyshlennik and Sourdough* (Norman: University of Oklahoma Press, 1945).

126. Ernest Gruening, "Alaska, Progress and Problems," *Scientific Monthly* 77 (July 1953): 7.

127. The authors of one recent comparative frontier history refer to Amazonia as the last terrestrial frontier, an increasingly common practice also among conservationists reacting to threats from agronomy and logging. See William W. Savage, Jr., and Stephen I. Thompson, eds, "The Comparative Study of the Frontier: An Introduction," *The Frontier: Comparative Studies* (Norman: University of Oklahoma Press, 1979), 2:10.

128. For a stimulating study of the Australian frontier, see Brian Fitzpatrick, "The Big Man's Frontier and Australian Farming," *Agricultural History* 21 (January 1947): 8–12.

129. Frederick Jackson Turner, "Contributions of the West to American Democracy," 1903, from *The Frontier in American History*, 245, 258.

130. George Wilson Pierson, "American Historians and the Frontier Hypothesis in 1941," in O. Lawrence Burnette, Jr., *Wisconsin Witness to Frederick Jackson Turner: A Collection of Essays on the Historian and the Thesis* (Madison: State Historical Society of Wisconsin, 1961), 138.

131. Vilhjalmur Stefansson, *Northwest to Fortune: The Search of Western Man for a Commercially Practical Route to the Far East* (London: George Allen and Unwin Ltd., 1958) 243–84; Melody Webb, *The Last Frontier*, 52–54.

132. Charles Vevier, "The Collins Overland Line and American Continentalism," *Pacific Historical Review* 28 (1959): 237–53.

133. Bancroft, *History of Alaska*, 577–78.

134. Sherwood, *Exploration of Alaska*, 15–44. For William H. Dall's account, see *Alaska and its Resources*, 355–58. Much of the information concerning Alaska's resources contained in Senator Sumner's proratification speech (1867) was supplied by members of the Western Union expedition such as Henry Bannister, assistant to Robert Kennicott, Dall's predecessor as chief of the expedition's Scientific Corps.

135. Stefansson, *Northwest to Fortune*, 294–330; Goetzmann and Sloan, *Looking Far North*, 176–78.

136. For useful details see Webb, *The Last Frontier*, 158–67.

137. U.S. Department of the Interior, Bureau of Land Management, *Eagle-Fort Egbert: a Remnant of the Past* (Washington, D.C.: GPO, ca. 1980), 18–20.

138. William L. Mitchell, *The Opening of Alaska*, edited by Lyman L. Woodman (Anchorage, Alaska: Cook Inlet Historical Society, 1982), 23.

139. The desire for an "All-American" telegraph line was realized when a submarine cable was laid between Valdez and Seattle in 1904.

140. Mitchell, *The Opening of Alaska*, introduction, 1.

141. Ibid., 100.

142. Olaus J. Murie, *Journeys to the Far North* (Palo Alto, Calif.: The Wilderness Society and Crown Publishers, 1973), p. 112; *The Alaska-Yukon Caribou*, North American Fauna, No. 54, Bureau of Biological Survey (Washington, D.C.: GPO, 1935), 4.

143. Mitchell, *The Opening of Alaska*, contains a photograph of a tripod located some 30 miles north of Valdez at the head of Keystone Canyon. Wireless telegraphy superseded WAMCATS in the 1920s. In 1935 Alaska Communications Systems took over WAMCATS. In the meantime, most of the line has been absorbed into microwave and satellite systems. In some places, the line still survives as part of the local telephone system.

144. For authoritative accounts of the Alaska Road Commission and its activities, see Claus-M. Naske, *Alaska Road Commission Historical Narrative: Final Report* (Fairbanks: State of Alaska, Department of Transportation and Public Facilities, June 1983); *Paving Alaska's Trails: The Work of the Alaska Road Commission* (Lanham, Md.: University Press of America, 1986).

145. The road was named for Major Wilds P. Richardson, the first commissioner of the Alaska Road Commission.

146. For a fictionalized account of these struggles, laden with anticonservationist anecdotes (p. 233, for example), see Rex Beach, *The Iron Trail* (New York: Harpers, 1913).

147. George E. Walsh, "Arctic Railroad Building," *Cassier's Magazine*, ca. 1900, 266; clipping, Scott Polar Research Institute, Cambridge. The article's

THE TRANS-ALASKA PIPELINE CONTROVERSY

title was a misnomer. Southeast Alaska does not qualify as arctic, or even subarctic. According to one account, construction claimed at least 32 lives. Goetzmann and Sloan, *Looking Far North*, 56.

148. The *Skagway Guide of 1899*, as quoted in William E. Brown, *This Last Treasure: Alaska National Parklands* (Anchorage: Alaska Natural History Association, 1982), 35. The railroad was shut down in 1982 and became part of a hiking trail in Klondike Gold Rush National Historic Park. In May 1988 it reopened for the tourist trade, offering round trips to the summit of White Pass 41 miles from Skagway.

149. Francis A. Walker, "The Indian Question," *The North American Review* 116 (April 1873): 349. For a historian's assessment of the importance of the railroads as an agent of frontier "recession," see Frederic Logan Paxson, "The Pacific railroads and the Disappearance of the Frontier in America," *Annual Report of the American Historical Association*, Washington, D.C., 1907, 105–18.

150. *Alaska-Yukon Magazine* 9 (February 1910): 186.

151. Ibid. 13 (February 1912): 20–27.

152. As quoted in Ernest Gruening, *The State of Alaska* (New York: Random House, 1954), 184–85.

153. For details see Webb, *The Last Frontier*, 247–55.

154. For a description of the reaction in Fairbanks when news was received of the passage of the railroad bill on 12 March 1914, see Margaret E. Murie, *Two In The Far North* (New York: Alfred A. Knopf, 1962), 67–68. Margaret E. Murie was the wife of naturalist Olaus J. Murie and the first woman graduate of the University of Alaska.

155. Margaret Murie recalls how, as a fifteen-year-old, she was a Seattle-bound passenger aboard the last horse-drawn sleigh to operate over the Valdez Trail on 4 May 1918. Ibid., 68–84. Although the entire line did not operate until 1921, enough track had been completed by 1919 to allow mail and passengers to be carried over the incomplete sections by other methods.

156. *Fairbanks Daily News-Miner*, 11 August 1928. Ohlson, who had previously been division superintendent of the Northern Pacific in Duluth, held this new position until 1945. See William H. Wilson, "The Alaska Railroad and the Agricultural Frontier," *Agricultural History* 52 (April 1978): 263–79; "The Alaska Engineering Commission and a New Agricultural Frontier," ibid. 42 (October 1968): 339–50. These are sober appraisals of the inflated hopes for an agrarian future in Alaska and of the efforts to introduce commercial agriculture (primarily the Matanuska and Tanana Valleys) between 1898, when the federal government established seven agricultural experimentation stations in the territory, and the Second World War. For a different view (and representative booster position), see C. C. Georgeson, "The Possibilities of Agricultural Settlement in Alaska," *Pioneer Settlement: Cooperative Studies*, Special Publication No. 14 (New York: American Geographical Society, 1932): 50–60. Georgeson was in charge of the experimentation stations established in 1898.

157. Scott C. Bone, *Alaska, Its Past, Present, and Future* (Juneau: Governor's Office, 1925), 1. For a useful general discussion of booster rhetoric and its characteristics ("confusion of present and future, fact and hope"), that is equally applicable to Alaska, see Daniel J. Boorstin, "Booster Talk: The Language of Anticipation," *The Americans: The National Experience* (New York: Random House, 1965), 296–98.

158. For Clark's account of his journey, see "A Bike Ride from Valdez to Fairbanks in 1906," Terrence Cole, ed., "Wheels On Ice," *Alaska Journal* 15 (Winter 1985): 27–47.

159. Clark to James Oliver Curwood, 29 January 1923, in response to a request for information about conditions in the interior of Alaska. Curwood, an author from Michigan, had set a number of travelogues and adventure tales in Alaska. Politically, Curwood was a Bull Moose Republican and he later became a national director of the Izaak Walton League. "I do not want to see that wonderful country [Alaska] completely and utterly raped as our own states have been," he wrote to Clark, 6 January 1923, John A. Clark Papers, UAF.

160. The first dredge actually arrived in the Yukon basin in 1899. For a useful discussion of developments in gold mining technology and methods in the Alaskan interior, upon which I have relied for these details, see Webb, *The Last Frontier*, 269–89.

161. The ditch was named for James M. Davidson, an engineer who first applied this method of providing high volume water pressure. On 18 April 1974, the *Fairbanks Daily News-Miner* carried a picture of lengths of steel pipe being hauled through Fairbanks. However, this was not Japanese pipe being trucked out from storage for the construction of the Trans-Alaska Pipeline. The pipe in the picture had been salvaged from the abandoned Davidson Ditch. The dredging operations of the Fairbanks Exploration Company had shut down in 1962 and ownership of the ditch had transferred to the Chatanika Power Company. In 1967, sections were destroyed by flooding and the ditch was shut down. The Bureau of Land Management halted the salvage operation where it affected pipe located on public lands to ensure compliance with environmental regulations governing land reclamation. Some sections were left *in situ* and in 1977 were nominated for inclusion in the National Register of Historic Places.

162. John De Novo, "Petroleum and the United States Navy," *Mississippi Valley Historical Review* 41 (March 1955): 647–53. In the early nineteenth century, the U.S. Navy had created live oak reserves along the Atlantic and Gulf seaboards from North Carolina to Louisiana to guarantee shipbuilding supplies. See Gates, *History of Public Land Law*, 553.

163. See Philip S. Smith, J. B. Mertie, and W. T. Foran, "Summary of Recent Surveys in Northern Alaska," *Mineral Resources of Alaska: Report on Progress of Investigations in 1924*, USGS Bulletin 783-E (Washington, D.C.: GPO, 1926), 151–68; John C. Reed, *Exploration of Naval Petroleum Reserve No. 4 and Adjacent Areas of Northern Alaska, 1944–53, Part 1, History of the Exploration*, USGS Professional Paper 301 (Washington, D.C.: GPO, 1958), 19–20.

164. Smith et al., "Summary of Recent Surveys," 165.

165. Stefansson, *The Northward Course of Empire*, introduction, xix.

166. Boats powered by engines were known as "gas launches" and appeared during the first decade of the twentieth century. See Webb, *The Last Frontier*, 224.

167. Ibid., 264.

168. For Dufresne's biographical details, see Sherwood, *Big Game*, 8–11.

169. Frank Dufresne, *Alaska's Animals and Fishes* (New York: A. S. Barnes and Company, 1946), xv.

170. The road was named for James Steese, commissioner of the Alaska Road Commission in the 1920s.

171. See Sherwood, *Big Game*, 73. Regulations passed in 1941 made it illegal to shoot game within a half mile of any road or the Alaska Railroad, ibid., 72.

172. Frank Dufresne, "What of Tomorrow?," *Alaska Sportsman* 3 (April 1937): 9. *Alaska Sportsman*, a local equivalent of *Forest and Stream*, began life in 1935.

173. Dufresne, *Alaska's Animals and Fishes*, 3.

174. Leopold and Darling reiterate this point in *Wildlife in Alaska*, 56–59.

175. For these and other biographical details see George Marshall's introduction to *Arctic Wilderness* (Berkeley and Los Angeles: University of California Press, 1956), ix–xx. This selection from Bob Marshall's Alaskan journals and letters was reissued as *Alaska Wilderness: Exploring the Central Brooks Range* (Berkeley and Los Angeles: University of California Press, 1970), with a fresh introduction by George, one of Bob's brothers. See also George Marshall, "Bob Marshall and the Alaska Arctic Wilderness," *Living Wilderness* 34 (Autumn 1970): 29–32. Also useful is Stephen R. Fox, *John Muir and his Legacy: The American Conservation Movement, 1890–1975* (Boston: Little, Brown and Company, 1981), 206–12.

176. Marshall, *Arctic Wilderness*, 1. Also useful on Marshall and Alaska is Nash, "Tourism, Parks and the Wilderness Idea," 15–16; *Wilderness and the American Mind*, 287.

177. Robert Marshall, "The Development of Alaska Resources Should Be Retarded," 2, undated (but post-1935), typewritten manuscript, sent by John Sieker, his successor as chief of the Forest Service's Division of Recreation and Lands, to Bob's brother, George, 12 September 1941, Robert Marshall Papers, Bancroft Library, University of California, Berkeley (UCB). (Hereafter referred to as RMP.)

178. Marshall to Helen Terrys, 12 November 1929, RMP.

179. Marshall to Al Cline, 15 July 1930, RMP.

180. Robert Marshall, *Arctic Village* (London: Jarrolds, 1934), 117–18. Wiseman's first wireless station preceded the arrival of the first airplane by a few months.

181. Ibid., 120–21. Marshall points out that there were only 6.25 miles of road on which this automobile could travel, and for only four months of the year.

182. Marshall, *Arctic Wilderness*, 113. Marshall made his fourth and final visit to Wiseman and the upper Koyukuk in June 1939. He died of a heart attack at the age of thirty-eight in November 1939.

183. Frank Dufresne, *My Way Was North: An Alaskan Autobiography* (New York: Holt, Rinehart and Winston, 1966), 193. Margaret Murie, who grew up in Fairbanks, was also struck by the similarity between airplane and insect, commenting wryly that before the airplane, "the hum of the mosquito was heard in the land." *Two in the Far North*, 58.

184. *Alaska Life: The Territorial Magazine* 3 (July 1940): 6.

185. Vilhjalmur Stefansson, "Routes to Alaska," *Foreign Affairs* 19 (July 1941): 863. In 1938 the 30,000 whites in Alaska bought 30,000 separate airline passages, ibid.

186. *Alaska Life* 3 (March 1940): 2.

187. U.S. Congress, House of Representatives, National Resources Committee (NRC), *Alaska: Its Resources and Development*, House Document no. 485, 75th Congress, 3d session (Washington, D.C.: GPO, 1938).

188. In this sense its sober realism was comparable to John Wesley Powell's famous *Report on the Arid Lands* (1878). For a discussion of the National Resources Committee (NRC) report in the context of federal attitudes toward the development and settlement of Alaska in the 1930s, and a discussion of various refugee schemes, see Miller, *The Frontier in Alaska*, 161–79.

189. NRC, *Alaska: Its Resources and Development*, 101.

190. Ibid., 40.

191. Ibid., 86.

192. Ibid., 129, 142.

193. Ibid., "Comments on the Report of Alaska's Recreational Resources Committee," appendix B, 213.

194. Nash, "Tourism, Parks and the Wilderness Idea," 17; *Wilderness and the American Mind*, 288.

195. Alfred Hulse Brooks, "The Future of Alaska," *Annals of the Association of American Geographers* 15 (December 1925): 163–79. Brooks compared southeast Alaska favorably with the New England coast settled by the first Europeans (163). He attacked the icebox image (164) and stressed that the government railroad would stimulate agricultural production for local consumption (173). After an elaborate comparison between Alaska and Scandinavia, he placed Alaska's eventual population at 10 million (178–89). Brooks's article "The Value of Alaska," *Geographical Review* 15 (January 1925), 25–50, was almost identical in tone and content. Brooks applies the term "last frontier" to Alaska (27).

196. John Hellenthal was the brother and legal partner of Simon Hellenthal, the Federal District Court Judge in Anchorage (Third Division) from 1935 to 1943.

197. John Hellenthal, *The Alaskan Melodrama* (New York: Liveright, 1936), 211, 303.

198. John Hellenthal, "Why Not Go Modern On Conservation?" *Alaska Life* 4 (January 1941), 3, 13, 17–19, 27–28, 30.

199. Ibid., 19.

200. One should note that Hellenthal's thinking (or, at least, his terminology) was rather muddled. By referring to preservationism as orthodoxy, he contradicted the claim that it was heresy.

201. Hellenthal, "Why Not Go Modern?," 13.

202. Frederick Jackson Turner, "Social Forces in American History," 1910, in *The Frontier in American History*, 320.

203. Louis R. Huber, "What Price Frontier? Alaska Has Inherited The Role Of Pioneer! What Keeps Her From Fulfilling It?" *Alaska Life* 4 (February 1941): 4–5, 15.

204. Ibid., 4. For Ickes's attitude to a sanctuary, see Sherwood, *Big Game*, 56.

205. Huber, "What Price Frontier?," 5.

206. A. E. Johann, *Pelzjäger, Prärien und Präsidenten; Fahrten und Erlebnisse Zwischen New York und Alaska* (Berlin: Deutschen Verlag, 1937). The relevant chapter is "Alaska—Zukunftsland der Weissen Rasse."

207. These references are to the extracts (6–7) in "Does Hitler's Lebensraum Include Alaska?," *Alaska Life* 4 (1941): 6–7, 23–25.

208. Ibid., 25.

CHAPTER 2. THE IMPACT OF WAR

1. Clarence L. Andrews, *The Story of Alaska* (Seattle: Lowman and Hanford Company, 1931).

2. Ibid. Andrews repeated this analogy in the postscript ("The Alaska of the Future") to a second edition: (Caldwell, Idaho: The Caxton Printers, 1938), 241. For Andrews's attitude toward Alaskan economic development, see 236–43, 1938 ed.

3. James C. Malin, *The Grassland of North America: Prolegomena to Its History* (Lawrence, Kans.: Author, 1947), 267–68.

4. Pan-American flew from Seattle to Ketchikan and Juneau. The first direct service from Seattle to Anchorage was introduced after the Second World War.

5. The commission's members were Herbert H. Rice, chairman; Ernest Sawyer, assistant to the secretary of the interior; and Major Malcolm Elliott of the U.S. Army Corps of Engineers, who was president of the Alaska Road Commission. MacDonald was appointed exploratory engineer. For these details, see U.S. Congress, House of Representatives, Committee on Roads, hearings, *H.R. 3095 (Proposed Highway to Alaska)*, 77th Congress, 2d session, 4–6 February 1942 (Washington, D.C.: GPO, 1942), 3.

6. For these details, see the British Columbia-Yukon-Alaska Highway Commission, *Report on the Proposed Highway Through British Columbia and the Yukon Territory to Alaska*, August 1941 (Ottawa: King's Printer, 1942), 8–9.

7. Hearings, *Proposed Highway to Alaska*, 35.

8. Dimond (1881–1953) had been a teacher, miner, and lawyer, as well as mayor of Valdez.

9. See Dimond's comments, hearings, *Proposed Highway to Alaska*, 4; David A. Remley, *Crooked Road: A History of the Alaska Highway* (New York: McGraw-Hill, 1976), 120–22. A useful collection of material (with a Canadian emphasis) on the politics and construction of the Alaska Highway is Kenneth Coates, ed., *The Alaska Highway: Papers of the 40th Anniversary Symposium* (Vancouver: University of British Columbia Press, 1985). The 40th anniversary symposium was held at Fort St. John, British Columbia, in 1982.

10. Dimond, "Empire Builders Comparable to Railroad Barons," transcript of radio address, 14 August 1943, 3, Legislative Bill Files, 1935, box 23, Anthony J. Dimond Papers, University of Alaska Archives, Fairbanks, Alaska (UAF). (Hereafter referred to as Dimond Papers.)

11. Dimond to Franklin D. Roosevelt, 11 June 1935, Dimond Papers.

12. Dimond to Hull, 24 February 1936, Dimond Papers.

13. Dimond to Rettie, 1 November 1941, Dimond Papers.

14. For references to Harriman's scheme, see the testimonies of Congressman Warren G. Magnuson (Democrat, Seattle, Washington), hearings, *Proposed Highway to Alaska*, 72; U.S. Congress, Senate, Committee on Foreign Relations, hearings (testimony of Donald MacDonald), *Subcommittee on Senate Resolution 253 (A Resolution Providing for an Inquiry into the Location of the Alaska Highway on the So-Called C or Prairie Route)*, 77th Congress, 2d session, 1, 12, 16 June 1942 (Washington, D.C.: GPO, 1942), 40. (Hereafter referred to as Hearings, *Senate Resolution 253*.)

15. Donald MacDonald, "Alaska-Yukon Highway," *The Alaska Yukon Gold Book*, edited and compiled by Fred N. Atwood (Seattle: Sourdough Stampede Association, Inc., 1930), 64.

16. International Highway Association, foreword, typescript, 1929, 7, 8, Alaska Highway, 1932–61, E. L. "Bob" Bartlett Papers, UAF. (Hereafter referred to as Bartlett Papers.)

17. MacDonald to FDR, transcript of telegraph wire, 11 June 1935, Legislative Bill Files, 1935, box 23, Dimond Papers.

18. MacDonald, draft speech, Joint House Memorial Session (undated), 3, Dimond Papers.

19. Congressman Warren G. Magnuson chaired AIHC. The other four presidentially appointed members were Donald MacDonald (Alaskan member); Ernest Gruening, the director of the Division of Territories and Island Possessions of the Department of the Interior; Thomas Riggs, a former governor of Alaska

and an engineer by profession; and James W. Carey, a civil engineer from Portland, Oregon. A comparable Canadian body, the British Columbia-Yukon-Alaska Highway Commission was also established in 1938.

20. Alaskan International Highway Commission, *Report of the International Highway Commission to the President*, House Document 711 (Washington, D.C.: GPO, 23 April 1940). The summary contained no reference to the military importance of an international highway. This omission was strongly criticized by Congressman William M. Whittington (Democrat, Mississippi), the chairman of the House Committee on Roads, at the hearings on the proposed highway to Alaska, which his committee conducted in February 1942. See Hearings, *Proposed Highway to Alaska*, 54.

21. AIHC, *Report of the International Highway Commission*, 1, 13.

22. Hearings, *Proposed Highway to Alaska*, 23.

23. Airbases were also constructed at Kamloops, Williams Lake, and Prince George (all in British Columbia); Whitehorse (Yukon Territory); and at Northway and Big Delta in Alaska.

24. U.S. Congress, House of Representatives, Committee on Roads, *The Alaska Highway—An Interim Report*, House Report 1705 (Washington, D.C.: GPO, 1946), 6.

25. For an extensive discussion of these routes, see Vilhjalmur Stefansson, "Routes to Alaska," *Foreign Affairs* 19 (July 1941): 861–69. The renowned Canadian explorer-anthropologist vigorously supported an international highway as a boost to his schemes for the colonization of the North American Arctic. Stefansson, who was keen to see the oilfields at Norman Wells developed, wanted Route D.

26. AIHC's Canadian counterpart, the British Columbia-Yukon-Alaska Highway Commission, wanted Route B. See *Report on the Proposed Highway Through British Columbia and the Yukon Territory to Alaska*, 18–36.

27. House Committee on Roads, *The Alaska Highway—An Interim Report*, 9.

28. Hearings, *Proposed Highway to Alaska*, 1.

29. Ibid., 6.

30. Ibid., 11, 42–44. Certain Midwesterners, such as John Moses, the governor of North Dakota, preferred Route C. So did the U.S.-Canada-Alaska Prairie Highway Association; ibid., 129, 93–108.

31. For the U.S. Army's reasons for choosing Route C, see the testimonies of Brigadier General C. L. Sturdevant, assistant chief of engineers, U.S. War Department, and Colonel J. K. Tully, General Staff, U.S. Army Corps of Engineers. Hearings, *Senate Resolution 253*, 3–6, B-11. The Northwest Staging Route was used by American pilots delivering planes destined for the Soviet Union as part of the lend lease program. Soviet pilots collected the planes in Fairbanks and flew them to the European front.

32. Sigirid Arne, "U.S. Army Engineers Speed Alaska Highway," *Washington Post*, 14 June 1942.

33. Harold W. Richardson, "Alcan—America's Glory Road: Part 1: Strategy and Location," *Engineering News-Record* 129 (17 December 1942): 81–96. Richardson, the *Engineering News-Record* western editor, claimed to be the first person to travel the entire length of the road from Edmonton to Fairbanks. For the history of construction, with data, maps, and related documents, see House Committee on Roads, *The Alaska Highway—An Interim Report*, 12, 21, 73–130. See also Stuart Ramsay Tompkins, *Alaska: Promyshlennik and Sourdough* (Norman: University of Oklahoma Press, 1945), 296–300.

34. For a summary, see Lyman L. Woodman, "Building the Alaska Highway:

A Saga of the Northland," *The Northern Engineer* 8 (Summer 1976): 11–28. Woodman was Public Affairs Officer for the Alaska District (Anchorage) of the U.S. Army Corps of Engineers.

35. Arne, *Washington Post,* 14 June 1942.

36. William Gilman, *Our Hidden Front* (New York: Reynal and Hitchcock, Inc., 1944), 75. Gilman's statement was quoted by the naturalist Olaus Murie, "Alaska," *Living Wilderness* 10 (12 February 1945): 6. For similar sentiments, see the report of Theodore Strauss, "Alcan Road Booms Vast Untamed Area," *New York Times,* 1 January 1943.

37. Philip H. Godsell, *The Romance of the Alaska Highway* (Toronto: Ryerson Press, 1946), 179.

38. Instructive are some of the chapter headings in Douglas Coe, *Road to Alaska: The Story of the Alaska Highway* (New York: Julian Messner, 1943); "The Battlefield;" "The Plan of Attack;" "The North's Secret Weapon;" and "The Enemy Surrenders."

39. Don Menzies, *The Alaska Highway: A Saga of the North* (Edmonton, Alberta: Stuart Douglas, 1943), 1.

40. Woodman, "Building the Alaska Highway," 19.

41. Menzies, *The Alaska Highway,* 19.

42. Lyman L. Woodman, *The Army's Role in the Building of Alaska,* U.S. Army Alaska, Headquarters, Anchorage: U.S. Army Pamphlet 360–65, 1 April 1969, 82.

43. Coe, *Road to Alaska,* 27.

44. Ibid., 143–48; Woodman, "Building the Alaska Highway," 23.

45. Elizabeth Parks Bright, *Alaska, Treasure Trove of Tomorrow* (New York: Exposition Press, 1956), 48. For similar appraisals, see William Gilman, "Colonists for Alaska," *Yale Review* 34 (Summer, 1945): 68; Froelich Rainey, *National Geographic* 83 (February 1943): 143. For details of construction tactics, see Harold W. Richardson, "Alcan—America's Road: Part 3, Construction Tactics," *Engineering News-Record* 130 (14 January 1943): 131–38; Gilman, *Our Hidden Front,* 62–75.

46. Godsell, *The Romance of the Alaska Highway,* 180, foreword, xi.

47. Hearings, *Senate Resolution 253,* 1. Donald MacDonald and Warren G. Magnuson, representing the Alaskan International Highway Commission, echoed Langer's views. Ibid., 36–64. See also Magnuson's protest in the *Congressional Record,* 77th Congress, 2d session, 88 (9 June 1942), A2310–11.

48. Hearings, *Senate Resolution 253,* 93–94.

49. Dimond, Speech, American Road Builders' Association, Washington, D.C., November 1942, draft typescript, 5, Legislative Bill Files, 1935, box 23, Dimond Papers.

50. For MacDonald's reaction, see "Highway . . . Hell!" *Alaska Life* 2 (September 1942): 3–12.

51. Remley, *Crooked Road,* 223.

52. Ibid., 42. Remley's assessment of the historical significance of the Alaska Highway seems based largely on the fact that "it runs nearly to America's land's end"; ibid. Wallace Stegner, the novelist and authority on the history of the West, has challenged Remley's appraisal of the road's importance: "What makes me unwilling to accept the Alaska Highway as an extension or reaffirmation of our national character and national destiny, is that it leads literally nowhere. It dead-ends against a row of rusty oil drums in the abandoned mining camps of Circle." See Stegner's review of *Crooked Road, Sierra Club Bulletin* 62 (Summer 1977): 61. Circle City on the Yukon River lay at the terminus

of the Steese Highway. Thanks to the Alaska Highway, it was possible to drive to Circle from the lower forty-eight. In 1942, Circle was the most northerly point accessible on the American public road system.

53. U.S. Bureau of the Census, *Current Population Reports,* Series P-25, no. 80 (Washington, D.C.: GPO), 7 October 1953. Figures quoted by George W. Rogers, *The Future of Alaska: The Economic Consequences of Statehood* (Baltimore: Resources for the Future and Johns Hopkins University Press, 1962), 95.

54. Alaska Employment Security Commission figures, as quoted by Rogers, *The Future of Alaska,* 93.

55. For a contemporary account of the wartime boom in Alaska see R. L. Duffus, "Alaska Hits Her Stride," *New York Times Magazine,* 2 November 1952. For historical assessments, see Claus-M. Naske, "The Alcan: Its Impact on Alaska," *The Northern Engineer* 8 (Spring 1976): 12–18; Rogers, *The Future of Alaska,* 93–102.

56. Ivan Bloch, "Alaska Power Resources in Relation to Mineral Development," *Journal of the Electrochemical Society* 103 (October 1956): 587.

57. U.S. Congress, Senate, Committee to Investigate the National Defense Program, hearings, *Investigation of the National Defense Program, Part 22; Canol Project,* 11 September, 26 October, 12–13, 16–19, 22–24 November, 20 December 1943 (Washington, D.C.: GPO, 1944), 9528, 9716. For a contemporary report on Canol, prepared for the Great Falls Chamber of Commerce, Great Falls, Montana, see W. F. Wright, "The Future of the Alaska Highway," 294–95. appendix P, exhibit 1, Committee on Roads, *The Alaska Highway—An Interim Report.*

58. Hearings, *Investigation of the National Defense Program,* Glen M. Ruby, exhibit 1099 (3 October 1942), 9854; see also exhibit 1137 (29 March 1943), H. LeRoy Whitney, chairman, War Production Board, 9882.

59. Ibid., 9391, 9514. For confirmation of these points, see a report by the Anchorage Chamber of Commerce, "Summary of Canol Project in Canada During WW2 (Alaska's 1957 Oil Discovery and its Strategic Importance)," Anchorage, (undated) 3. For a general historical discussion see Theodore J. Karamanski, "The Canol Project: A Poorly-Planned Pipeline," *Alaska Journal* 9 (Autumn 1979), 17–21.

60. Hearings, *Investigation of the National Defense Program,* 9599–9600, 9632–33, 9646, exhibit 1098 (15 April 1942), 9846–49. Not even technical studies of matters such as the number of pump stations required, or optimum pipe size, were conducted. There were no cost projections for materials, supplies, or work force. See Anchorage Chamber of Commerce, "Summary of Canol Project," 3.

61. Hearings, *Investigation of the National Defense Program,* 9545, 9935.

62. Riggs to Frederic A. Delano, 14 May 1943. Legislative Bill File, box 23 (4), Dimond Papers. Delano was chairman of the National Resources Planning Board in Washington, D.C.

63. Hearings, *Investigation of the National Defense Program,* 9528.

64. In 1953 Canol Four, the only part of the project that remained in operation, was replaced by a 626-mile, 8-inch line from Haines to Fairbanks. Petroleum products shipped to Haines were pumped via eleven stations north to the Canadian border. Here the line emerged above ground until it reached Big Delta, where it disappeared underground again to complete the final 96 miles.

65. For a detailed account of all aspects of this operation, see John C. Reed, *Exploration of Naval Petroleum Reserve No. 4 and Adjacent Areas of*

Northern Alaska, 1944–53; Part 1, History of the Exploration, USGS Professional Paper 301 (Washington, D.C.: GPO, 1958), 19–20. Reed was commander of the U.S. Naval Reserve. The explorations were directed by Commodore William G. Greenman, chief of the Office of Naval Petroleum and Oil Shale Reserves from 1944 until 1950.

66. Reed, *Exploration,* 22, 28. The author of the report was a veteran of the 1920s explorations, Lieutenant William T. Foran, chief geologist, Technical Subcommittee, U.S. Army-Navy Fuel Oil Board for Alaska. Foran subsequently played an important part in attracting commercial oil companies to Alaska through the Northern Development Company and the Yakutat Development Company.

67. The U.S. Navy did discover a reasonable quantity of gas just outside Pet 4 but was not greatly interested in this fuel.

68. Reed, *Exploration,* 184.

69. For an account of a trial installation at Barter Island, Alaska, prior to construction of the fully-fledged system, see V. B. Bagnall, "Operation DEW line," *Journal of the Franklin Institute* 259 (June 1955): 481–90. Bagnall was project manager, Defense Projects Division, Western Electric Company.

70. For a brief history and account of DEWline logistics, see "Canada's First Arctic Coast Mega-Project," *Beaufort* 2 (August 1982): 6–9. Although the title only refers to Canada, the article does cover the Alaskan aspects. *Beaufort* is published by Dome Petroleum Ltd., Esso, Canada. For an account by a writer of popular arctic stories, laden with references to the "Godforsaken" Arctic, see Richard Morenus, *DEWline, Distant Early Warning: The Miracle of America's First Line of Defense* (New York: Rand McNally and Company, 1957).

71. Reed, *Exploration,* 169, 184.

72. H. F. Flanders (Western Electric Company), "Contributions of the DEWline in Arctic Engineering," Proceedings of the 8th Alaska Science Conference, Anchorage, 10–13 September 1957: Albert W. Johnson, ed., *Science in Alaska,* 1957, 54.

73. Some 818,000 fuel drums were either flown in or shipped north for the construction phase alone. Laid end-to-end, these would have stretched 180 miles. At any given time, some 7,500 men were at work on the line. See "Canada's First Arctic Coast Mega-Project," 6.

74. Bagnall, "Operation DEW line," 488–90.

75. Morenus, *DEWline,* 138.

76. *Alaska Life* 5 (August 1942): 66.

77. This aspect of the Alaska Highway, its impact on the lifestyles of Native Alaskan and bush whites ("rural residents" in contemporary parlance), has yet to be addressed by scholars. Remley's *Crooked Road,* though based in large part on the oral testimonies of local residents, does not engage this topic in any depth. The impact of the road on Canadians who lived in its path has been more thoroughly investigated, particularly in the 1970s as a precedent for a possible Mackenzie Valley or Alaska Highway gas pipeline from the North Slope oilfields to the American Midwest. See *Alaska Highway 1942–82, Sources of Information in the Yukon Archives: A Partial Listing of Highway Literature* (Edmonton, Alberta: Boreal Institute, June 1982).

78. "Phooey—on the Highway, 'By An Alaskan Businessman,'" *Alaska Life* 4 (July 1941): 3, 14–15, 30. This anonymous businessman claimed the road would bring no economic benefits. He also argued that it would be too expensive to maintain in winter and that it would be better to spend money building roads in Alaska instead.

79. Bright, *Alaska, Treasure Trove of Tomorrow,* 48.

80. Robert Marshall, "The Development of Alaska Resources Should Be Retarded," 3, Robert Marshall Papers, Bancroft Library, University of California, Berkeley (UCB). (Hereafter referred to as RMP.) For the same views, see Marshall's letter to Mordecai Ezekiel, 9 March 1939, ibid. Marshall's other objections were the deflection of road building and maintenance resources away from needier, more densely populated areas, and, fourthly, the misappropriation of national recreational resources. He believed that the federal government should concentrate on providing access for the "low income majority" to recreational areas close to large population centers, rather than cater to the affluent minority who could afford to vacation in the far north. His fifth point was the increased dangers of forest fires, and his final one was the impact of an invasion from the south on the Alaskan wage and price structure. "The Development of Alaska Resources Should Be Retarded," 3–4.

81. Marshall, "The Development of Alaska Resources Should Be Retarded," 3, 4.

82. Marshall to Emery Tobin, editor, *Alaska Sportsman,* 22 March 1939, RMP.

83. Robert Marshall, "Opportunities for Refugees in Alaska," 19 April 1937, 2; Marshall to Ezekiel, 9 March 1939, ibid. See also, Robert Marshall, "Should We Settle Alaska?" *New Republic* 102 (8 January 1940): 49–50. For details of various refugee resettlement schemes for Alaska in the 1930s see Orlando W. Miller, *The Frontier in Alaska* (New Haven: Yale University Press, 1975), 168–76; 170–71 for Marshall's attitude to the resettlement of refugees in Alaska.

84. Dave Bohn, *Glacier Bay: The Land and the Silence* (San Francisco: Sierra Club, 1967), 94–106.

85. "Admiralty, Island in Contention," *Alaska Geographic* 1 (Summer 1973): 59.

86. Marshall to the *Journal of Forestry,* 30 March 1936, RMP.

87. Marshall and Althea Dobbins, "Last Vestiges of the Frontier," typescript, undated, without page numbers, RMP.

88. Frank Dufresne, *Mammals and Birds of Alaska,* U.S. Fish and Wildlife Service, Circular no. 3, (Washington, D.C.: GPO, 1942), 17. For a sober appraisal of Alaska's chances of attracting postwar settlers, see William Gilman, "Colonists for Alaska," *Yale Review* 34 (Summer 1945): 666–82. Describing Alaska as a frontier that "simply has not grown up" (666), Gilman attacked both the icebox image and booster propaganda as a disservice to Alaska. Alaska, he stressed, while not uninhabitable, was no haven for the land-hungry unemployed.

89. As quoted by Olaus Murie, *Living Wilderness,* February 1945, 6. For the full text of Gabrielson's remarks, see Ira N. Gabrielson, "Alaska Wildlife," *Audubon* 45 (November-December 1943): 329–35.

90. Gabrielson, "Alaska Wildlife," 335.

91. But they were not entirely inconspicuous and caused considerable debate in Alaska. See Morgan B. Sherwood, *Big Game in Alaska: A History of Wildlife and People* (New Haven: Yale University Press, 1981), 137–40, for information on game law violations by the military and the U.S. Army's response. According to Sherwood, Gabrielson's informant on these matters was Kermit Roosevelt, the grandson of Theodore Roosevelt, who was stationed in Anchorage during the war.

92. Lois Crisler to George Marshall, 1 May 1957, Sierra Club Office Files (San Francisco): Arctic Wilderness File, Sierra Club Papers, Bancroft Library, University of California, Berkeley (UCB).

93. Lois Crisler, *Arctic Wild* (New York: Harper and Brothers, 1958), 64,

205, 300.

94. Crisler, *Living Wilderness* 24 (Summer 1960): 19.

95. Robert Marshall, *Arctic Wilderness*, edited by George Marshall (Berkeley and Los Angeles: University of California Press, 1956), vi. A. Starker Leopold, Professor of Zoology and Forestry at the University of California, Berkeley, was the son of the pioneering American ecologist, Aldo Leopold (1887–1948). He died in 1983.

96. Flanders, "Contributions of the DEWline in Arctic Engineering," 53–4; Morenus, *DEWline*, 83.

CHAPTER 3. BOOSTERS AND CONSERVATIONISTS IN THE POSTWAR ERA

1. Daniel J. Boorstin, *The Americans: The Democratic Experience* (New York: Random House, 1973), 273–74.

2. See Timothy C. Weiskel, "Agents of Empire: Steps Toward an Ecology of Imperialism," *Environmental Review* 11 (Winter 1987): 275–89.

3. For a review that does these frontier figures justice, see Bradford Luckingham, "The Urban Dimension of Western History," in Michael P. Malone, ed., *Historians and the American West* (Lincoln: University of Nebraska Press, 1983), 323–43.

4. Richard C. Wade, *The Urban Frontier: The Rise of Western Cities, 1790–1830* (Cambridge: Harvard University Press, 1959); Earl Pomeroy, *The Pacific Slope: A History of California, Oregon, Washington, Idaho, Utah and Nevada* (New York: Alfred Knopf, 1966), especially the chapter "The Power of the Metropolis." Another useful study by Pomeroy is "The Urban Frontier of the Far West," in John G. Clark, ed., *The Frontier Challenge: Responses to the Trans-Mississippi West* (Lawrence: University of Kansas, Regents Press, 1971), 7–29.

5. Daniel J. Boorstin, *The Americans: The National Experience* (New York: Random House, 1965), 113; *The Democratic Experience*, 69.

6. Alaska Development Board, *Charting Alaska's Progress, 1951–52,* Biennial Report, Juneau, 14.

7. Boorstin, *The National Experience,* 124. For the role of the press in promoting communities and attracting settlers in the nineteenth century, see chapter 17, "The Booster Press," ibid., 124–34.

8. Evangeline Atwood, letter to the author, 11 March 1987.

9. The existing study of the role of the Alaskan press in the statehood campaign is disappointing. See Carroll V. Glines, Jr., "Alaska's Press and the Battle for Statehood" (Master's thesis, American University, 1969).

10. See Hickel's autobiographical *Who Owns America?* (Englewood Cliffs, N.J.: Prentice-Hall, 1971), 67–68; Herb and Mirriam Hilscher, *Alaska, USA* (Boston: Little, Brown and Company, 1959); chap. 16, "Alaska's Number One Young Businessman."

11. The first American mall was built in Kansas City in 1922. See Boorstin, *The Democratic Experience,* 272–73.

12. Boorstin, *The National Experience,* chap. 18, "Palaces of the Public," 134–47, examines the role of hotels in developing the frontier.

13. Operation Statehood, "A Plea From Alaskans for Help," 1957; as quoted in George W. Rogers, *The Future of Alaska: The Economic Consequences of Statehood* (Baltimore: Resources for the Future and Johns Hopkins University Press, 1962), 170.

14. Claus-M. Naske, *E. L. "Bob" Bartlett of Alaska: A Life in Politics* (College: University of Alaska Press, 1979), 122. The definitive political history of statehood is Naske, *An Interpretative History of Alaska Statehood* (Anchorage: Alaska Northwest Publishing Company, 1973). Also useful is Ernest Gruening, *The Battle for Alaska Statehood* (College: University of Alaska Press, 1967).

15. U.S. Congress, House of Representatives, Committee on Public Lands, Subcommittee on Territorial and Insular Possessions, hearings, *Statehood for Alaska, H.R. 206*, 80th Congress, 1st session, 16–24 April 1947 (Washington, D.C.: GPO, 1947), 2.

16. Ibid., 324–29.

17. The salmon industry paid no taxes on wages because its employees were paid at the place of hire (usually Seattle) after the canning season. The Merchant Marine Act (Jones Act) of 1920, which prohibited the use of foreign-built and foreign-flag ships in interstate commerce, eliminated competition out of Canadian ports such as Vancouver and Prince Rupert, which were closer, thereby encouraging monopoly rates. The steamship companies charged the canneries preferential rates.

18. For a good example of the icebox image's deployment, see the comments of Senator John C. Stennis (Democrat, Mississippi), *Congressional Record*, 81st Congress, 2d session, 96 (27 November 1950), S15935–36.

19. The Alaska Statehood Committee's biennial reports placed great emphasis on the need to gain more control over Alaskan resources. See *Biennial Report*, Anchorage, 15 February 1951, 11–12; *Biennial Report*, Anchorage, 1 August 1953, 21, 40.

20. See Frank E. Buske, "Rex Beach: A Frustrated Goldseeker Lobbies to 'Win the Wilderness over to Order,'" *Alaska Journal* 10 (1980): 37–42.

21. U.S. Congress, Senate, Committee on Interior and Insular Affairs, hearings, *Alaska Statehood*, 81st Congress, 2d session, 24–29 April 1950 (Washington, D.C.: GPO, 1950), 233–41.

22. These political considerations had precedents. In 1864 Nevada was admitted to the Union after only three and a half years as a territory, despite its sparse population, to strengthen the Republicans in the Senate.

23. Gruening's book, *The State of Alaska* (New York: Random House, 1954), is a restatement of the classic traditionalist interpretation of Alaskan history. Gruening's approach is clearly indicated by the chapter headings; the period 1867–84 he calls "The Era of Total Neglect;" 1884–98 is "The Era of Flagrant Neglect;" 1898–1912 is "The Era of Mild but Unenlightened Interest;" and the period 1912–33 is "The Era of Indifference and Unconcern."

24. Naske, *An Interpretative History*, 115.

25. Delegates of the People of Alaska, *The Constitution of the State of Alaska*, Alaska Constitutional Convention, University of Alaska, Fairbanks, 5 February 1956, 24. For further details of the natural resource policies adopted, see Victor Fischer, *Alaska's Constitutional Convention* (College, Alaska: Institute of Social, Economic and Government Research, 1975), 129–37.

26. The Wildlife Management Institute (1911) superseded the American Wildlife Institute, which, in turn, had its origins in the American Game Association. Gutermuth was the first secretary of the Natural Resources Council of America (1946). He was operative in founding the World Wildlife Fund and served as

its secretary, treasurer, and director in the ensuing years. In 1963 he was appointed director of the National Rifle Association. He was a full member of the Boone and Crockett Club.

27. U.S. Congress, House of Representatives, Subcommittee on Territorial and Insular Possessions, hearings, *Statehood for Alaska, S.49 and S.35,* 85th Congress, 1st session, 11–15, 25–29 March 1957 (Washington, D.C.: GPO, 1957), 379.

28. Ibid., 379; Joseph Penfold, director of the Izaak Walton League of America; ibid., 418. Charles H. Callison, conservation director of the NWF, also stressed the national recreational interest in Alaskan wildlife: "Every sportsman dreams of making a trip to Alaska;" statement submitted by Gutermuth, ibid., 484–85.

29. Ibid., 380.

30. Ibid., 260.

31. George Marshall to Gruening, 9 August 1957, Correspondence File (1916–74), Ernest Gruening Papers, University of Alaska Archives, Fairbanks, Alaska (UAF). (Hereafter referred to as Gruening Papers.)

32. George Marshall to Michael Nadel, 3 November 1969, George Marshall Papers, Bancroft Library, University of California, Berkeley (UCB).

33. George Marshall, letter to author, 7 May 1986.

34. U.S. Congress, Senate, Committee on Interstate and Foreign Commerce, Subcommittee on Merchant Marine and Fisheries, hearings, *S. 1899, A Bill to Authorize the Establishment of the Arctic Wildlife Range, Alaska,* 86th Congress, 1st session, part 1, 30 June 1959 (Washington, D.C.: GPO, 1960), 7. (Hereafter referred to as Hearings, *AWR.*) At the same hearings, Gutermuth denied being "in bed" with the fish trusts to frustrate Alaskan efforts to conserve salmon stocks by perpetuating federal management of Alaskan fisheries, ibid., 47.

35. Ernest Patty, *North Country Challenge* (New York: McKay Company, 1969), 249. Patty states that a meeting took place in Washington, D.C., but he does not provide names or dates. George Marshall later denied that such a meeting had occurred. George Marshall to Michael Nadel, 3 November 1969, Correspondence File (1964–73), George Marshall Papers, UCB.

36. The State of Alaska also received unprecedented freedom of choice in selecting these lands. Mineral policy marked a break with the past too since the federal government did not retain ownership of subsurface minerals in the lands selected by the state. For details, see Richard A. Cooley, "State Land Policy in Alaska: Progress and Prospects," *Natural Resources Journal* 4 (January 1965): 455–67.

37. "Alaska—Land of Beauty and Swat," *Time* 71 (9 June 1958): 20.

38. Hearings, *Alaska Statehood,* April 1947, 334.

39. *Muktuk* (or *maktak*) is an Eskimo delicacy made from whale skin and blubber.

40. U.S. Congress, House of Representatives, Committee on Public Lands, Subcommittee on Territories and Insular Possessions, hearings, *Alaska Statehood, (House Resolution 83),* 80th Congress, 1st session, 30 August–12 September 1947, 83. Marston became a member of Gruening's "committee of distinguished Americans" and served as a delegate to the 1955 constitutional convention in Fairbanks.

41. Ernest Gruening, "Go North, Young Man!" *Reader's Digest* 44 (January 1944): 53–57.

42. Ibid., 53.

43. Hearings, *Alaska Statehood (House Resolution 83)*, 389–90; Gruening, *The Battle for Alaska Statehood*, 41.

44. Ernest Gruening, "Alaska: Progress and Problems," *Science in Alaska*, 1952, abstract, 13. For his full remarks, see *Scientific Monthly* 77 (1953): 3–12. Gruening also quotes this remark in *The Battle for Alaska Statehood*, 41.

45. Walter Prescott Webb, *The Great Frontier* (Cambridge, Mass.: Houghton Mifflin, 1951), 284–87.

46. Ibid., 284. Webb adopted the term *fringe* from a study of contemporary world frontiers by the geographer, Isaiah Bowman: *The Pioneer Fringe*, Special Publication no. 13 (New York: American Geographical Society, 1931). For brief references to limited Alaskan agricultural opportunities, see 4, 226–27, 248, 253.

47. As quoted in John Sherman Long, "Webb's Frontier and Alaska," *Southwest Review* 56 (Autumn 1971): 301. Long focuses on twentieth-century efforts to introduce agriculture into Alaska, such as stock raising on Kodiak Island and the Matanuska Valley colonization scheme. He presents a rather distorted picture of the Matanuska experiment in the 1930s and overestimates its success. Long's sympathies are with those responsible for these heroic efforts and he is critical of the federal government and conservation policies, which, he believes, hindered the process of agricultural development in Alaska. Prominent among the author's sources are boosterist works by C. L. Andrews, the historian; Rex Beach, novelist; Scott C. Bone, the one-time territorial governor of Alaska; and John J. Underwood, the journalist whose writings promoted Alaskan economic development.

48. Hearings, *Alaska Statehood*, 75. Dimond was then judge for the 3d Judicial District of Alaska.

49. Among other things, the Alaska Development Board (ADB) promoted agriculture and settlement, particularly by Second World War veterans who had served in the territory. See George Sundborg, *Opportunity in Alaska* (New York: Macmillan, 1945).

50. Hearings, *Alaska Statehood*, 1950, Washington, D.C., 195. One of Frederick Jackson Turner's books was called *The Rise of the New West, 1819–29* (New York: Harper and Brothers Publishers, 1906).

51. U.S. Congress, House of Representatives, Interior and Insular Affairs Committee, Subcommittee on Territories and Insular Possessions, hearings, *Alaska Statehood*, 83d Congress, 1st session, 14–17 April 1953 (Washington, D.C.: GPO, 1953), 80–86.

52. Frederick Jackson Turner, "Pioneer Ideals and the State University," 1910, in *The Frontier in American History* (Huntington, N.Y.: Robert E. Krieger, reprint, 1976), 269.

53. U.S. Congress, Senate, Committee on Interior and Insular Affairs, Subcommittee on Territories and Insular Affairs, hearings, *Alaska Statehood*, 85th Congress, 1st session, 11–15, 25–29 March 1957 (Washington, D.C.: GPO, 1957), 151, 156. The *Fairbanks Daily News-Miner* had changed its stance on statehood in 1954, after Snedden acquired the newspaper from Lathrop; see editorial, 27 February 1954.

54. Patrick O'Donovan, "The 49th Star on the U.S. Flag," (London) *Observer*, 13 July 1958.

55. BLM figures, as quoted by Rogers, *The Future of Alaska*, 137; For the inducements offered to postwar military homesteaders, see Orlando W. Miller, *The Frontier in Alaska* (New Haven: Yale University Press, 1975), 187–89; Bureau

of Land Management, *Alaska: Information Relative to the disposal and leasing of Public Lands in Alaska* (Washington, D.C.: GPO, 1948), 18.

56. William E. Warne, Assistant Secretary of the Interior, *Papers Presented at the Seminars on Alaska, 2–3, 6–7 February 1950,* 2; as quoted in Rogers, *The Future of Alaska,* 137.

57. H. Johnson and R. Coffman, *Land Occupancy, Ownership, and Use on Homesteads in the Kenai Peninsula, Alaska, 1955,* Bulletin 21, (Palmer, Alaska: U.S. Department of Agriculture, November 1956); as quoted in Rogers, *The Future of Alaska,* ibid. For further details, see Miller, *The Frontier in Alaska,* 209–10.

58. Rogers, *The Future of Alaska,* 41–42; Miller, *The Frontier in Alaska,* 192, 286.

59. Ibid., 207; Karl E. Francis, "Outpost Agriculture, The Case of Alaska," *Geographical Review* 57 (October 1967): 496–505. Francis emphasized the difference between agriculture that is technically possible and agriculture that is economically viable. He argued that Alaskan agriculture will always be restricted to the provision of a few fresh products to the local market to supplement imports.

60. Oil discovered at Katalla on the southcentral coast in 1902 had produced 154,000 barrels of crude, which was refined locally and sold to fishing boats over a 31-year period. This was equivalent to one week's yield from the Swanson River field on the Kenai Moose Range, and an afternoon's output from Prudhoe Bay; Arnold and Helen Nelson, "The Bubble of Oil at Katalla," *Alaska Journal (A 1981 Collection):* 19–27. The Alaska Petroleum and Coal Company advertised Katalla as "The New Pennsylvania" in 1905, ibid., 21.

61. *Anchorage Daily Times,* 31 July 1957.

62. Ibid., 29 July 1957.

63. Hearings, *Alaska Statehood (House Resolution 83),* 83.

64. In 1935—urged by Jay N. Darling, head of the Bureau of Biological Survey—DuPont, the Hercules Powder Co., the Remington Arms Co., and other gun and ammunition manufacturers donated funds to federate the organized sportsmen of the United States; to organize a national wildlife conference; and to fund wildlife management training courses at land grant colleges. This resulted in the creation that year of the National Wildlife Federation (NWF), still the nation's largest conservation organization; the annual North American Wildlife Conference; and the foundation of the Cooperative Wildlife Research Unit.

65. Charles H. Callison, hearings, House Committee on Merchant Marine and Fisheries, 1956. Copy of otherwise unidentified 7-page testimony in the Gruening Consolidated File, box 4 (45). See also *Conservation News* 20 (15 December 1955): 1–3; 21 (1 March 1956): 1–3; 21 (15 July): 5; 21 (10 October 1956): 7–9. *Conservation News* is a bulletin published by the NWF.

66. *Anchorage Daily Times,* 19 November 1957.

67. Ibid.

68. *Anchorage Daily Times,* 18 December 1957.

69. Hearings, Proposed Regulations Relating to Oil and Gas Leasings on Federal Wildlife Lands. House Committee on Merchant Marine and Fisheries, Washington, D.C., 9–10 December 1957, 3. The version referred to here is a transcript in Alaska General: Consolidated Files, box 4 (45), Moose Range, Gruening Papers.

70. Ibid., 4.

71. Ibid., 4, 7, 14–16.

72. Ibid., 1–3, 7, 12.

73. Gruening to the editor of *True*, 27 December 1957, Consolidated Files, Alaska General, box 4 (45), Gruening Papers, UAF.

74. *Anchorage Daily Times*, 20 December 1957.

75. Burton Atwood, "Oil on the Moose Range," 1. This report was part of "Conservation Problems in Alaska," Observations by Burton H. Atwood, 27 September 1957, Consolidated Files, Alaska General, box 4 (45), Gruening Papers.

76. Gruening, testimony, hearings, Proposed Regulations Relating to Oil and Gas Leasings, December 1957, 10, Consolidated Files, Alaska General, box 4 (45), Gruening Papers.

77. See A. Starker Leopold and Frank Fraser Darling, *Wildlife in Alaska: An Ecological Reconnaissance* (New York: The Ronald Press, 1953), 57–58.

78. See the testimony of Albert M. Day, representing the Anchorage Chamber of Commerce, hearings, Proposed Regulations Relating to Oil and Gas Leasings, 4. Copy in Consolidated Files, Alaska General, box 4 (45), Gruening Papers. Also, Robert Atwood to General Nathan F. Twining, Chairman, Joint Chiefs of Staff, Washington, D.C., 25 October 1957, 9–10, ibid. This is not substantiated by biologists and environmental historians. Sherwood relates that moose were already there and that the settlers, soon after their arrival, began to kill moose and other wildlife, whether it affected their agriculture or not. See Morgan B. Sherwood, *Big Game in Alaska: A History of Wildlife and People* (New Haven: Yale University Press, 1981), 7–8. Moose had certainly caused trouble for the operators of the Alaska Railroad. On 10 January 1947, for example, their habit of walking on the tracks, which afforded the easiest passage through heavy snow, resulted in one derailment and fourteen delays in schedule. Gruening (Hearings, Proposed Regulations Relating to Oil and Gas Leasings, 11) and Atwood to Twining, 25 October 1975, 10, emphasized this point. According to statistics quoted by Atwood, the Alaska Railroad killed between 150 and 250 moose per annum (Atwood to Twining, ibid., 10).

79. Collins was Chief of the Park Service's State and Territorial Recreation Division (Region Four). Sumner was head of its Biological Survey. For the final report see William J. Stanton, *Alaska Recreational Survey*, 2 vols, National Park Service, Department of the Interior (Washington, D.C.: GPO), 1953.

80. Adolph Murie, *A Naturalist in Alaska* (Old Greenwich, Conn.: Devin Adair Company, 1961), 4.

81. For a summary of Murie's research, see *The Alaska-Yukon Caribou*, North American Fauna, no. 54, Bureau of Biological Survey (Washington, D.C.: GPO, 1935).

82. For biographical details, see Murie's obituary in *Tundra Times*, 4 November 1963.

83. Olaus Murie, "Planning for Alaska's Big Game," Alaska Science Conference, National Academy of Sciences, National Research Council, Washington, D.C., 9–11 November 1950, *Science in Alaska*, edited by Henry B. Collins (Washington, D.C.: Arctic Institute of North America, 1952): 258–68.

84. Ibid., 258. In 1947 Murie had chaired a Committee on Aircraft and Wilderness, which recommended a landing ban in federally protected areas and a minimum altitude of 2,700 feet for planes passing over them. See *Living Wilderness* 12 (Autumn 1947): 2–6. According to Murie, the airplane, as a means of travel in wilderness, was "foreign to the whole wilderness idea" (ibid). This investigation, the first of its kind, occurred with specific reference to the Quetico-

Superior region of northern Minnesota, an area whose wild values conservationists believed were under mounting pressure, not only from would-be resource extractors but also technologically powered recreationists. In 1949 President Truman prohibited flights within the Boundary Waters Canoe Area of the Quetico-Superior at altitudes below 4,000 feet above sea level.

85. In 1932 a federal public works project completed the road from Anchorage to Matanuska begun by the Anchorage Chamber of Commerce. The Glenn Highway, which connected Palmer with the Richardson Highway, was constructed in 1941.

86. Murie, "Planning for Alaska's Big Game," 258, 267.

87. Arctic Wilderness Folder, Sierra Club Office Files (San Francisco), Bancroft Library, University of California, Berkeley (UCB).

88. Sierra Club Resolution, 3 May 1952, Sierra Club Office Files, UCB.

89. Robert O. Beatty to Richard Leonard, secretary of the Sierra Club, 3 October 1952. Sierra Club Office Files, UCB.

90. WFOC, Annual Meeting, Echo Summit, Calif., Sierra Club Office Files, UCB.

91. Collins to Al Anderson (manager of the Alaska Development Board), 2 October 1953. Edgar Wayburn, Private Papers, San Francisco, Calif. Wayburn was the club's vice president and the official with chief responsibility for Alaska throughout the 1960s and 1970s.

92. George Collins and Lowell Sumner, "The Northeast Arctic: The Last Great Wilderness," *Sierra Club Bulletin* 38 (October 1953): 13–26. This was an updated version of Sumner's address, "Alaska's Biological Wealth: Why Let History Repeat Itself?" proceedings, 2d Alaska Science Conference, 4–8 September 1951, *Science in Alaska* (Washington, D.C.: Arctic Institute of North America, 1951), 337–39. A further report by Sumner, "Arctic Wilderness," appeared in the *Living Wilderness* 18 (Winter 1953–54): 5–16.

93. Collins and Sumner, "The Northeast Arctic," 25.

94. Collins and Sumner to Richard Leonard, the secretary of the Sierra Club, 12 August 1953. *La Foule* (The Crowd) was the name French-Canadian fur traders had given to the phenomenon of caribou migration.

95. Editorial, *Living Wilderness* 18 (Autumn 1953): 2.

96. The Conservation Foundation (1948) is an international, nonmembership, privately-financed research, education, and informational organization.

97. Other expedition members were Brina Kessel of the University of Alaska's zoology department, and University of Alaska graduate students George Schaller and H. Robert Krear.

98. *Living Wilderness* 21 (Autumn/Winter 1956–57): 30. This quotation can be located in the *Collected Verse of Robert Service* (London: Ernest Benn Ltd., 1930), 22.

99. Lowell Sumner, "Your Stake in Alaska's Wildlife and Wilderness," *Sierra Club Bulletin* 41 (December 1956): 68.

100. William H. Hackett, *Alaska's Vanishing Frontier: A Progress Report*. House Subcommittee on Territories and Insular Possessions (Washington, D.C.: GPO, 1951), 1.

101. Sumner, "Your Stake in Alaska's Wildlife and Wilderness," 63–66.

102. See "Background Summary of Proposed Modification of Public Land Order No. 82," undated typescript, Legislative Bill File, 87th Congress, 1961–2, box 12, Bartlett Papers, UAF, 3.

103. Murie to Charles E. Wilson (Secretary of Defense), 20 May 1955. As quoted in the *Living Wilderness* 20 (Spring/Summer 1955): 22.

104. Murie to BLM (Anchorage), 17 December 1956), Military Land Withdrawals, box 13, Alaska Conservation Society Papers, UAF. (Hereafter referred to as ACS Papers.)

105. Wilderness Society, Annual Meeting, Moose, Wyoming, 17–21 August 1957, ACS Papers.

106. See Murie's report to the executive committee of the Wilderness Society, 27 June 1957, Arctic Wilderness File, George Marshall Papers, UCB.

107. TVSA, Arctic Wildlife Range Resolution, 14 May 1957, included as part of hearings, AWR, part 2, Fairbanks, 30 October 1959, 295–96. (Unless Part 1 is specified, subsequent references are to Part 2 of the hearings.) Forty-three TVSA members voted for the resolution, two abstained and five voted against it.

108. Lake to Ross L. Leffler, 18 October 1957, hearings, AWR, Fairbanks, supporting documents, 294.

109. Janzen to the BLM, 7 November 1957, BLM File (ANWR), box 13, ACS Papers.

110. For the "worthless lands" thesis—that only economically-valueless lands receive protection as wilderness and national parks—see Alfred Runte, National Parks: The American Experience (Lincoln: University of Nebraska Press, 1979), 48–64.

111. Fairbanks Chamber of Commerce to BLM (Anchorage), 30 January 1958, BLM File (ANWR), box 13, ACS Papers.

112. Bartlett to the Department of the Interior, 14 February 1958, Legislative Bill File, 87th Congress, 1961-2, box 12, Bartlett Papers, UAF.

113. See hearings, AWR, part 1, 61.

114. Living Wilderness 24 (Spring 1959): 29.

115. Bartlett to Clarence Anderson, 23 July 1959, General Subject File, 1959–63, box 63, Gruening Papers, UAF. Bartlett's subcommittee conducted hearings on the bill in Washington, D.C. on 30 June and in Ketchikan, Juneau, Anchorage, Seward, Cordova, Valdez, and Fairbanks on 20, 22, 24, 26–31 October 1959.

116. Hearings, AWR, part 1, 63, 8–9, 14.

117. See Celia Hunter's recollections in Maxine E. McCloskey, ed., Wilderness: The Edge of Knowledge (San Francisco: Sierra Club, 1970), 186–87. These details were confirmed during an interview with Celia Hunter, conducted by William S. Schneider and the author in Fairbanks, 15 August 1985. Those who testified in Fairbanks were Virginia (Ginny) and Morton Wood, Leslie and Teri Viereck, William O. Pruitt, Celia Hunter, John Thompson, Gerald Vogelsang, and Frederick Dean. Bartlett later remarked that the best testimony for the range was delivered in Fairbanks. (Bartlett to Frank Griffin, 23 February 1960, box 12, Bartlett Papers.)

118. Alaska Conservation Society News Bulletin 2 (Summer 1961), back cover.

119. Hearings, AWR, 127.

120. Ibid., 336, 407–8.

121. See editorial, Alaska Sportsman 25 (July 1959), 7.

122. Peter Bading, vice president of Jonas Brothers, taxidermists, hearings, AWR, 149–50; Rick Houston of Alaska Air Adventures, letter to the Anchorage Daily News, 8 December 1960.

123. Glenn DeSpain (TVSA), hearings, AWR, 291. For an assessment of the value of wildlife to Alaska in direct cash terms, see John Buckley, Wildlife in the Economy of Alaska, Alaska Cooperative Wildlife Research Unit, Biological Papers of the University of Alaska, no. 1, College, Alaska, 1955.

124. Hearings, AWR, 326.

125. Ibid., 351, 354. Thompson also referred to Lowell Sumner's writings.

126. Anderson to Gruening, 21 May 1959, box 63, Gruening Papers; hearings, *AWR*, Part 1, 62.

127. Gruening to Doug Bullard, 16 July 1959, General Subject File, 1959–63, box 63, Gruening Papers.

128. Harry R. Geron, hearings, *AWR*, 397.

129. Ibid., part 1, 49.

130. Ibid., 52.

131. Ibid., part 2, 413–14.

132. Ibid., 337.

133. Wolff to BLM (Anchorage), 5 March 1958, BLM File (ANWR), box 12, ACS Papers.

134. Hearings, *AWR*, 362, 364.

135. Ibid., 232, 229. Vogler had been one of those who thought FDR was a communist and who believed that the socioeconomic planning of the 1930s (based on the false and dangerous belief that there was no more physical frontier) had disastrously tried to replace individual opportunity, historically provided by the frontier, with social security. See John McPhee, *Coming Into The Country* (New York: Farrar, Straus and Giroux, 1976), 315.

136. Hearings, *AWR*, 183–84. Miners felt that the provisions in the bill guaranteeing access to minerals were meaningless since they precluded appropriation of surface title to claims and prohibited mechanical access by tracked vehicle and airplane.

137. Editorial, *Alaska Sportsman* 25 (July 1959): 7.

138. "Too Big? Too Soon?—No!" editorial, *Fairbanks Daily News-Miner*, 30 October 1959.

139. *Anchorage Daily Times*, 6 December 1959.

140. Murie to Bartlett, 15 April 1960, BLM File (ANWR), box 13, ACS Papers.

141. Alaska Sportsmen's Council, Special Bulletin, 27 April 1960, box 13, ACS Papers; Hearings, *AWR*, 92.

142. Bartlett to Murie, 13 August 1960, Legislative Bill File, 1961–62, box 12, Bartlett Papers.

143. One range advocate who lobbied Bartlett hard was Louis R. Huber, a resident of Seattle who was in the business of making nature films. In the early 1940s, Huber had been the author of a number of anticonservationist articles (see chapter 1). He was now a member of the Alaska Conservation Society. See his letters to Bartlett, dated 7, 10 and 18 June 1960, box 12, Bartlett Papers.

144. Bartlett, statement, 29 June 1960, Bartlett Papers.

145. *Congressional Record*, 87th Congress, 1st session, 107 (21 September 1961), 20601–3.

146. Gruening to Leslie Viereck, 4 November 1959, BLM File (ANWR), box 13, ACS Papers.

147. *Anchorage Daily Times*, 12 December 1960; *Fairbanks Daily News-Miner*, 8 December 1960.

148. Ernest Gruening, news release, 12 December 1960, box 13, ACS Papers.

149. Musk oxen were reintroduced into the northeast Arctic in 1969 from Canadian stocks and now number some 500 in this area. Like the buffalo, the musk ox was one of the few large animals to have survived the ice ages in North America. Interestingly, many early twentieth-century observers, such as Ernest Thompson Seton, thought it was related to the buffalo. See Barry Lopez,

Arctic Dreams: Imagination and Desire in a Northern Landscape (New York: Bantam, 1987), 49.

150. William O. Pruitt, "Animal Ecology and the Arctic National Wildlife Range," 12th Alaska Science Conference, College, Alaska, 28 August–1 September 1961, *Science in Alaska* (College: Alaska Division, American Association for the Advancement of Science, 1961), 44.

151. Murie to David Spencer, 6 January 1961, BLM Files (ANWR), box 13, ACS Papers.

152. Bartlett to Eugene D. Smith, 27 March 1961, box 11, Bartlett Papers.

153. Gruening to Egan, 27 June 1961, box 11, Bartlett Papers.

154. Bartlett to Egan, 29 June 1961, folder H. R. 3155, ANWR, box 11, Bartlett Papers. Accordingly, the Alaskan senators adopted new tactics in a bid to defeat the purpose of Seaton's order. In May 1961 Bartlett asked the Senate Appropriations Committee to withhold funds for the administration and management of ANWR and the Izembek National Wildlife Range. The Izembek National Wildlife Range (415,000 acres) had also been withdrawn as part of Public Land Order 2214. The Kuskokwim National Wildlife Range was also created by this order (1.8 million acres). For Bartlett's explanation of his action, see the *Congressional Record,* 87th Congress, 1st Session, 107 (21 September 1961), S1962–64. Throughout the 1960s, Gruening and Bartlett were successful in blocking appropriations for ANWR. In 1969, after Bartlett's death in 1968 and the defeat of Gruening's reelection bid, the Senate provided funds for the management of ANWR. See Claus-M. Naske, "The Arctic National Wildlife Range," in David L. Spencer, Claus-M. Naske, and John Carnahan, *National Wildlife Refuges of Alaska; An Historical Perspective* (Anchorage, Alaska: U.S. Fish and Wildlife Service and the Arctic Environmental Information and Data Center), 1979, 111.

155. Robert Marshall, *Arctic Wilderness* (Berkeley and Los Angeles: University of California Press, 1956), 68, 103.

156. Philip S. Smith and J. B. Mertie, *Geology and Mineral Resources of Northwest Alaska,* USGS Bulletin 815 (Washington, D.C.: GPO, 1930), 48.

157. Robert Marshall, *Arctic Village* (London: Jarrolds, 1934), 115.

158. National Resources Committee, *Alaska: Its Resources and Development,* House Document No. 485, 75th Congress, 3d Session 1938 (Washington, D.C.: GPO, 1938), 8.

159. Richard L. Neuberger, the future senator from Oregon, explored this shift from what he called the "breathtaking" to "resource protection" in a pre environmentalist era article, "How Much Conservation?" *Saturday Evening Post* 212 (15 June 1940): 12–13, 89–90, 92, 94–96. Neuberger, who had been a military aide to General O'Connor during the construction of the Alaska Highway, focused on the attempt during the 1930s to extend and expand the Olympic National Monument (1909) into a fully fledged Olympic National Park that would protect significant tracts of commercially valuable timber among its 187,000 additional acres.

160. Raymond Dasmann, *The Destruction of California* (New York: Collier, 1969), 73.

161. Stewart Brandborg, hearings, *AWR,* part 1, 50. Leslie Viereck, ibid., part 2, 407.

162. For a general discussion, with examples, of arguments for wilderness preservation based on the cultural and historical value of wilderness, see Roderick Nash, *Wilderness and the American Mind* (New Haven: Yale University

Press, 1982), 260–62. For many useful reflections on the importance of landscape as history, see Donald W. Meinig, ed., *The Interpretation of Ordinary Landscapes: Geographical Essays* (New York: Oxford University Press, 1979).

163. Lois Crisler, "Where Wilderness is Complete," *Living Wilderness* 22 (Spring 1957): 4.

164. *Alaska Sportsman* 27 (June 1961): 7.

CHAPTER 4. PROJECT CHARIOT, 1958–1963

1. Alaska Employment Security Commission, *Financing Alaska's Employment Security Program*, vol. 2, 1 October 1958, 19. As quoted in George W. Rogers, *The Future of Alaska: Economic Consequences of Statehood* (Baltimore: Johns Hopkins University Press, 1962), 94. For corroboration and further data see Brent R. Bowen, *Defense Spending in Alaska*, Institute of Social, Economic and Government Research, College, Alaska, July 1971.

2. Alaska Employment Security Commission figures, cited by Rogers, *The Future of Alaska*, 94.

3. Frederick Jackson Turner, "The Significance of the Frontier in American History," from *The Frontier in American History*, 1920 (Huntington, N.Y.: Robert E. Krieger, reprint, 1976), 4.

4. The phrase is Barry Commoner's. See his book *Science and Survival* (London: Victor Gollancz, Ltd, 1966), 15.

5. Project Plowshare was also referred to as the Plowshare Program and Operation Plowshare. The AEC's Division of Peaceful Nuclear Explosives assumed control over Plowshare when this division was established in 1961.

6. For a short, general discussion of the origins, purposes, and scope of Project Plowshare, see Richard T. Sylves, "U.S. Nuclear Exotica: Peaceful Use of Nuclear Explosives." This paper was delivered as part of a panel discussion on "Science, Technology, and Public Policy" at the Midwest Political Science Association's annual conference, April 1985. Useful for details of other Plowshare projects is Ralph Sanders, "Nuclear Dynamite: A New Dimension in Foreign Policy," *Orbis—A Quarterly Journal of World Affairs* 4 (Fall 1960): 307–22. Sanders, a political scientist, was an enthusiastic advocate of Plowshare.

7. AEC, Division of Peaceful Nuclear Explosives, "Plowshare Program, Fact Sheet, Project Chariot," revised, Washington, D.C., April 1962, 1–2, (hereafter referred to as "Fact Sheet.") The original "Fact Sheet" is undated. All subsequent references are to the revised copy, box 13 (AEC File, 1957–68), Don C. Foote Papers, University of Alaska, Fairbanks (UAF). (Hereafter referred to as Foote Papers.) The AEC Files in the Foote Papers (boxes 10, 11, 12, 13, and 19–26) are the most accessible source for the study of Project Chariot. Don Foote (who died in an automobile accident in Fairbanks in 1969) was a geographer who took part in the AEC's bioenvironmental studies for Chariot.

8. T. L. Péwé, D. M. Hopkins and A. H. Lachenbruch, *Engineering Geology Bearing on Harbor Site Selection Along the Northwest Coast of Alaska from Nome to Point Barrow*. USAEC Report TEI-668. USGS Open File Report, Washington, D.C., 14 October 1959. See also, R. Kachadoorian, "Engineering Geology of the Chariot Site," from John N. Wolfe and Norman J. Wilimovsky, eds., *The Environment of the Cape Thompson Region, Alaska*, Division of Technical Information, Springfield, Va., 1966, 86–95.

9. AEC, "Fact Sheet," 3. One kiloton is the equivalent of 1,000 tons of TNT.

10. See E. J. Longyear, *Report to the University of California Radiation Laboratory on the mineral potential and proposed harbor locations in Northwest Alaska*, Minneapolis, 18 April 1958, box 11 (13), Foote Papers. Longyear submitted its report to the AEC in August 1958 but it was not made public until October 1959. See Joseph Foote, "Report from Tigara" (undated), box 10 (16), Foote Papers.

11. AEC Press Release no. 131, San Francisco Operations Office, Oakland, Calif., 9 June 1958, box 11 (23), ibid.

12. *Juneau Daily Empire*, 15 July 1958.

13. *Fairbanks Daily News-Miner*, 15 July 1958.

14. Ibid., 24 July 1958. For the direct quotes, see "Project Chariot: Foote, Joe, Background Materials," brief, 1 March 1961, 12, Foote Papers, box 10 (13).

15. In Juneau, Teller asked for other suggestions as to how the AEC could assist the development of the young state. Projects suggested to him were a seven-mile canal across the Alaska Peninsula between Herendeen Bay and Balboa Bay to shorten the distance from Bristol Bay to the Pacific, a deep-water port on Norton Sound to serve Nome, the excavation of a shipping basin near Katalla on the southcentral coast and a dam on the Yukon River at Rampart Canyon. (See *Juneau Daily Empire*, 15 July 1958. This report did not identify who made these suggestions.) Though Teller gave the impression that the AEC was open to ideas, there is little indication that it seriously considered any earth-moving scheme in Alaska apart from Chariot.

16. *Fairbanks Daily News-Miner*, 24 July 1958.

17. See the transcript of an interview with Rogers conducted and transcribed by Ronald K. Inouye for the University of Alaska, Alaska and Polar Regions Department, Oral History Program, April 1986, 43–45. Rogers believed that Longyear's economic feasibility report on Chariot contained "a strong element of predetermination" and was naive because it failed to examine the demand for minerals from Cape Thompson and did not calculate the disadvantages of trying to compete with more favorably located regions of the world. See his review of the Longyear report, sent to Don Foote on 15 February 1961, 2, 4, 8, box 11 (21), Foote Papers.

18. The president of the University of Alaska, Ernest Patty, conveyed these concerns to the AEC. Patty to AEC, 7 October 1958, box 21 (Correspondence), Foote Papers.

19. Reed to the *Fairbanks Daily News-Miner*, letter dated 27 January 1959, printed 3 February. Reed had written his first letter criticizing Chariot the previous summer. He had suggested that St. Michael, near the mouth of the Yukon and ice-free almost five months of the year, would be a better spot for a harbor, ibid., 22 July 1958.

20. Ibid., 19 January 1959.

21. Johnson to the *Fairbanks Daily News-Miner*, letter dated 13 January, printed 23 January. Johnson had written his first critical letter to the *Fairbanks Daily News-Miner* the previous summer (30 July issue).

22. St. Louis was one of six American cities chosen in 1957 by the U.S. Public Health Service for tests to determine the amount of strontium-90 (Sr-90), a radioactive isotope, in milk supplies. Strontium-90 moves through the environment in concert with calcium, a chemically similar element. Calcium in soil

is a major source of plant food. CNI was formed to answer questions from citizens about the effects of Sr-90. For details, see Barry Commoner, *Science and Survival*, 101–9; *The Closing Circle: Confronting the Environmental Crisis* (London: Jonathan Cape, 1972), 55–56.

23. For a brief discussion of nuclear engineering as a threat to the negotiations for a ban on weapons tests, see Frederick Reines, "The Peaceful Nuclear Explosion," *Bulletin of the Atomic Scientists* 15 (March 1959): 123.

24. *New York Times*, 12 September 1958.

25. As quoted in *Bulletin of the Atomic Scientists* 15 (January 1959): 47.

26. U.S. Congress, Joint Committee on Atomic Energy, hearings, *Development, Growth, and State of the Atomic Energy Industry*, 86th Congress, 1st session, Washington, D.C., 17 February 1959 (Washington, D.C.: GPO, 1959), 13, 28. McCone gave the impression that a specific site at Cape Thompson had not yet been selected.

27. Reuben Kachadoorian, R. H. Campbell, C. L. Sainsbury, and D. W. Scholl, *Geology of the Ogotoruk Creek Area, Northwestern Alaska*, Report TEM-976, 1 November 1958. (Marked "Official Use Only.") This report, the product of field studies carried out in the summer of 1958, was made public on 14 October 1959. The AEC's "Fact Sheet" (1962) referred to this report and conceded the doubtful mineral potential of the Cape Thompson region (3). The geological survey also discussed geological and meterological factors to which critics had drawn attention. However, it did not consider them evidence that the site was physically unsuitable for the blast. Critics stressed the problems posed by high and unpredictable winds, silting, slumping, and longshore drift, which, aggravated by the heavy storms characteristic of the area, would, they claimed, soon seal off a harbor's mouth. See Don Foote, History Notes (C), box 11 (19), 15, Foote Papers. The AEC's consultant geologists described the mouth of Ogotoruk Creek as topographically and geologically "well-suited" for a harbor and claimed that "little maintenance" would be required. They considered slumping (from thawed ground) a minor problem: "it is doubtful that large-scale slumping into the excavation will occur as a result of the blasting and subsequent thawing of the permafrost." Likewise, they recorded that the effects of silting from suspended material carried by Ogotoruk Creek would be "negligible." The geologists did report that longshore transportation of sand and gravel by currents "will tend to form spits or bars across the harbor entrance," but did not view this as a serious problem; "only minor intermittent dredging would be required to keep the entrance open." See Kachadoorian et. al., *Geology of the Ogotoruk Creek Area*, 5–6.

28. AEC, "Fact Sheet," 3.

29. In December 1959 the AEC reduced the blast further to its final size of 280 kilotons; one 200 kiloton bomb and four 20 kiloton bombs. The small ones (400 feet underground) would create a channel 900 feet wide, 200 feet deep and 1,900 feet long. The large one (800 feet underground) would produce a turning basin 400 feet deep and with a diameter of 1,800 feet. According to the AEC, this further modification was necessary for technical and financial reasons. AEC to Bartlett, 13 January 1961, 3, box 11 (23), Foote Papers. See also AEC, "Fact Sheet," 3–4.

30. C. L. Anderson (director of the Alaska State Department of Fish and Game), to John N. Wolfe, 15 June 1959, Case Files, Cape Thompson, Ralph J. Rivers Papers, University of Alaska, Fairbanks (UAF). (Hereafter referred to as Rivers Papers.)

31. Press Release, Governor's Office, Juneau, 20 August 1959, box 10 (9), Foote Papers.

32. Press Release, Governor's Office, 27 August 1959, box 10 (9), Foote Papers.

33. *Anchorage Daily Times*, 29 August 1959. Finally, in November 1959, the Alaska Legislature appointed the Commissioner of Natural Resources, Phil R. Holdsworth, to head the state's Atomic Energy Board as Coordinator of Atomic Development in Alaska. This body, set up in March, had been dormant until now.

34. Bartlett, *Newsletter,* 20 February 1959, box 10 (9), Foote Papers.

35. AEC, "Fact Sheet," 2. The BLM's announcement of the withdrawal appeared in the *Federal Register* 23 (5 September 1958) and was also publicized by press release to the *Fairbanks Daily News-Miner, Jessen's Weekly,* the two Anchorage papers, and various radio and television stations. In addition, it was posted in certain post offices around Alaska. Point Hope's post office was not one of them. The full size of the withdrawal was announced in the *Federal Register* 24 (30 April 1959), 3385, box 10 (2), Foote Papers.

36. Bartlett to Richard Cooley, 28 July 1959, Rivers Papers.

37. *Fairbanks Daily News-Miner,* 10 January 1959.

38. See Don Foote, "Cape Thompson: The People," from Committee on Nuclear Information (CNI), *Project Chariot: A complete report on the probable gains and risks of the AEC's Plowshare Project in Alaska, Nuclear Information* 3 (4–7), St. Louis, 3 June 1961, 14.

39. Committee for the Study of Atomic Testing in Alaska to AEC (via Rivers), 18 February 1959, Rivers Papers.

40. Luedecke to Rivers, 27 March 1959, Rivers Papers.

41. The AEC claimed that this study was an integral part of the original proposal for Project Chariot. Critics believed that the AEC was responding to pressure.

42. Atomic Energy Commission, San Francisco Operations Office, *Chariot Environmental Program,* Oakland, Calif., July 1959, 2–3. Copy courtesy of the Lawrence Livermore National Laboratory Archives, Livermore, Calif.

43. The only Alaskan member of the eight-member Committee on Environmental Studies was Robert L. Rausch of the Arctic Health Research Center, U.S. Public Health Service, Anchorage. Rausch was on the Alaska Conservation Society's board of directors.

44. Other participating institutions were the U.S. Public Health Service, General Electric's Hanford (Washington) Laboratories, the University of Washington's Laboratory of Radiation Biology and Department of Oceanography, and the Environmental Radiation Division of the University of Caliifornia, Los Angeles.

45. Libby to Gruening, 13 March 1959, box 20 (60), Foote Papers. Libby was a veteran of the Manhattan Project and the only AEC Commissioner who was a scientist.

46. McIntyre to McCone, 11 April 1959, box 25 (3), Foote Papers.

47. Due to ill-health, Murie did not play a very active role in the fight against Chariot. He died in 1963.

48. University of Alaska, Fairbanks, 18 May 1959. AEC Press Release. Transcribed excerpts from a tape recording of this speech are contained in box 12 (51) of the Foote Papers. For the original recording, see box 59 (18) (Audio-Visual), ibid. One of Teller's favorite quips was, "If your mountain is not in

the right place, just drop us a card." As quoted in *Anchorage Daily News,* 27 June 1959.

49. Teller, commencement address, University of Alaska, Fairbanks, 18 May 1959. According to George Rogers, Teller talked about turning Eskimos into coal miners during their meeting in July 1958 (Inouye interview, 44).

50. *Nome Nugget,* 29 June 1959.

51. *Anchorage Daily Times,* 26 June 1959.

52. Ibid., 10 February 1959.

53. *Fairbanks Daily News-Miner,* 15 August 1959.

54. *Fairbanks Daily News-Miner,* 23 and 20 August 1959. See also the 18, 22, and 24 August editions of the *News-Miner* for the rest of the series.

55. Committee on Environmental Studies, AEC-SAN, *Statement,* 7 January 1960, 2, box 25 (2), Foote Papers.

56. Don Foote, History Notes (C), 37, box 11 (19), Foote Papers.

57. *Jessen's Weekly,* 18 August 1960.

58. *New York Times,* 17 August 1960.

59. See for example, *New York Times,* 17 and 18 August 1960.

60. According to the AEC, it had intended to send representatives to the coastal villages in the vicinity of the blast site in the summer of 1959. It explained that this visit had to be postponed because many villagers were absent on hunting trips at this time of year. AEC, San Francisco Operations Office, news release, Oakland, Calif., 4 March 1960. Courtesy of Lawrence Livermore National Laboratory Archives.

61. See the transcript of a recording of the meeting on 14 March made by the Reverend Keith Lawton, the Episcopalian missionary at Point Hope, 9, box 12 (29), Foote Papers.

62. Ibid., 22–23, 21. The third AEC representative was Robert Rausch.

63. The "Ranger" tests at Frenchman's Flat in 1951 had broken many storefront windows in downtown Las Vegas.

64. Lawton tape, transcript, 25, 35.

65. E. C. Shute, Manager, AEC-SAN, to David Frankson, President, Point Hope Village Council, 11 June 1960, box 25 (1), Foote Papers.

66. Wilderness Society, Governing Council Resolution, annual meeting, Pine Creek Camp, Salmon, Idaho, 31 August 1960. As reported in *Living Wilderness* 25 (Autumn-Winter 1960–1): 43.

67. AEC, TID-12439, *Bioenvironmental Features of the Ogotoruk Creek Area, Cape Thompson, Alaska: A First Summary by the Committee on Environmental Studies for Project Chariot,* 20 December 1960, 52–53, 55–56, 57. (The report was not available to the public until May 1961.) See also AEC, "Fact Sheet," 5.

68. *Fairbanks Daily News-Miner,* 5 December 1960.

69. Viereck to Crisler, 27 August 1960, Alaska Wilderness File, George Marshall Papers.

70. Viereck to Wood, 29 December 1960, 1, box 22 (155), Foote Papers. According to Richard Cooley, Wood dismissed Alaskan opponents of Chariot as "crackpots with ulterior motives." Cooley to Don Foote, 19 April 1961, box 19 (34) (Correspondence), Foote Papers.

71. Viereck to Wood, 29 December 1960, 2, 3, ibid. The ethics of research and political control of scientific "truth" are topics that lie beyond the scope of this chapter. In May 1961 Viereck's contract with the University of Alaska was not renewed. According to a report in *Time* (13 September 1963, 63), Pruitt

was fired from his job at the University of Alaska following his protest that the blast might harm local Natives. In a letter to the AEC, he charged that the final report of his findings as a member of the environmental research team was modified by Brina Kessel, the head of the university's Zoology Department, prior to submission to the AEC. (Kessel was in charge of the university's contract with the AEC.) Pruitt argued that in this way his three years' work was "diluted and aborted." (Pruitt, "Statement to Project Chariot Environmental Committee and All Concerned," 25 April 1962, Foote Papers, box 25.) Furthermore, the American Civil Liberties Union, investigating an alleged breach of academic freedom, questioned University of Alaska president Wood about accusations that his intervention eventually led to the denial of a one year appointment of Pruitt to the faculty of Montana State University. (See William B. Stern, "ACLU Probes Pruitt Case, *Polar Star,* 18 October 1963. Accompanying this feature was a copy of a letter from Louis M. Hacker, chairman of the ACLU's Academic Freedom Committee, to Wood.) However, the ACLU was unable to obtain sufficient evidence to proceed with legal action. Albert Johnson later left the university due to dissatisfaction with Kessel and Wood. (Albert Johnson, interview by Dan O'Neill, San Diego, 11 April 1988, transcript, 41–42.

72. Don Foote, History Notes (C), 25, box 11 (19), Foote Papers.

73. Foote to Arthur H. Lachenbruch, 25 November 1960, box 21 (84) (Correspondence), Foote Papers. Lachenbruch was chief of the Alaska branch of the USGS.

74. Foote to Wolfe, 11 March 1961, box 25, Foote Papers. According to Foote, his "affiliation" with the AEC ended on 31 May 1961. However, he remained at Point Hope until the end of August 1962. In the fall of 1962, he taught a graduate seminar in geography at McGill University entitled "Project Chariot: A Unique Incident in Arctic History?"

75. Arthur Grahame, "A-Test Alaska Threat?" *Outdoor Life* 127 (January 1961): 10–11, 37; *National Wildlands News* 2 (February 1961): 2, 5; *Defenders of Wildlife News* 36 (Spring 1961): 2; "Project Chariot—The Long Look," *Sierra Club Bulletin* 46 (May 1961): 5–9, 12–13.

76. CNI, *Project Chariot: A complete report on the probable gains and risks of the AEC's Plowshare project in Alaska, Nuclear Information,* (3) (Nos. 4–7), St. Louis, 3 June 1961. A number of these reports (by Foote, Viereck and Pruitt) had appeared in the ACS's Spring 1961 *Bulletin.* For a review of the report, see *Science* 133 (23 June 1961): 2000–2001.

77. Commoner, "Biological Risks from Project Chariot," *Project Chariot: a complete report,* 9–12.

78. Friedlander, "Predictions of Fallout from Project Chariot," ibid., 8.

79. *Fairbanks Daily News-Miner,* 19, 21, 22, and 23 August 1961. Snapp's probing series did not indicate an editorial shift. As late as 4 January 1962, C. W. Snedden remained a staunch supporter, insisting that abundant supplies of coal, iron, and tin were waiting for a decent harbor. Snedden also fondly recalled Teller's vision of mining jobs for local Natives.

80. As quoted in Paul Brooks and Joseph Foote, "The Disturbing Story of Project Chariot," *Harper's* 224 (April 1962): 61. William O. Pruitt recalled: "Every village has several tape recorders. . . . There is a constant traffic in tapes from one village to another. . . . The Chariot information and the material contained in the Alaska Conservation Society and CNI bulletins got into the tapes, and literally swept the Arctic coast from Kakhtovik all the way down to Nome and below. I recall, also, meeting an Eskimo driving a dog team on

the trail one time, and by golly, he had a copy of the CNI bulletin tucked inside his parka." Presentation, National Conference for Scientific Information, Scientists' Institute for Public Information, New York, 16 February 1963. As quoted in Commoner, *Science and Survival,* 107.

81. *Fairbanks Daily News-Miner,* 18 November 1961. The Association on American Indian Affairs (AAIA) sponsored the conference, which the assistant Secretary of the Interior, John Carver, and Jim Hawkins, Alaska director of the Bureau of Indian Affairs, attended. The other event that precipitated the *Inupiat Paitot* was the U.S. Fish and Wildlife Service's sudden enforcement of a seasonal ban on Native hunting of migratory ducks. These regulations (1916 Migratory Bird Treaty) were usually ignored by Native hunters and federal officials alike. The arrest of a Native hunter led to civil disobedience in the shape of the so-called Barrow "Duck-in."

82. Rock became editor and publisher. Henry S. Forbes, a wealthy Massachusetts physician, who was in charge of the Alaska policy committee of the New York City-based Association on American Indian Affairs, provided the financial backing. For Rock's recollection of these events, see Ronald K. Inouye's transcript (October 1985) of an interview with Rock by Levi Lott, 3–4. The first issue of *Tundra Times* appeared in the fall of 1962. For discussion of Chariot, see issues 2 (15 October 1962), and 3 (1 November 1962). For an account by two journalists of Chariot's role in the establishment of *Tundra Times,* see Patrick Daley and Beverly James, "An Authentic Voice in the Technocratic Wilderness: Alaskan Natives and the *Tundra Times,*" *Journal of Communication* 36 (Summer 1986): 10–30. Also useful is Lael Morgan, *Art and Eskimo Power: The Life and Times of Alaskan Howard Rock* (Fairbanks, Alaska: Epicenter Press, 1988), 182–200.

83. Brooks and Foote, "The Disturbing Story of Project Chariot," *Harper's* 224 (April 1962): 60–67. Brooks was editor-in-chief at Houghton Mifflin of Boston. Joseph Foote was a lawyer and journalist in Rhode Island.

84. AEC-SAN Press Release, Oakland, Calif., 24 August 1962. Foote Papers, AEC File, box 25 (183). The second summary report of the Committee on Environmental Studies repeated that the chance of damage to plants, animals, and people "appears exceedingly remote." Committee on Environmental Studies for Project Chariot, Plowshare Program (Bette Weichold, ed.), *Bioenvironmental Features of the Ogotoruk Creek Area, Cape Thompson, Alaska (Including Assessments of the Effects of Nuclear Detonations Proposed for an Excavation Experiment): A Second Summary Report,* TID-17226, Washington, D.C., October 1962, v. The report is prefaced (Guidelines, iii) by a lengthy quotation from Thomas Jefferson's instructions of a scientific nature to Captain Meriwether Lewis prior to the Lewis and Clark Expedition (1803–6).

85. A. R. Luedecke to Ralph Rivers, 22 August 1962, Rivers Papers; *New York Times,* 25 August 1962. One LRL official had recommended that Chariot be cancelled in April 1962. See John S. Foster to John S. Kelly, 30 April 1962. Copy courtesy of Lawrence Livermore National Laboratory Archives.

86. For a similar opinion, see "Eskimos Win Fight With Science," *Washington Post,* 25 August 1962.

87. Sharon Francis to Don Foote, 1 February and 8 May 1962; Francis to Foote, 29 August 1962, box 22 (3), Foote Papers.

88. Don Foote to Howard Rock, 12 November 1962, box 22 (122), Foote Papers.

89. Don Foote to Howard Rock, 5 March 1963, Ibid.

90. John S. Kelly to Rivers, 2 May 1963, Rivers Papers. Kelly was the director of the AEC's Division of Peaceful Nuclear Explosives.

91. *Newsletter,* 20 February 1959, box 10 (9), Foote Papers.

92. Samuel P. Hays, *Conservation and the Gospel of Efficiency: The Progressive Conservation Movement, 1890–1920* (Cambridge: Harvard University Press, 1959), chap. 1, 1. Hays makes the same point in "From Conservation to Environment: Environmental Politics in the United States since the Second World War," in Kendall E. Bailes, ed., *Environmental History: Critical Issues in Comparative Perspective* (Lanham, Md.: University Press of America, 1985), 202. Conservation in general, dominated as it was by an "elite"—usually professional and expert—whether located inside federal agencies or in private groups, never constituted or produced a "movement" in the way environmentalism later did.

93. Hunter to Don Foote, 27 November 1961, Correspondence, box 21 (71), Foote Papers. Hunter was secretary of the Alaska Conservation Society.

94. In 1956 James Marshall's wife, Lenore Marshall, cofounded the National Committee for a Sane Nuclear Policy (SANE) with Norman Cousins, editor of *Saturday Review of Literature.*

95. George Marshall to Richard Leonard, 29 April 1960. Marshall wrote to Leonard, the secretary of the Sierra Club, urging a resolution against Chariot. Alaska Wilderness File, George Marshall Papers, UCB.

96. Viereck to Lois Crisler, 27 August 1960, 3, Alaska Wilderness File, George Marshall Papers.

97. Viereck to Crisler, 27 August 1960, Alaska Wilderness File, George Marshall Papers, UCB. For details of damage to terrain and subsequent melting of the permafrost, see Reuben Kachadoorian, "Engineering Geology of the Chariot Site," Wolfe and Wilimovsky, eds., *The Environment of the Cape Thompson Region,* 94–95.

98. Leonard, memorandum to Brower, 17 April 1961, Alaska Wilderness File, George Marshall Papers.

99. See Donald Worster, *Nature's Economy: A History of Ecological Ideas* (Cambridge: Cambridge University Press, reprint, 1985), 340.

100. Thomas R. Dunlap, *DDT: Scientists, Citizens, and Public Policy* (Princeton: Princeton University Press, 1981), 102–4; Ralph H. Lutts, "Chemical Fallout: Rachel Carson's 'Silent Spring,' Radioactive Fallout, and the Environmental Movement," *Environmental Review* 9 (Fall 1985): 211–25.

101. Interview with Barry Commoner, conducted by Dan O'Neill, New York City, 27 April 1988. However, in his book *The Closing Circle* (1972), Commoner traced his ecological awakening further back to 1953, when Troy, New York, was showered with rain containing fallout from AEC tests in Nevada (49–65).

102. Viereck to Wood, 29 December 1960, 3, box 22 (155), Foote Papers.

103. *Juneau Daily Empire,* 17 March 1960.

104. *Outdoor Life,* January 1961, 137.

105. John C. Reed, review of *Environment of the Cape Thompson Region, Alaska, Science* 154 (October 1966): 372.

106. For the final report, see John N. Wolfe and Norman J. Wilimovsky, eds., *Environment of the Cape Thompson Region, Alaska,* 1966, especially chaps. 14, "Vegetation and Flora" by Johnson and Viereck, 277–355; 20, "Ecology of Terrestrial Mammals," by Pruitt, 519–64; 23, "Sea-Cliff Birds," by Swartz, 611–79; and 36, "A Human Geographical Study," by Foote, 1041–1107. For a praiseworthy review of the 1266-page report by John C. Reed, see *Science* 154 (October 1966): 372.

107. Committee on Environmental Studies for Project Chariot, *Second Summary Report*, October 1962, 3, 165. This statement was repeated in the final report (1966).

108. Brooks and Foote, "The Disturbing Story of Project Chariot," 67.

109. Don Foote to Joseph Foote, 20 December 1961, box 20 (Correspondence), Foote Papers.

110. James Marshall to Harvey Broome, 17 March 1960, Arctic Wilderness File, George Marshall Papers.

111. Undated, unidentified news clipping, Rivers Papers.

112. It is worth noting that although Nevada was eventually chosen as the site for the first nuclear tests on American soil, the AEC and the Department of Defense initially recommended (October 1950) Amchitka. See the official history of the AEC by Richard G. Hewlett and Francis Duncan, *Atomic Shield: 1947–1952: Volume II, A History of the United States Atomic Energy Commission* (University Park: Pennsylvania State University Press, 1969), 535.

113. See R. Glen Fuller and Melvin L. Merritt, eds., TID-26712, *The Environment of Amchitka Island, Alaska*, Division of Military Applications, Energy Research and Development Administration (ERDA) (Washington, D.C.: GPO, 1977), iii-iv, ix, 1, 3, 6, 647–48. The AEC had been disbanded in 1975 and its functions were divided between the Nuclear Regulatory Commission and ERDA.

114. *Fairbanks Daily News-Miner*, 10 June 1987.

115. Speech (question and answer session), Commonwealth North, Anchorage, 9 June 1987. Tape recording, courtesy of Christopher Toal, director, SANE/Alaska.

CHAPTER 5. RAMPART DAM, 1959–1967

1. Alaska was not a traditional "reclamation state." In other words, it was not covered by the Reclamation Act of 1902, which authorized the Reclamation Service (renamed Bureau of Reclamation in 1907) to operate in the western states. The bureau added hydro development to its responsibilities for irrigation and land reclamation in 1939.

2. U.S. Department of the Interior, Bureau of Reclamation (Michael W. Straus and Joseph M. Morgan), *Alaska: A Reconnaissance Report on the Potential Development of Water Resources in the Territory of Alaska, for Irrigation, Power Production and Other Beneficial Uses*, 2 May 1949, House Document 197, 82d Congress, 2d session (Washington, D.C.: GPO, January 1952).

3. By 1960 the corps had spent over $1 million locating 169 sites and issuing seven reports. See U.S. Army Corps of Engineers, U.S. Army Engineer District, Alaska, *Yukon and Kuskokwim River Basins, Harbors and Rivers in Alaska Survey Report*, Interim Report No. 7, Anchorage, 1 December 1959 (revised 1 June 1962), 113–17.

4. Rampart's dimensions were 4,700 feet by 530 feet. Coulee was 4,173 feet long and 550 feet high. Rampart's capacity was 5.04 million kilowatts compared to Coulee's 1.944 million. For these and other statistics, see Ivan Bloch and Associates (Industrial Consultants), *Background Facts on Rampart Canyon Project, Yukon River, Alaska*, Portland, Oregon, October 1963, 1, Public Works

File, Rampart Dam. Ernest Gruening Papers, University of Alaska Archives, Fairbanks (UAF). (Hereafter referred to as Gruening Papers.)

5. U.S. Congress, Senate, Committee on Interior and Insular Affairs, Subcommittee on Irrigation and Reclamation, hearings, *Eklutna Project, Alaska, S. 3222, A Bill to Authorize Modification and Reconstruction,* 11 May 1960, 86th Congress, 2d session (Washington, D.C.: GPO, 1960).

6. *Tundra Times,* 4 November 1963. See also Gruening's autobiography. *Many Battles: The Autobiography of Ernest Gruening* (New York: Liveright, 1973), 496.

7. U.S. Congress, Senate, Committee on Interior and Insular Affairs, Subcommittee on Irrigation and Reclamation, hearings, *Hydroelectric Power Requirements and Resources in Alaska,* 7, 13, 15 September 1960, 86th Congress, 2d session (Washington, D.C.: GPO, 1961). (Hereafter cited as hearings, *Hydro Requirements.*) The hearings were held in Anchorage, Fairbanks, and Juneau on 7, 13, and 15 September 1960. Gruening, not an official member of the Subcommittee on Irrigation and Reclamation, stood in as chairman for Clinton P. Anderson and was the sole senator in attendance.

8. Don Eyinck, chairman, Resources and Industrial Development Committee, Fairbanks Chamber of Commerce. Eyinck also thought the lake would be an ideal site for submarines and for launching intercontinental ballistic missiles. Ibid., 62.

9. Ibid. (Juneau, 15 September 1960), 151. Bloch served as chief of the Bonneville Power Administration's Industrial and Resources Administration from 1938 until 1947.

10. Ibid., xvi–xvii.

11. Ibid., vii. Scientists (especially oceanographers) denied that there would be a noticeable change in temperature. See Bruce W. McAlister and Wayne V. Burt, "Predicted Water Temperatures for the Rampart Dam Reservoir on the Yukon River," proceedings, 14th Alaska Science Conference, Anchorage, 27–30 August 1963, *Science in Alaska,* edited by George Dahlgren (College: Alaska Branch, American Association for the Advancement of Science, 1964), 249–65.

12. Hearings, *Hydro Requirements, xix.*

13. The Fish and Wildlife Coordination Act of 1934 authorized the U.S. Fish and Wildlife Service (then the Bureau of Biological Survey) to prepare reports for the Bureau of Reclamation and the Corps of Engineers that assessed the impact of their proposed projects on fish and wildlife. These reports were to include suggestions for mitigating any damage that might occur. Though purely advisory, they were to be submitted to Congress as an integral part of any feasibility report. (The Fish and Wildlife Service was created in 1940 when the Bureau of Commercial Fisheries [Department of Commerce] and the Biological Survey [Department of Agriculture] were transferred to Interior.)

14. See U.S. Army Corps of Engineers, *Proceedings, Conferences of the Rampart Economic Advisory Board,* Anchorage, 27 May 1961. See also the *Proceedings* of the second (10–12 January 1962) and third conferences (30–31 March 1962). These were attended by REAB and representatives of the DRC, the Corps of Engineers, the USFWS (represented by Gordon Watson), and Alaska's congressional delegation.

15. Development and Resources Corporation (Gordon R. Clapp), *The Market for Rampart Power: Working Papers pertaining to the report to the U.S. Army Corps of Engineers, Alaska District,* New York, 1962. Soon afterward, the

Senate Committee on Public Works published the report: *The Market for Rampart Power, Yukon River, Alaska,* (Development and Resources Corporation), 87th Congress, 2d session (Washington, D.C.: GPO, 23 April 1962). Subsequent references are to the committee print.
    16. Ibid., 9, 20–24, 79–96.
    17. Ibid., 96, 100, 32.
    18. Ibid., 10, 74–77, 97, 124–29.
    19. Ibid., 102–3.
    20. Ibid., 16–17. The statistics for this national growth were based on the U.S. Federal Power Commission's figures for 1960.
    21. Ibid., 41.
    22. Ibid., 40, 29.
    23. George Sundborg, *Hail Columbia: The Thirty Year Struggle for Grand Coulee Dam* (New York: Macmillan, 1954), 17, 9.
    24. The term "puffing" is taken from Ray Allen Billington, *Land of Savagery, Land of Promise: The European Image of the American Frontier in the Nineteenth Century* (New York: W. W. Norton, 1981), 215.
    25. George Sundborg, "The 'Biggest Dam' on the Mighty Yukon," *Rural Electrification* 17 (July 1959): 15. This was included as part of the hearings record, *Hydro Requirements,* appendix, 195–96.
    26. Gruening, "A Rampart on the Yukon," (speech, rough draft, undated), Gruening Papers. The verse quoted and the poetic extracts from Gruening's speech can be located in the *Collected Verse of Robert Service* (London: Ernest Benn, 1930), 8, 12, 13.
    27. Ernest Gruening, report to the Senate Public Works Appropriations Subcommittee, (rough draft), 9 May 1959. Public Works, Rampart Dam, Gruening Papers.
    28. John F. Kennedy, address, Alaska State Fair, Palmer, 3 September 1960. "Résumé of Findings," hearings, *Hydro Requirements,* xix. Eisenhower's Interior Secretary, Fred Seaton, retorted "that's a hell of a way to get a dam" (*Anchorage Daily Times,* 7 September 1960). The Russian dam that Kennedy referred to was at Bratsk on the Angara River. When completed in 1966, this was the world's largest. Bratsk has since lost this distinction to the Krasnoyarsk project on the Yenisey (1971). Both these dams are at a considerably lower latitude than Rampart would have been.
    29. Kennedy, Speech, Edgewater Hotel, Anchorage, 3 September 1960. As reported in the *Anchorage Daily Times* of that date.
    30. FDR, Commonwealth Club, San Francisco, Calif., 23 September 1932. See Samuel I. Rosenman, ed., *The Public Papers and Addresses of FDR,* Vol. 1, *The Genesis of the New Deal, 1928–32* (New York: Random House, 1938), 750.
    31. Hearings, *Hydro Requirements,* 10–11.
    32. The authors of a political and administrative case study of Rampart and Devil Canyon stress this interagency conflict between the Bureau of Reclamation and the Corps of Engineers. See Claus-M. Naske and William R. Hunt, *The Politics of Hydroelectric Power in Alaska: Rampart and Devil Canyon—A Case Study,* Institute of Water Resources, University of Alaska, Fairbanks, 1978, 6.
    33. Dominy, hearings, *Hydro Requirements,* 10, 66–67.
    34. Ibid., 154.
    35. Floyd E. Dominy, Report to the Secretary of the Interior, Bureau of

Reclamation, Department of the Interior, *Devil Canyon Project, Alaska*, Washington, D.C., 6 March 1961.
36. Furthermore, flooding here would not involve the loss of any archaeological or historical sites. For the ACS's views on the relative merits of Rampart and Devil Canyon, see Robert B. Weeden, "Conservation and Kilowatts," *Alaska Conservation Society News Bulletin* 2 (May 1961): 3–5. Weeden, a biologist with the Alaska Department of Fish and Game, was editor of the *News Bulletin*.
37. *Anchorage Daily News*, 17 January 1961. The *Daily News* approached Devil Canyon as a supplement to Rampart, not a long-term alternative.
38. Joe Vogler to William Wood, 12 June 1961, Gruening Papers. The subject of Vogler's letter was the ACS article, "Conservation and Kilowatts." (*News Bulletin*, May 1961), comparing Devil Canyon and Rampart. Naske and Hunt argue that Rampart's size was one of its major attractions (*The Politics of Hydroelectric Power*, 25). They also argue that Rampart advocacy thwarted the Devil Canyon proposal at a time when the latter's authorization could have been possible (ibid., 39). They emphasize that Alaskans did not have much of an opportunity to compare the two projects. The Bureau of Reclamation could not advance its case for Devil Canyon as Rampart proponents monopolized the field (ibid., p. 10).
39. Vogler to Wood, 5 January 1961, Gruening Papers.
40. See the report on the ASC's annual meeting in Juneau, *Alaska Conservation Society News Bulletin* 2 (January 1961): 6.
41. *Alaska Sportsman* 27 (July 1961): 7.
42. Garrison Dam (1945) had displaced 1,544 members of the Three Tribes of the Fort Berthold Reservation. See Arthur E. Morgan, *Dams And Other Disasters* (Boston: Porter Sargent, 1971); Michael J. Lawson, *Dammed Indians, The Pick Sloan Plan and the Missouri River Sioux, 1944–80* (Norman: University of Oklahoma Press, 1982).
43. Joint Resolution No. 11, 2d Legislature, 1st session, 16 February 1961. The amendment was passed on 24 February by a margin of 20-18. See Jay Hammond, "Other Side of Rampart," *Anchorage Daily Times*, 11 November 1961.
44. This agreement was not unprecedented. In 1949 the rivals signed the Newell-Weaver agreement that divided dam sites on the Columbia.
45. Gruening to Judge Harry O. Arend, 12 June 1961, Gruening Papers. Arend was an associate justice of the Supreme Court of Alaska.
46. Sharrock to Austin Ward of the Fairbanks Chamber of Commerce, 21 August 1963, series 1, box 33, Fairbanks Chamber of Commerce Collection, UAF. (Hereafter, FCCC.)
47. "Railbelt" is a term for the corridor along the Alaska Railroad between the state's two premier cities, Anchorage and Fairbanks, where over two-thirds of Alaska's total population is located.
48. The conference was held in the McKinley Park Hotel, Mount McKinley National Park, 7–8 September 1963. It is clear from Sharrock's memorandum to "Participants in the Rampart Dam Conference" (28 August 1963) that Bartlett was among "persons or organizations expected to participate." Series 1, box 33, FCCC. Bartlett's absence caused surprise and some annoyance in the Rampart inner circle.
49. Yukon Power for America (YPA), *Addresses Presented at Rampart Dam Conference*, Fairbanks, 1963, 61, 55.
50. Ibid., 39–41.

51. Udall, who had no intention of making an official pronouncement on the project before his department's studies were published, was in Nairobi, Kenya, attending the 8th Assembly of the International Union of the Conservation of Nature and Natural Resources.

52. YPA, *Addresses Presented,* 45.

53. Ibid., 51. For Gabrielson's comments on Rampart, see James B. Trefethen, ed., *Transactions,* 28th North American Wildlife and Natural Resources Conference, Detroit, 4–6 March 1963 (Washington, D.C.: Wildlife Management Institute, 1964), 25.

54. YPA, *Addresses Presented,* 52.

55. *Anchorage Daily News,* 23 October 1963.

56. *Anchorage Daily News,* 17 October 1963. In Anchorage, the grant from the city was severely criticized for similar reasons by the Alaska Methodist University's Young Republican Club (10 October 1963). The Golden Valley Electrical Association of Fairbanks made a $5,000 contribution.

57. Series 1, box 33, FCCC.

58. Bartlett to Ed Merdes, 27 November 1963, Gruening Papers.

59. Sundborg to Howard Zahniser (the editor of *Living Wilderness*), 4 October 1963, Gruening Papers.

60. Ibid. For a full statement of this case, including the project's recreational value, see YPA, *The Rampart Canyon Project,* Fairbanks, 3 September 1964 (fisheries, 17; Natives, 14–15; industry, 5–6). YPA envisaged boating, sailing, marinas, hunting lodges, and float plane facilities—all accessible by road, rail, and air. See also, YPA, *The Rampart Story,* undated brochure, which estimated the number of year-round jobs that would be created at 57,000. Series 1, box 33, FCCC.

61. *Anchorage Daily News,* 13 September 1963.

62. Arthur D. Little, Inc. (Cambridge, Mass.), *Industrial Opportunities in Alaska: A Summary of Reports to the State of Alaska,* State of Alaska Division of Industrial Development, Department of Economic Development and Planning, Juneau, September 1962. The report in question was "Potential for Use of Alaska's Energy Resources."

63. KFAR-TV, Fairbanks, 25 September 1963, box 6 (Tape 6), Alaska Conservation Society Papers Papers, University of Alaska. Fairbanks (UAF). (Hereafter, ACS Papers.)

64. Egan to Johnson, 8 December 1963, Gruening Papers.

65. *Alaska Conservation Society News Bulletin* 4 (February 1963): 8–9.

66. Richard Starnes, "The Rampart We Watch," *Field and Stream* 68 (August 1963): 12, 64–66.

67. *Fairbanks Daily News-Miner,* 9 December 1963.

68. *Jessen's Weekly,* 19 February 1964.

69. Pierce, letter to the *Fairbanks Daily News-Miner* (13 December 1963), in reply to an accusation by Warren Taylor (ibid., 3 December).

70. Hammond to Robert Weeden, 22 February 1964, box 4 (Correspondence), ACS Papers.

71. Wood to Win Noyes, 28 October 1963, box 4, ibid. Win Noyes was the wife of Richard M. Noyes, chairman of the Sierra Club's Pacific Northwest Chapter, Eugene, Oregon. The Northwest Chapter had responsibility for Alaska.

72. Gruening to the editor of *Young Citizen,* 17 February 1964. (The offending article appeared in the 14 January issue.) Clipping, box 7, ACS Papers.

73. Merdes to Major General James C. Jensen, 27 February 1964. The letter

concerned a Mr. Bellringer, whom Merdes suspected of preparing anti-Rampart material for the North American Wildlife Conference that was to be held in March in Las Vegas. Gruening Papers. In Las Vegas, Gabrielson continued his criticism of the project. See James B. Trefethen, ed., *Transactions* (Washington, D.C.: Wildlife Management Institute, 1964), 25–66. The conference passed a resolution calling for more thorough wildlife studies and the consideration of alternatives.

74. The BLM held sessions in Anchorage, Fairbanks, and Fort Yukon on 13, 14, and 15 February 1964 respectively. A session was also held in Washington, D.C. on 24 March at which the Alaska Conservation Society (Celia Hunter), the Wildlife Management Institute (C. R. Gutermuth), and the Wilderness Society (Stewart Brandborg) testified. Interior approved the application for withdrawal in February 1965.

75. Tobin to Roger R. Robinson, state director, BLM (Anchorage), 21 February 1964; see also, Glenn DeSpain to Robinson, 17 February, ACS Papers.

76. *Alaska Sportsman* 30 (December 1964): 7.

77. *Fairbanks Daily News-Miner,* 16 February 1964.

78. Trappers later went on record as the largest single group of Alaskans opposed to Rampart. Ten thousand members strong, the Alaska Trappers Association denounced Rampart as an alien idea imported by outsiders, and "an atrocity to our source of income and our way of life." Alaska Trappers Association to Gruening, 4 March 1968, Gruening Papers.

79. Gruening, *Many Battles,* 497.

80. *Tundra Times,* 17 February 1964. Celia Hunter, interviewed in Fairbanks on 15 August 1985 by William S. Schneider and the author, recalls Young's stance on Rampart as the only issue on which he and the Alaska Conservation Society ever saw eye-to-eye.

81. *Anchorage Daily Times,* 31 October 1964.

82. Harry L. Rietze, J. T. Barnaby, and Gordon Watson, U.S. Fish and Wildlife Service, *A Report to the Secretary of the Interior, Rampart Canyon, Dam and Reservoir Project Committee,* Washington, D.C., April 1964. Rietze was regional director, Bureau of Commercial Fisheries; Barnaby was regional director, Bureau of Sport Fisheries and Wildlife; and Watson was Alaska Administrator, Branch of River Basin Studies. As reported at the 30th North American Wildlife and Natural Resources Conference, Mexico City, 1965. See James B. Trefethen, ed., *Transactions,* (Washington, D.C.: Wildlife Management Institute, 1965), 13.

83. Taylor to Gruening, 27 June 1964, Gruening Papers

84. Frank Dufresne, "Rampart Roulette," *Field and Stream* 69 (August 1964): 77. But neither veteran conservationist played a significant role in the fight against Rampart.

85. Inter-College Colloquium on Human Behavior, University of Alaska, Fairbanks. As reported in the *Polar Star* 20 (15 November 1964): 4.

86. Department of the Interior, *Rampart Project, Alaska: Market for Power and Effect of Project on Natural Resources,* Field Study Report 998, Washington, D.C., January 1965, vol 3., 850–934 (fish and wildlife), 935–78 (Natives), 981–84 (recreation). This three-volume, 1,000-page report included contributions from the Bureau of Reclamation, the Bureau of Indian Affairs (BIA), the BLM, the USFWS, the Bureau of Outdoor Recreation (BOR), the Bureau of Mines, the Alaska Railroad, and the USGS.

87. The members were Henry Caulfield, director, Resources Program; Jo-

seph M. Morgan, director, Division of Water and Power Development; James T. McBroom (USFWS); Joseph C. McCatskill of Mineral Resources; Roderick Riley (BIA); and Eugene Zumwalt (BLM).

88. *New York Times,* 28 February 1965.
89. Gruening to Merdes, 19 April 1965, Gruening Papers.
90. Gruening to Merdes, 23 April 1965, ibid.
91. 12 July 1965, ibid.
92. Terris Moore, "Alaska's first American century: the view ahead," *Polar Record* 14 (1968), 12.
93. Here are the actual numbers involved:

	Pro		Anti	
	U.S.	Alaska	U.S.	Alaska
1963	6	4	9	3
1964	10	6	25	8
1965	20	13	61	23
1966	5	4	34	13
Total	41	27	129	47

94. Arthur Laing, "Waterfowl—A Resource in Danger," 37th annual convention, Ontario Federation of Anglers and Hunters, Ottawa, 19 February 1965, box 4, ACS Papers.
95. Laing to Gruening, 12 October 1965, Gruening Papers.
96. For a further expression of economic faith in nuclear power by an Alaskan opponent of Rampart, see Dan Swift, "The Economics of Rampart Dam," *Alaska Conservation Society News Bulletin* 5 (December 1964): 4–7. Rampart advocates often expressed surprise that conservationists were so enthusiastic about the nuclear alternative, whose cost competitiveness they denied and of whose dangers they warned. This faith in nuclear power as a clean and cheap alternative to damming wild rivers was characteristic of a number of conservationists at this time. David Brower, for example, extolled its virtues in his elegy for Glen Canyon, the film *The Place No One Knew.*
97. Box 6 (tape 5), ACS Papers.
98. Paul Brooks, "The Plot to Drown Alaska," *Atlantic Monthly* 215 (May 1965): 53–70. The article was based on a week-long exploration of the Yukon Flats by boat and plane in the summer of 1964. *Atlantic Monthly,* one of the nation's premier literary magazines, had been sympathetically inclined toward preservationism since the early years of the century when it had carried a series of articles by John Muir against the damming of Hetch Hetchy.
99. Ibid., 57–59. For the reaction of the Fairbanks Chamber of Commerce, see Randy Acord, "The Plot to Let Alaska Breathe," (undated), series 1, box 33, FCCC. Acord was chairman of the Fairbanks Chamber of Commerce's Rampart Power Committee until 1966.
100. Ernest Gruening, "The Plot to Strangle Alaska," *Atlantic Monthly* 216 (July 1965): 56–59.
101. Gruening to Sidney L. James, 18 February 1965, Gruening Papers. James was the publisher of *Sports Illustrated.* Gruening wrote in response to an article by Robert H. Boyle, "America Down The Drain," *Sports Illustrated* 21 (16 November 1964): 78–90. This polemical piece denouncing progress for its own sake was especially damning of what the author, a keen sportsman, dubbed

"wrecklamation" projects such as Rampart, which he derided as "a cement contractor's Eskimo pie" (85).

102. Gruening, Speech (draft), Committee on Projects, National Rivers and Harbors Congress, 52d Annual Convention, 9 June 1965, Gruening Papers.

103. Gruening, *Congressional Record*, 88th Congress, 1st session, 108 (4 October 1963), 17729. In this issue, Gruening inserted an article from the *Seattle Post-Intelligencer* (27 September 1963) on the duck increase around Coulee.

104. Bloch to Gruening, 3 December 1963, Gruening Papers.

105. *Fairbanks Daily News-Miner*, 30 November 1963; Sundborg to Fred Dimmer of Pennsylvania, 21 October 1965, Gruening Papers.

106. Walt Pederson to Gruening, 31 August 1965, Gruening Papers.

107. Gruening to Udall, 3 September 1965, Gruening Papers.

108. Gottschalk to Gruening, 16 September 1965, Gruening Papers.

109. Rampart Canyon Dam and Reservoir Project Committee, Division of Biology and Agriculture, National Academy of Sciences, National Research Council, *A Report to the Secretary of the Interior*, Washington, D.C., 1 November 1965, 1.

110. This was part of the American Association for the Advancement of Science (AAAS) conference at the University of California, Berkeley, in late December 1965. For a report, see the *Los Angeles Times*, 12 January 1966.

111. Gruening to Sundborg, 8 November 1965, Gruening Papers.

112. Stephen H. Spurr, *Rampart Dam and the Economic Development of Alaska*, Ann Arbor, Mich., March 1966, 3. Steven H. Spurr, its author, was professor of Natural Resource Studies and dean of the Horace H. Rackham School of Graduate Studies, University of Michigan, Ann Arbor. The other members of the team were Ernest Brater and Justin W. Leonard, also of the University of Michigan; A. Starker Leopold of the University of California, Berkeley; William A. Spurr of Stanford; and Michael F. Brewer, Center for Natural Resource Policy Studies, George Washington University, Washington, D.C. The sixteen-month study, commissioned in July 1964 after a review of the USFWS report, involved five trips to Alaska. It was funded by the Boone and Crockett Club; the Conservation Foundation; Defenders of Wildlife; the Duck Hunters Association of California; the Izaak Walton League of America; the National Audubon Society; the National Wildlife Federation; the Nature Conservancy; the New York Zoological Society; the Sierra Club; the Sport Fishing Institute; the Wilderness Society, and the Wildlife Management Institute. The NRCA (1946) was an umbrella organization consisting of the groups listed above. Spurr presented a condensed version of the 62-page report, of which only six pages were devoted to fish and wildlife, at the 31st North American Wildlife Conference, Pittsburgh, Pa., 14–16 March 1966. See "Alaska's Economic Rampart," James B. Trefethen, ed., *Transactions* (Washington, D.C.: Wildlife Management Institute, 1966), 448–53. See also, A. Starker Leopold (with Justin W. Leonard), "Alaska dam would be resources disaster;" Stephen H. Spurr, "Rampart Dam—A Costly Gamble," *Audubon* 68 (May–June 1966): 173–78.

113. Spurr, *Rampart Dam and the Economic Development of Alaska*, 16–19, 43, 46, 44, 48, 54.

114. Gruening to the *New York Times*, 20 April 1966, Gruening Papers. For YPA's response to the Spurr report see Gus Norwood, "Rampart Dam in Perspective," *Alaska Construction*, July-August 1966: 18–20.

115. Ward, address (undated), series 1, box 33, FCCC.

116. "Big Money Rides Against Consideration of Rampart," *Fairbanks Daily News-Miner*, 13 April 1966.

117. Department of the Interior, *Alaska Natural Resources and the Rampart Project*, Washington, D.C., 18 June 1967, iv, 36.

118. Gruening to Udall, 3 May 1967, Gruening Papers.

119. Gruening claimed that Udall had made this remark during a telephone conversation, as related in a letter to Walt Pederson, 4 May 1967, Gruening Papers. See also *Rural Electrification Newsletter* 25 (23 June 1967): 1. Shortly before his death, Gruening asserted that the energy crisis would not have existed had Rampart been built. (*Ketchikan Daily News*, 13 December 1973. Gruening was touring Alaska to promote his autobiography, *Many Battles*.) Nor had he changed his mind about the reasons for its defeat. "Rampart was knocked off by extremists," he recalled bitterly, "conservationists headed by Stewart Udall . . . for no good reason whatsoever." (*Anchorage Daily Times*, 15 December 1973.)

120. David S. Black, address, Pacific Northwest Trade Association (Seattle), 58th annual conference, Edmonton, Alberta, Canada; as reported in the *Pacific Northwesterner*, October 1967, 7.

121. Gruening, telegram, minutes, YPA, 5th annual meeting, 14 January 1968, Fairbanks, series 1, box 33, FCCC.

122. *Fairbanks Daily News-Miner*, 9 December 1969.

123. U.S. Army Corps of Engineers, *Notice of Completion of Report on the Rampart Canyon Project*, Washington, D.C., 25 June 1971, 1.

124. Ibid., 23, 32, 24.

125. YPA, 8th annual meeting, Anchorage, as reported in the *Anchorage Daily News*, 4 September 1971.

126. Henry Raymond to Gruening, 15 August 1965, Alaska Public Works File, box 1, Correspondence (1965), Ralph Rivers Papers. (Hereafter referred to as Rivers Papers.)

127. Thomas Johnson to Gruening, 7 April 1965, Gruening Papers.

128. Alaska Public Works File, box 4, Rivers Papers.

129. Dean Hurliman to Gruening, 6 August 1965, Gruening Papers.

130. Wes and Eleanor Webb to Gruening, 1 August 1965, Gruening Papers.

131. Anon to William Wood, 11 May 1963, Gruening Papers.

132. Hallet Morse to Gruening, 6 May 1965, Gruening Papers.

133. Norwood, McKinley Park Conference, 1963, from YPA, *Addresses Presented*, 16, 30.

134. Phil Holland to Gruening, 20 February 1964, Gruening Papers.

135. *Fairbanks Daily News-Miner*, 22 October 1963.

136. Sundborg to John Johnson of Indiana, 17 November 1965, Gruening Papers.

137. Randy Acord, "The Plot to Let Alaska Breathe," ca. 1965, 2, series 1, box 33, FCCC.

138. Gus Norwood, "The Rivers of Alaska," Third Annual Convention of the Alaska Association of Soil Conservation Subdistricts, Fairbanks, Alaska, 7 November 1968, Case Files, box 4, Rivers Papers.

139. Carroll Pursell, "Conservation, Environmentalism, and the Engineers: The Progressive Era and the Recent Past," in Kendall E. Bailes, ed., *Environmental History: Critical Issues in Comparative Perspective* (Lanham, Md.: University Press of America, 1985), 176–92.

140. Gruening, speech (draft), National Electric Week Dinner, Anchorage, 13 February 1963, Gruening Papers.

141. For details of McGee's concept, see Michael J. Lacey, "The Mysteries of Earth-Making Dissolve: A Study of Washington's Intellectual Community

and the Origins of American Environmentalism in the Late Nineteenth Century" (Ph.D. diss., George Washington University, 1979), 105–30.

142. Clayton R. Koppes, "Environmental Policy and American Liberalism: The Department of the Interior, 1933–53," in Bailes, *Environmental History*, 442–47; J. Leonard Bates, "Fulfilling American Democracy: The Conservation Movement," *Mississippi Valley Historical Review* 44 (June 1957): 29–57. Bates argues that conservation was essentially a social, as opposed to a scientific reform movement, concerned with economic justice and the extension of democracy.

143. Senate speech, 24 June 1960, from "Résumé of Findings," hearings, *Hydro Requirements*, xviii.

144. Gruening to Udall, 11 April 1966, Gruening Papers.

145. For details, see Owen Stratton and Phillip Sirotkin, "The Echo Park Controversy," Inter-University Case Program, Cases in Public Administration and Policy Formation no. 46, University of Alabama, Birmingham, Alabama, 1959. Roderick Nash has pinpointed the Echo Park controversy as the decisive event in shaping the modern conservation movement. See *Wilderness and the American Mind* (New Haven: Yale University Press, 3d ed., rev., 1982), 209–19.

146. Samuel P. Hays, "From Conservation to Environment: Environmental Politics in the United States since World War Two," in Bailes, *Environmental History*, 203.

147. Radin to the *New York Times*, 22 March 1966, in reply to its editorial, "The World's Biggest Boondoggle," 8 March 1966, Gruening Papers.

148. Virginia Hill Wood, "Rampart—Foolish Dam," *Alaska Conservation Society News Bulletin* 6 (Spring 1965): 7; reprinted in *Living Wilderness* 29 (1965): 3–7.

149. Unidentified letter to Gruening, 1965, Gruening Papers. Returning for a moment to the reasons for Rampart's defeat, it is clear that President Johnson, other considerations aside, was unwilling to back a huge project whose chief advocate was one of the loudest critics of his Vietnam policies.

150. Frederick Ryan to Gruening, 15 March 1966, Gruening Papers.

151. This is not to suggest that most of the Alaskans who supported Chariot also backed Rampart. The latter obviously attracted more support from business interests and politicians. An assertion that the same broad forces had been aligned for and against the two ventures (Alan Cooke, *Polar Record* 12 [1964]: 277–80) drew fire from Ivan Bloch and George Sundborg, who resented the comparison and the inference that they had approved of Chariot. (Sundborg and Bloch to Cooke, 4 November 1964, Gruening Papers.) Cooke had worked as Don Foote's assistant on the human geographical studies for the AEC at Cape Thompson.

152. Anthony Netboy, "Fish Versus Dams—A New Look At River Development," YPA Annual Meeting, Fairbanks, 20 June 1964, 3, series 1, box 33, FCCC. Netboy, a salmon expert, was a biologist at Portland State College, Oregon.

CHAPTER 6. OIL: THE FORCES GATHER

1. Both companies were subsidiaries of Alaska Interstate Company of Houston, Texas.

2. *Tundra Times,* 4 November 1966. Hicken reemphasized this in a recent interview on KUAC-FM radio station, University of Alaska, Fairbanks, 22 May 1988. Tape recording courtesy of Ronald K. Inouye.

3. Tape recording, KUAC-FM interview, ibid.

4. *Tundra Times,* 20 January 1967.

5. Civil Action No. A-21-67. On 27 September 1968 the judge in Anchorage ruled that the secretary of the interior could not interfere with the selection of lands by the State of Alaska under Section 6(B) of the Statehood Act because of pending Native land claims. Udall appealed this decision and a stay order was issued pending a final adjudication. See U.S. Congress, Senate, Committee on Interior and Insular Affairs, hearings, *Nomination of Governor Walter J. Hickel, of Alaska, to be Secretary of the Interior,* 91st Congress, 1st session, January 1969 (Washington, D.C.: GPO, 1969), 96–108.

6. *Tundra Times,* 20 January 1967.

7. The NORTH Commission consisted of ten members appointed by the governor. All served on a voluntary basis. Five were Alaskans; its chairman, Al Swalling; a contractor from Anchorage, C. W. Snedden, and Jack White (both of whom had been leading members of Yukon Power for America); John Manley, the general manager of the Alaska Railroad; and John B. Coghill, a businessman who was mayor of Nenana. The other members were Reginald Whitman, former vice president of the Great Northern Railroad; Samuel Pryor, former vice president of Pan American World Airways; Everett Hutchinson, President Johnson's undersecretary of transportation; Russell G. Smith, the chairman of Asiatic Development of San Francisco; and Donald G. Smith, a former manager of the Alaska Railroad. The famous aviator, Charles A. Lindbergh, was an ex-officio member.

8. *Tundra Times,* 28 July 1967; also quoted in "Now Alaska starts to 'open up' on its own," *U.S. News and World Report* 64 (10 June 1968): 88.

9. Interview with Hickel, *Tundra Times,* 6 October 1967.

10. Institute of Business, Economic and Government Research (IBEGR), "The Petroleum Industry in Alaska," *Monthly Review of Alaska Business and Economic Conditions* 1 (August 1964), University of Alaska, Fairbanks, 1.

11. Brent R. Bowen, *Defense Spending in Alaska,* Institute of Social, Economic and Government Research (ISEGR), University of Alaska, Fairbanks, July 1971, 1.

12. IBEGR, "Alaska's Economy in 1967," *Monthly Review of Alaska Business and Economic Conditions* 5 (1967): 2.

13. "Alaska Strikes it Rich: With Interview with Governor W. J. Hickel," *U.S. News and World Report* 65 (10 December 1968): 52; also quoted by Michael Frome, "Hickel and the Arctic," *Field and Stream* 74 (November 1969): 12. For the NORTH Commission's full report, see *North Alaska—How to realize its promise and potential; Report of the NORTH Commission* (Washington, D.C.: GPO, 1970). The NORTH Commission's plans came to nothing but Hickel recalls the organization fondly as "a beautiful idea" and "still a good thought." See interview with Hickel, KUAC-TV, Fairbanks, 22 May 1988.

14. "Alaska Strikes it Rich," *U.S. News and World Report,* 51–53.

15. For these details, see "History of Walter J. Hickel Highway," 2–3; no author or date, but probably written by a member of the Alaska State Department of Highways in 1969. Courtesy of India Spartz, Alaska Historical Library, Juneau. This typescript accompanies Alaska Photograph Album 82 at the Alaska Historical Library.

16. In 1970 the office of secretary in Alaska was changed by constitutional amendment to lieutenant governor of Alaska.

17. "History of Walter J. Hickel Highway," 4.

18. See Tom Brown, *Oil on Ice: Alaskan Wilderness at the Crossroads*, edited by Richard Pollak. (San Francisco: Sierra Club, 1971), 42–46.

19. The Oral History Collection at the University of Alaska Archives contains an unprocessed tape recording of a talk given by Johansen about the construction of the ice road at the monthly meeting of the American Institute of Mining, Metallurgical and Petroleum Engineers, Fairbanks, 26 May 1969.

20. Jane Pender, an Alaskan journalist, traveled the ice road during its second season and contributed a four-part series to the *Anchorage Daily News*, 22–30 March 1970.

21. With development in Cook Inlet came marine degradation. The U.S. Army Corps of Engineers reported slicks in December 1965, and again in March and July 1966. Complaints from commercial fishermen about damage to equipment were persistent. In response, Charles D. Evans (Bureau of Commercial Fisheries) organized a meeting of state and federal representatives in Anchorage in September 1966, to consider pollution control measures. In April 1967 Secretary Udall referred to seventy-five pollution incidents reported by state and federal agencies. In spite of mounting evidence and continuous expressions of state and federal concern, national conservation organizations displayed little interest. Sponsored by the American Petroleum Institute (API), Thomas Kimball, the executive director of the National Wildlife Federation, paid a five-day visit in the fall of 1967 to inspect drilling operations in Cook Inlet and the pipelines and refineries of Pan American, Shell, and Marathon oil companies. His subsequent press interviews left the impression that there was no cause for alarm. His blithe statements angered many Alaskans, including Senator Bartlett, a champion of the fishing industry. A few weeks after his laudatory comments, a more critical Kimball warned the oil industry that it would have "a growing public relations problem unless most responsible citizens are convinced that every reasonable precaution is being undertaken to protect the quality of the Alaskan environment." Kimball to Carl R. Arnold (API), 6 October 1967, Edgar Wayburn, Private Papers, San Francisco, Calif. (Hereafter referred to as Wayburn Papers.) In February 1968 Governor Hickel ordered the arrest of the captain of the Atlantic Richfield-chartered tanker, *Rebecca*, for dumping oily ballast in Cook Inlet.

22. "Gates of the Arctic" was a name coined by Bob Marshall. Udall's proposal to create a Gates of the Arctic National Monument by executive order was turned down by President Johnson during the closing days of his administration.

23. Robert Marshall, *Alaska Wilderness: Exploring the Central Brooks Range*, edited by George Marshall (Berkeley and Los Angeles: University of California Press, 1970), introduction, ix–x. *Alaska Wilderness* was originally called *Arctic Wilderness* (1956). In 1956, John Buckley, professor of wildlife management at the University of Alaska, reported the construction during the past winter of a winter road from Circle on the Yukon to the Arctic Ocean near Blow River. However, this road was only used by Caterpillar tractors. See *Sierra Club Bulletin* 41 (December 1956): 54.

24. Samuel Wright, review of *Alaska Wilderness, American West* 7 (November 1970): 56.

25. For a representative article see "Last Frontier," *American West*, November 1970, 32–37.

THE TRANS-ALASKA PIPELINE CONTROVERSY

. For his wife's account of this sojourn at Big Lake, see Billie Wright, *Four Seasons North: A Journal of Life in the Alaskan Wilderness* (New York: Harper & Row, 1973).

27. Samuel Wright, "A Letter from the Arctic," (15 February), *Living Wilderness* 33 (Spring 1969): 5–6.

28. Richard Starnes, "Can the Arctic be Saved?" *Field and Stream* 75 (July 1970): 16.

29. As quoted in Luther J. Carter, "North Slope: Oil Rush—Alaska May or May Not Become a Kuwait of the Arctic But the Oil Men's Arrival Is Changing a Wilderness," *Science* 166 (3 October 1969): 91.

30. *U.S. News and World Report* 67 (10 November 1969): 64.

31. *New York Times*, 18 December 1968. Hickel says he was quoted out of context and claims to have made this remark with specific reference to the failure to harvest "over-ripe timber in Southeast [Alaska]." See Walter J. Hickel, *Who Owns America?* (Englewood Cliffs, N.J.: Prentice-Hall, 1971), 21; hearings, *Hickel Nomination*, 16.

32. Drew Pearson and Jack Anderson, "Washington Merry-Go-Round," *Washington Post*, 23 December 1968. These allegations persisted throughout December 1968 and January 1969. See David Sanford, "Hickel's Pickle," *New Republic* 160 (18 January 1969): 11–12.

33. Steven Pavish to Joseph Penfold, national director, IWLA, 22 December 1968, box 35, M. Brock Evans Papers, Suzzallo Library, University of Washington, Seattle; TVSA to Senator Henry M. Jackson, 29 December 1968, hearings, *Hickel Nomination*, 276.

34. Ibid., 54.

35. Ibid., 241–42, 244.

36. Ibid., 14, 62.

37. Ibid., 130, 207.

38. Hearings, *Hickel Nomination*, 13, 207. For an example in the press, see *Newsweek* 73 (13 January): 35–36. Fall had favored the private leasing of naval reserves such as Teapot Dome (Wyoming) and to this end sought their transfer to the Department of the Interior. For the story of Teapot Dome, see Burl Noggle, *Teapot Dome: Oil and Politics in the 1920s* (Baton Rouge: Louisiana State University Press, 1962); J. Leonard Bates, *The Origins of Teapot Dome: Progressives, Parties, and Petroleum, 1909–21* (Urbana: University of Illinois Press, 1963). Fall was a great advocate of the development of Alaskan lumber and wanted Interior to take the Forest Service away from the Department of Agriculture.

39. Hearings, *Hickel Nomination*, 311, 360, 440. He had also made numerous (unsuccessful) applications for gas leases on the Kenai Peninsula. For Hickel's own account of the hearings, see *Who Owns America?* 11–37.

40. Senators Nelson (Democrat, Wisconsin), Moss (Democrat, Utah), and McGovern (Democrat, North Dakota) voted against his confirmation. For news coverage of the hearings, which were the longest for the nomination of an interior secretary in the history of the Interior Department, see *Time* 93 (17 January 1969): 13; ibid., 93 (24 January 1969): 22.

41. Rogers, keynote address, "Alaska in Transition: Wilderness and Development," in Maxine E. McCloskey, ed., *Wilderness: The Edge of Knowledge* (San Francisco: Sierra Club, 1970), 149. For the full text of Rogers's remarks, see 143–53 of this record of the conference proceedings. Extracts from Rogers's address appear in Gordon Scott Harrison, ed., *Alaska Public Policy: Current Problems and Issues* (College, Alaska: Institute of Social, Economic and Government Research, 1970), 223–35.

42. Brock Evans, "A Conservationist Views Alaska's Wilderness," in Mc-Closkey, *Wilderness: The Edge of Knowledge*, 88.

43. Robert Weeden, "Arctic Oil: Its Impact on Wilderness and Wildlife," ibid., 165 (full text, 157–67).

44. John Muir, *Steep Trails*, edited by William Frederic Badè (Boston: N. S. Berg, 1918), 348.

45. From the transcript of a tape sent by Harmon "Bud" Helmericks to Edgar Wayburn, vice president of the Sierra Club, 12 March 1971, Wayburn Papers.

46. White to the Sierra Club, 3 March 1969, Sierra Club Office Files, Alaska: Oil (folder 14), Bancroft Library, University of California, Berkeley (UCB).

47. Transcript, Helmericks tape, 12 March 1971, Wayburn Papers.

48. Ibid. These were so ubiquitous along the Arctic coast that a prominent Alaskan engineer once quipped that the drum should be designated the official state flower. The official was Colonel E. L. Hardin, chief, U.S. Army Corps of Engineers, Alaska Branch, as quoted in Barry Weisberg, "Alaska—The Ecology of Oil," *Ramparts* 8 (January 1970): 28; and Ben East, "Is It TAPS for Wild Alaska?" *Outdoor Life* 145 (May 1970): 45.

49. Transcript, Helmericks tape.

50. John P. Milton, "Arctic Walk," *Natural History* 78 (May 1969): 45–52. See also Milton's book, *Nameless Valleys, Shining Mountains: The Record of an Expedition into the Vanishing Wilderness of Alaska's Brooks Range* (New York: Walker and Co., 1970). Milton was deputy director of the Conservation Foundation's International Programs Division.

51. Milton, "Arctic Walk," 49, 51.

52. See Ray Allen Billington, "The American Frontiersman: A Case Study in Reversion to the Primitive," *America's Frontier Culture: Three Essays* (College Station: Texas A&M University Press, 1977), 19–50. This essay was delivered as Billington's Harmsworth Inaugural Lecture at the University of Oxford, England, 2 February 1954. It was first published in 1956 by the Clarendon Press (Oxford).

53. Kenneth Brower, *Earth and the Great Weather* (San Francisco: Friends of the Earth, 1971), 165.

54. William Tucker, "Is Nature Too Good For Us?" *Harper's* 264 (March 1982): 31. Tucker is the author of *Progress and Privilege: America in the Age of Environmentalism* (Garden City, N.Y.: Doubleday, 1982).

55. Here is a selection: R. L. Schueler, "Ecology: the new religion," *America* 122 (21 March 1970): 292–95; J. Margolis, "Our Country 'tis of thee, land of ecology," *Esquire* 73 (March 1970): 124–25; "Special Issue on the Environment: A national mission for the seventies," *Fortune*, February 1970; John Pekkanen, "Ecology: a cause becomes a mass movement," *Life* 68 (January 1970): 22–32; "Special Issue on America the Beautiful," *Look* 4 November 1969; "This Ecology Craze," *New Republic* 162 (7 March 1970): 8–9; Special Issue, "The Ravaged Environment," *Newsweek* 75 (26 January 1970): 33–45; Paul Friggens, "Last Chance for Mother Earth," *Reader's Digest* 96 (May 1970): 63–67; "Special Issue: The Last Chance—Now," *Sports Illustrated*, 2 February 1970; Special Issue, "The Emerging Science of Survival," *Time* 96 (2 February 1970); "Issue of the year: the environment," ibid., 97 (4 January 1971): 21–22. The July 1970 issue of *Current History* (59) was devoted to the American ecological crisis. See especially Carroll Quigley, "Our Ecological Crisis," 3–12, 49.

56. Clay S. Schoenfeld, "The Ecology of the New Conservation," 35th North American Wildlife Conference, Chicago, 1970, *Proceedings* (Washington, D.C.: Wildlife Management Institute, 1970), 357.

57. Beginning with his handling of this event, Hickel surprised some of his critics in the press and the conservationist community. See "Apprentice Noah," *Time* 93 (21 March 1969): 22. Some commentators saw this as evidence of a "conversion." See "The Education of Wally Hickel," ibid., 94 (1 August 1969): 13. (These references are to the European edition of *Time*.) Hickel shut down Union Oil's operations in the Santa Barbara Channel for a while and campaigned for the Outer Continental Shelf Act, which enforced absolute liability for clean-up (without cause) of spills.

58. Roderick Nash, reviewing Robert Easton's *Black Tide* (1972), *Living Wilderness* 37 (Autumn 1973): 34.

59. Tape recording, interview, KUAC-TV, Fairbanks, 22 May 1988. See also, Hickel, *Who Owns America?* 90–100.

60. Useful are Donald Fleming, "Roots of the New Conservation Movement," *Perspectives in American History* 6 (1972): 7–91; Schoenfeld, "The Ecology of the New Conservation," 351–67; Richard L. Means, "The New Conservation," *Natural History* 78 (August-September 1969): 16–25; Stephen R. Fox, *John Muir and his Legacy* (Boston: Little, Brown and Company, 1981), 302–23.

61. For an analysis of the postwar changes in the structure of the American economy and American society conducive to the rise to prominence of environmentalism as a quality of life issue in the late 1960s, see Samuel P. Hays, "From Conservation to Environment: Environmental Politics in the United States since World War Two," *Environmental Review* 6 (Fall 1982): 14–41. For a more extended discussion of these ideas see Hays, *Beauty, Health, and Permanence: Environmental Politics in the United States, 1955–85* (Cambridge: Cambridge University Press, 1987).

62. Bill Voigt, *Outdoor America* 36 (March 1971): 5.

63. Michael J. McCloskey, "Wilderness Movement at the Crossroads: 1945–70," *Pacific Historical Review* 5 (August 1972): 351.

64. The other founder members of the EDF were Carol Yannacone; Charles F. Wurster, a biology professor at the State University of New York, Stony Brook; and George M. Woodwell and Dennis Puleston, two ecologists at Brookhaven National Laboratory, Long Island.

65. Rachel Carson, *Silent Spring* (Boston: Houghton Mifflin, 1962). The silent spring that Carson envisaged was one stripped of the sounds of birds, which had all succumbed to the pesticide.

66. *Science* 158 (22 December 1967): 1552–56. The EDF did not win but its legal action created a delay that meant that spraying plans had to be postponed for a year. For an account of the origins, purpose, and early activities of the EDF, see Thomas R. Dunlap, *DDT: Scientists, Citizens, and Public Policy* (Princeton: Princeton University Press, 1981), 143–54. The federal government banned DDT in 1972.

CHAPTER 7. THE TRANS-ALASKA PIPELINE CONTROVERSY: 1, 1969–1971

1. The "big three" were ARCO, Humble (a subsidiary of Standard Oil of New Jersey, now Exxon), and British Petroleum (BP). ARCO was the outcome of a merger (1966) between Atlantic Richfield and Atlantic Refining. BP was in the process of merging with Standard Oil of Ohio (SOHIO). The merger took place on 1 January 1970. BP had been the first company to begin North

Slope exploration following statehood. This was BP's first entry into the American market. TAPS was enlarged in the fall of 1969 to include Mobil Oil, Phillips Petroleum, Union Oil (California), Amerada Hess, and Home Oil (Canada).

2. Atlantic Pipeline Company, et al., *Transcontinental Pipeline Project: Transportation of Alaskan Crude Oil, Economic Feasibility Study: Seattle to Chicago, Prudhoe Bay to Chicago, Chicago to East Coast*, Houston, Tex., 31 September 1968. In June 1969 several oil companies established Mackenzie Valley Pipeline Research Ltd. (MVPR) to continue investigations of a Canadian line. In January 1970 MVPR published a report confirming the feasibility of a Canadian line. See Bechtel Summary Report, *Technological Feasibility and Cost Study; Mackenzie Valley Pipeline Research, Ltd.*, Houston, Tex., January 1970.

3. TAPS decided against building a cold oil pipeline because crude oil is hard to pump when the oil is at a low temperature. Extensive mechanical refrigeration facilities would have been needed to chill the oil when it came out of the ground and then to keep it chilled along the line since moving oil acquired heat due to friction. An added disadvantage was that in the process of chilling hot oil, large quantities of wax are produced, creating a considerable disposal problem.

4. Train, a tax judge, had been president of the Conservation Foundation from 1965 to 1969.

5. The Secretaries of Commerce; Defense; Health, Education and Welfare (HEW); Transportation; and Housing and Urban Affairs were also members.

6. U.S. Congress, Senate, Interior and Insular Affairs Committee, hearings, *The Status of the Trans-Alaska Pipeline Proposal, Part 3*, 91st Congress, 1st session, 25 November 1969, Enclosure (3) (Washington, D.C.: GPO, 1969), 99. (Hereafter referred to as Hearings, *TAPS Proposal*.)

7. In February 1970 the federal task force established the interdepartmental Technical Advisory Board, whose job was to review the work of the Menlo Park group.

8. Krumm, memorandum to Boyd L. Rasmussen, U.S. Congress, Senate, Committee on Interior and Insular Affairs, hearings, *Trans-Alaska Pipeline, Draft Environmental Impact Statement*, 92d Congress, 1st session, (Washington, D.C.: GPO), vol. 10, exhibit 124, 1971. (Hereafter referred to as Hearings, *Draft EIS*.)

9. Department of the Interior, "Pipeline Study Group Field Trip Report," 20 June 1969, Alaska Conservation Society Papers, University of Alaska Archives, Fairbanks (UAF). (Hereafter referred to as ACS Papers.)

10. Hearings, *TAPS Proposal, part 2*, Washington, D.C., 16 October 1969, 103.

11. According to Mary Clay Berry, TAPS justified these foreign contracts on the grounds that no American company made pipe of this size. See Mary Clay Berry, *The Alaska Pipeline: The Politics of Oil and Native Land Claims* (Bloomington: Indiana University Press, 1975), 104–05.

12. Train to Richard G. Dulaney, 10 June 1969, hearings, *TAPS Proposal, part 2*, 83. According to John C. Reed of the Arctic Institute of North America (AINA), the Department of the Interior's rigorous scrutiny of the TAPS proposal was largely due to Train's concerns. See John C. Reed, *Oil and Gas Development in Arctic North America Through 2000*, Arctic Institute of North America, Research Paper no. 62, Washington, D.C., September 1973, 5.

13. Hearings, *TAPS Proposal, part 2*, 88.

14. A few days later, on 12 August, the Senate Interior Committee held hearings in Washington, D.C., to assess Interior's capability to oversee oil devel-

opment on Alaska's public lands, i.e., funding and manpower requirements. Undersecretary of the Interior Russell Train estimated that 142 new positions needed to be created and that the task of overseeing the project would cost $4.235 million. (U.S. Congress, Senate, Committee on Interior and Insular Affairs, hearings, *Federal Oversight of Oil Development on Public Lands in Alaska*, 91st Congress, 1st session, 12 August 1969 (Washington, D.C.: GPO), 3.

15. *Jessen's Weekly*, 19 August 1969.

16. "Richest Auction in History," *Time* 94 (19 September 1969): 54.

17. Ibid. This quote is from Turner's commencement address at the University of Washington, Seattle, 17 June 1914. See Turner, "The West and American Ideals," *The Frontier in American History* 1920 (Huntington, N.Y.: Robert E. Krieger, reprint, 1976), 296.

18. "Richest Auction in History," 54.

19. "Alaskan Prospect," *National Parks Magazine* 43 (September 1969): 2.

20. For the book that grew out of the conference see George W. Rogers, ed., *Change in Alaska: People, Petroleum, and Politics* (College: University of Alaska Press, 1970).

21. Robert Engler, *The Politics of Oil: A Study of Private Power and Democratic Directions* (Chicago: University of Chicago Press, 1961). Engler's classic study of the impact of "the private government of oil" on the principles and institutions of a democratic society was written in advance of the environmentalist attack on the oil industry. Towards the end of the book, Engler wonders which groups "will retain the will and the capacity to raise the larger points about the proper climate for a responsible and democratic society" (482). He did not anticipate that conservationists would be the most critical of these groups during the coming decade.

22. *Anchorage Daily News*, 28 August 1969. Stevens had been appointed by Governor Hickel to succeed Senator Bartlett, who died on 11 December 1968. Stevens was then elected in 1972.

23. Department of the Interior, hearings, *An Application for a Trans-Alaska Pipeline before the Department of the Interior*, Fairbanks, Alaska, 29–30 August 1969.

24. Siemon W. Muller of the USGS proposed this classic definition in 1943. See Louis L. Ray, *Permafrost*, Popular Publications of the USGS (Washington, D.C.: GPO, 1983), 6.

25. Transcript (photocopy), Prepared Direct Testimony of Henry W. Coulter, (Docket No. OR 78-1) Presented on Behalf of The State of Alaska, Federal Energy Regulatory Commission, hearings, *Trans-Alaska Pipeline System (In the Matter of Phase II of Setting Tariff Rates for the Intrastate Transportation of Petroleum over the Trans-Alaska Pipeline System)*, Washington, D.C., 16 December 1981, 5–6. (Photocopy courtesy of Henry W. Coulter.) The State of Alaska took legal action because it believed that mismanagement in planning for the pipeline and in constructing it had resulted in at least $1.6 billion of "imprudent construction costs," which it claimed should be excluded from the rate base. The state won the case.

26. As reported in *Anchorage Daily Times*, 30 August 1969.

27. Henry W. Coulter, testimony, hearings, *TAPS (In the Matter of Phase II)*, 6. Coulter became a member of the Technical Advisory Board.

28. The Department of the Interior also outlined measures to protect archaeological artifacts and called for the implementation of equal opportunity employment practices regarding Native Alaskans. Hearings, *TAPS Proposal, part 1*, Washington, D.C., 9 September 1969, 39–77.

29. Hearings, *TAPS Proposal, part 2,* 109. Further stipulations, which also covered technical matters, were issued in November 1970.

30. *Fairbanks Daily News-Miner,* 24 October 1969.

31. *Nome Nugget,* 9 December 1969.

32. ARCO press release (12 August 1969), hearings, *TAPS Proposal, part 1,* 206.

33. Federal Task Force on Alaskan Oil Development, *A Preliminary Report to the President,* Washington, D.C., 15 September 1969, 13.

34. For a professional exposition of the technical problems presented by permafrost, see Oscar J. Ferrians, Reuben Kachadoorian, and Gordon W. Greene, *Permafrost and Related Engineering Problems of Alaska,* USGS Professional Paper 678 (Washington, D.C., GPO, 1969).

35. Federal Task Force, *A Preliminary Report,* 11.

36. Hearings, *TAPS Proposal, part 2,* 79.

37. Hearings, *TAPS Proposal, part 1,* 1.

38. Hearings, *TAPS Proposal, part 2,* 114.

39. Ibid., 111.

40. Ibid., 195.

41. Ibid., 173.

42. Ibid., 177; also, Edward Ernest Clebsch, representing the Association of Southeastern Biologists, ibid., 218; Michael Frome, conservation editor, *Field and Stream,* ibid., 222.

43. Ibid., 228.

44. Ibid., 125, 127, 112.

45. Ibid., 138–39. See also, Ronald H. Sundt to Gravel, 12 November 1969, General Correspondence 1969–80, series 3, subseries 2, Mike Gravel Papers, UAF. (Hereafter referred to as Gravel Papers.)

46. *Fairbanks Daily News-Miner,* 24 October 1969.

47. *Anchorage Daily News,* 20 November 1969.

48. Hearings, *TAPS Proposal, part 3,* 285, 294.

49. Federal Task Force, *A Preliminary Report,* 7.

50. Hearings, *TAPS Proposal, part 3,* 285–93.

51. NEPA was approved by a vote of 372 to 15 in the House of Representatives and by voice vote, without debate, in the Senate. See *Congressional Quarterly Almanac* 25 (1969): 525–27.

52. "'America the Beautiful' Doomed? (Interview with Walter J. Hickel)," *U.S. News and World Report* 67 (10 November 1969): 62.

53. *Fairbanks Daily News-Miner,* 3 February 1970.

54. Department of the Interior, *Environmental Statement, Yukon River-North Slope Road,* Washington, D.C., 20 March 1970.

55. *Fairbanks Daily News-Miner,* 30 March 1970.

56. Tanana Chiefs' Executive Committee Resolution, 27 July 1969. As reported in the *Fairbanks Daily News-Miner,* 13 April 1970.

57. *Wilderness Society et al. v. Hickel,* 325 F. Supp. 422–41, ERC 1335 (D.D.C. 1970), "Alaska," Carton 19, Friends of the Earth Papers, Bancroft Library, University of California, Berkeley (UCB). (Hereafter referred to as FOE Papers.)

58. For details, see Peter M. Hoffman, "Evolving Judicial Standards under the National Environmental Policy Act and the Challenge of the Alaska Pipeline," *Yale Law Journal* 81 (1972): 1610–12.

59. NEPA unleashed a plethora of environmentalist lawsuits. Another Alaskan project to come under NEPA's scrutiny at this time was the AEC's proposed

"Cannikin" blast on Amchitka. Amchitka was part of the Aleutian Islands National Wildlife Refuge and the blast was conducted in late 1971. (See chapter four.) The environmental impact statement submitted for "Cannikin" was also legally contested by conservationists. A majority of Alaskan politicians supported the injunction initiated by the Committee for Nuclear Responsibility and supported by SANE, FOE, the Sierra Club, the National Parks and Conservation Association, the Wilderness Society, and the American Association on Indian Affairs. These politicians attacked the AEC for what they considered token and sham compliance with NEPA. Together with conservationists they also questioned the validity of the national security argument that was used to justify the blast, and the insincerity of AEC efforts to assuage public fears concerning environmental impact. In addition, critics were angry at what they saw as a glaring discrepancy between the exhaustive scrutiny to which the TAPS project was being subjected and the AEC's efforts to prevent public input to the debate over whether Cannikin should proceed. They also believed that the AEC should have considered alternative sites. However, the courts ruled that national security was at stake and refused to issue an injunction.

60. *Fairbanks Daily News-Miner,* 13 April 1970. Some Natives did worry about the impact of the project on subsistence lifestyles based on hunting, fishing, and trapping. Though there was often a good deal of common ground between them, the relationship of Natives and environmentalists was frequently tense. Relations between Alaska Natives and conservationists during the TAPS controversy are not discussed in Mary Clay Berry's *The Alaska Pipeline.* Little has been written on this subject.

61. Earth Day events stretched over a week. For a report which described Earth Day as the biggest street festival since the Japanese surrender in 1945, see *Newsweek* 75 (4 May 1970): 22–24.

62. *Fairbanks Daily News-Miner,* 23 April 1970; Pat Ryan, "The Earth As Seen From Alaska; Hickel In Earth Day Observances," *Sports Illustrated* 32 (4 May 1970): 26–82, 31. A week after the pipeline authorization act was signed in 1973, Hickel reflected: "In 1969 and 1970 industry, government and the scientific community didn't have the engineering knowledge to build it right. Now they have it. We've learned a lot, and its [the delay] been beneficial in many respects." *Anchorage Daily News,* 25 November 1973.

63. Joe Vogler, "To All Unemployed And To All Who May Become Unemployed," Open Letter, *Fairbanks Daily News-Miner,* 26 April 1970.

64. Bill Tobin and Bill Arnold, "Alaskans and the Pipeline; A Position Paper," April 1970, Series 4 (Oil and Gas), box 6, Fairbanks Chamber of Commerce Collection (FCCC), UAF.

65. *Anchorage Daily News,* 27 April 1970.

66. Ibid., 28 April 1970. Russell Train, Pecora's predecessor, had left Interior to assume the position of chairman of the President's Council on Environmental Quality.

67. *Alaska Industry* 2 (June 1970): 27. The nation's press and magazines also emphasized the problems. For representative coverage, see "Great Land: boom or doom? (Last Chance for the Last Frontier)," *Time* 102 (27 July 1970): 26–40; Lewis Lapham, "Alaska: Politicians and Natives, Money and Oil (Once upon a Time There Was a Frontier)," *Harper's* 240 (April 1970): 85–102.

68. "TAPS Not Tundra's First Line—Not By 45 Years," *Oil and Gas Journal* 68 (10 August 1970): 104. The Davidson Ditch operated until 1967, when flooding destroyed a section.

69. *Anchorage Daily News,* 19 and 20 April 1970.

70. Robert Knox, "Will The TAPS Pipeline Really Go?" *Alaska Industry* 2 (June 1970): 26–28.

71. Arlon Tussing, "The Trans-Alaska Pipeline: Political Leadership and the News Media in Alaska," *Anchorage Daily News,* 22 July 1970.

72. As quoted in Robert Sherrill, "Unsafe At Any Width," *The Nation* 218 (11 June 1973): 747.

73. Hugh Gallagher wrote a book about his dealings with one particular Native politician, perhaps the most militant. See *Etok: A Story of Eskimo Power* (New York: G. P. Putnam, 1974). "Etok" was Charles W. Edwardsen, Jr., executive director of the Arctic Slope Native Association, many of whose members opposed oil development in the Arctic.

74. Holbert to *Fairbanks Daily News-Miner,* 12 August 1970.

75. Federal Task Force, *A Preliminary Report,* 20.

76. This incident is related in Daniel Jack Chasan, *Klondike '70: The Alaskan Oil Boom* (New York: Praeger Publishers, 1971), 33–34. Confirmation and further details were provided by Geoffrey Larminie of BP, interviewed by the author in London, 15 November 1985.

77. Robert Belous, "Unsolved Problems of Alaska's North Slope," *National Parks and Conservation Magazine* (Special Issue: Alaska) 44 (1970): 16–17, 20–21. (The National Parks Association changed its name to National Parks and Conservation Association, and the name of its journal, in 1970.) Belous refers specifically to the aforementioned example of damage by GSI; so does Derek Lambert, *The Great Land* (London: Bantam, 1978), 275–76.

78. See Robert Weeden and David Klein, "Wildlife and Oil: A Survey of Critical Issues in Alaska," *Polar Record* 15 (1971): 479–94.

79. Ibid., 479.

80. *Alaska Industry* 2 (October 1970): 2.

81. Ibid., 2 (August 1970): 2.

82. Ibid., 2 (July 1970): 7.

83. Ibid., 2 (June 1970): 11. To collect these drums and other refuse a vehicle called the rolligon was developed. The rolligon could be used in summer because it traveled on inflated rubber bags (which exerted less surface pressure) instead of tires or cleated tracks. Advertisements for the rolligon showed a man lying happily between the balloon tires and the tundra.

84. Ibid., 3 (September 1971): 2.

85. Ibid., August 1970: 68.

86. Ibid., September 1971: 4.

87. Ibid., 2 (November 1970): 2.

88. U.S. Department of the Interior, *Draft Environmental Impact Statement for the Trans-Alaska Pipeline, Section 102(2)C of the National Environmental Policy Act of 1969,* Washington, D.C., 15 January 1971, 119, 144–45. (Hereafter referred to as *Draft EIS.*)

89. Ibid., 142.

90. Ibid., 152–54.

91. Stanley B. Haas, "Marine Transport-The Northwest Passage," proceedings, 21st Alaska Science Conference, *Science in Alaska* (College: Alaska Division, AAAS, 1970): 76–81. Haas was project manager, Arctic Marine Task Force, Humble Oil and Refining Company, Houston. For another account of the voyage, see Tom Brown, *Oil on Ice: Alaskan Wilderness at the Crossroads* (San Francisco: Sierra Club, 1971), 86–94.

92. Hearings, *TAPS Proposal, part 1,* 2.

93. *Draft EIS,* 160–62.

94. Ibid., 163.

95. *New York Times,* 5 February 1971. John McPhee noted Brower's fondness for railroads in *Encounters with the Archdruid,* where he referred to Brower's failure to disapprove of the railroad that crosses the Sierra Nevada at Donner Summit. McPhee explained that Brower "developed an extraordinary affection for trains" during his childhood. "Malapropos as it may seem at this point in his career," argued McPhee, "he still has it. The force of nostalgia in Brower is such that it can in some instances bend logic." See John McPhee, *Encounters with the Archdruid: Narratives about a Conservationist and Three of His Natural Enemies* (New York: Farrar, Straus and Giroux, 1971), 29.

96. M. Brock Evans Papers, University of Washington Libraries, Seattle.

97. Marshall to Wright, 1 April 1971, Alaska (Oil), George Marshall Papers, Bancroft Library, University of California, Berkeley (UCB).

98. Hickel officially left office on 25 November 1970. His celebrated letter to Nixon in the wake of the Kent State killings in May 1970, on the subject of America's alienated youth, which was leaked to the press, provided the immediate circumstances for his dismissal.

99. U.S. Congress, Senate, Committee on Interior and Insular Affairs, hearings, *On the Nomination of Rogers C. B. Morton to be Secretary of the Interior.* 92nd Congress, 1st session, 25–26 January 1971 (Washington, D.C.: GPO, 1971), 73.

100. Ibid., 14.

101. *Fairbanks Daily News-Miner,* 20 February 1971.

102. Hearings, *Draft EIS,* vol. 11, exhibit 180, 48–53.

103. Arthur H. Lachenbruch, *Some Estimates of the Thermal Effects of a Heated Pipeline in Permafrost,* USGS Circular no. 632, Menlo Park, Calif., December 1970. For the most reliable popular discussion of the Lachenbruch Report and the technical problems presented by the TAPS proposal, especially relating to permafrost, see Ron Moxness, "The Long Pipe," *Environment* 12 (September 1970): 12–23, 36. *Environment* was the former *Scientist and Citizen* and was published by the St. Louis Committee for Environmental Information (formerly the Committee for Nuclear Information.) The name of the bulletin was changed in January 1969.

104. The report in question is referred to by D. C. Alverson (USGS) in a memorandum to the Federal Task Force, 5 August 1969 (as quoted in Moxness, "The Long Pipe," 23).

105. I. G. Black, "USSR Oil Leak: A Warning To Alaska's Pipeliners," *Iron Age* 208 (July 1971): 41.

106. Hearings, *Draft EIS,* Washington, D.C., 17 February 1971, vol. 1, 153. (Unless Washington, D.C. is specified, all subsequent references are to the sessions in Anchorage.)

107. The term can be attributed to Peter R. Janssen, the education editor of *Newsweek.* See his article "The Age of Ecology," *Ecotactics* (San Francisco: Sierra Club, 1970), 53–63.

108. Mrs. O'Meara, (Sierra Club member), hearings, *Draft EIS,* vol. 3, 783. See also, Sandra Dauenhauer, ibid., 25 February, vol. 2, 630.

109. Mrs. O'Meara, ibid.; William H. Babcock, ibid., 618. Babcock was a sociologist at the University of Alaska, Anchorage.

110. Those who hold biocentric views are often called "deep ecologists." For the authoritative works on "deep ecology," see Arne Naess, "The Shallow and the Deep, Long Range Ecology Movement," *Inquiry* 16 (1973): 95–100.

Bill Devall, "The Deep Ecology Movement," *Natural Resources Journal* 20 (April 1980): 299–322.

111. The philosopher Peter Singer has coined the term "speciesism," which he equates with racism and sexism, to denote human domination over other forms of life. See *Animal Liberation, A New Ethics For Our Treatment of Animals* (New York: Avon Books, 1975), 19.

112. "Saving Alaska," *Not Man Apart* 1 (September 1971). *Not Man Apart* is the monthly journal of FOE.

113. Hearings, *Draft EIS,* 25 February, vol. 2, 400. For a discussion of the extension of ethical and legal-political rights to both animate and inanimate natural objects as the logical culmination of the American radical tradition, see Roderick Nash, "The Significance of the Arrangement of the Deck Chairs on the Titanic (The Extension of Ethics and Humanity)," *Not Man Apart* 5 (October 1975): 7–9; "Do Rocks Have Rights?" *Center Magazine* 10 (November-December 1977): 2–12; "Rounding Out the American Revolution: Ethical Extension and the New Environmentalism," in *Environmental History: Critical Issues in Comparative Perspective,* edited by Kendall E. Bailes (Lanham, Md.: University Press of America, 1985), 242–57. For the seminal contributions to the legal debate, see Christopher D. Stone, "Should Trees Have Standing?—Towards Legal Rights For Natural Objects," *Southern California Law Review* 45 (1972): 450–501; Lawrence H. Tribe, "Ways Not To Think About Plastic Trees: New Foundations For Environmental Law," *Yale Law Journal* 83 (7 June 1974): 1315–48.

114. Hearings, *Draft EIS,* vol. 2, 400.

115. Ibid., vol. 3, 756.

116. Ibid., vol. 2, 401.

117. Ibid., vol. 3, 756.

118. Charles S. Collins, "A Quick Look at the Alaskan Oil Problems," hearings, *Draft EIS,* Anchorage, vol. 3, exhibit 13, 830.

119. Ibid., vol. 2, exhibit 4, 3.

120. Ibid., vol. 1, 41.

121. *Alaska Industry* 3 (October 1971): 2.

122. Alfred J. Loman, hearings, *Draft EIS,* vol. 5, exhibit 103, 2.

123. Ibid., George Dickson, 26 February, vol. 3, exhibit 19, 973.

124. Bill Walters, Prudhoe Bay Field Coordinator, ARCO, as quoted in *Oil Daily,* 17 September 1971, 91.

125. *Draft EIS,* 110, 112, 135, 140–41, 190, 193.

126. *Wilderness Society et al. versus Hickel,* 9, (Civil Action no. 92870), "Alaska" (Carton 19), FOE Papers.

127. Hearings, *Draft EIS,* Washington, D.C., vol. 1, 206.

128. Ibid., Anchorage, Supplemental Testimony, vol. 9, exhibit 91.

129. Ibid., vol. 1, 162–63. The references that follow are taken from Rice's "Prepared Statement" entitled "Arguments in Opposition To The Alaska Pipeline," submitted as part of the record. Wood confirmed having made the remarks that Rice attributed to him during an interview with the author in Fairbanks, 4 September 1984.

130. Some commentators have singled out the "Judaeo-Christian ethic" as the root cause of the ecological crisis. See, for instance, Lynn White, Jr., "The Historical Roots of Our Ecological Crisis," *Science* 155 (10 March 1967): 1203–7.

131. Rice, Prepared Statement, *Draft EIS,* vol. 1, 5–9.

132. Ibid., vol. 3, 687.

133. Ibid., vol. 1, 59, 115; vol. 3, 720.

134. Rice, Prepared Statement, 3–4.

135. *New York Times,* 24 March 1971.

136. Hearings, *Draft EIS,* Prepared Statement, 10.

137. Gravel, *Newsletter* 5 (January-April), 1971, 3, series 31, Public Relations, Mike Gravel Collection, UAF.

138. Havelock, "Should The Trans-Alaska Pipeline Be Constructed?" "The Advocates", PBS TV, Los Angeles, Calif., 11 May 1971, unofficial public service transcript, ACS Papers.

139. Frank C. Daniel, NRA's conservation director, contributed to the official hearings record. See *The American Rifleman* 119 (April 1971): 27.

140. Hearings, *Draft EIS,* vol. 6, exhibit 47.

141. *Christian Science Monitor,* 23 February 1971.

142. Fortier, "Should The Trans-Alaska Pipeline Be Constructed?" "The Advocates," ACS Papers.

143. Ruckelshaus to Morton, 12 March 1971. See *Final Environmental Impact Statement, Trans-Alaska Pipeline,* Department of the Interior, Washington, D.C., March 1971, vol. 6, A-41-52. (Hereafter referred to as *Final EIS.*)

144. Harthon L. Bill, ibid., A-105 (A-104-06).

145. Volpe to Morton, 24 March 1971, ibid., A-57 (A-53-8). The Department of Agriculture's Coordinator of Environmental Quality Activities, T. C. Byerly, was equally concerned about these environmental problems. Byerly to Morton, ibid., 5 March 1971, A-28-32.

146. Army Corps of Engineers, "Review Comments—Interior Department Environmental Impact Statement for the Trans-Alaska Pipeline." The 22-page report (dated 5 February) was submitted to the *Congressional Record* by Les Aspin (Democrat, Wisconsin). See *Congressional Record,* 92d Congress, 1st session, 117 (10 March 1971), E5950-3. Unlike all the aforementioned federal reports, this particular one was not included in the *Final EIS.*

147. M. G. Patton, "Department of Defense, Comments on the Department of the Interior Draft Environmental Impact Statement on the Trans-Alaska Pipeline," 3 March 1971, *Final EIS,* vol. 6, A-15 (A-11-27). Patton was Acting Deputy Assistant Secretary, Environmental Quality.

148. Stans to Horton, 16 April 1971, ibid., vol. 6, A-63-8.

149. Galler, ibid., A-89 (69-99).

150. Unspecified HEW official to Morton, 6 July 1971, ibid., A-102-03.

151. Tupling, "Washington Report," *Sierra Club Bulletin* 54 (June 1969), back cover.

152. Craig W. Allin, *The Politics of Wilderness Preservation* (Westport, Conn.: Greenwood Press, 1982), 19; Paul W. Gates, *History of Public Land Law Development* (Washington, D.C.: GPO, 1968), 356–86.

153. Don Dedera, "Alaska Pipeline in Limbo," San Francisco *Sunday Chronicle and Examiner,* 27 February 1972.

154. John P. Milton, "The Web of Wildness," *Living Wilderness* (Special Alaska Issue, in lieu of Winter 1971–72): 35; (1972): 16.

155. Leaflet, dated 1 June 1971, Oil and Gas File, box 42, Fairbanks Environmental Center Collection, UAF. For similar comments, see Barry Weisberg, "The Rape of Alaska," *Ramparts* 8 (January 1970): 29.

156. Glen A. Settle, "A Chronological Review of Dates Important in Antelope Valley History," Kern-Antelope History Society, Rosamund, California, (undated), 62. Sierra Club Office Files, S. F., Bancroft Library, UCB. These figures were quoted in a letter to *Not Man Apart* 3 (6 August 1973): 4.

157. The advertisement was dated 5 July 1973. The picture had previously appeared in *Living Wilderness* 32 (July 1972).

158. *Fairbanks Daily News-Miner,* 9 July 1973.

159. David Williams, Hearings, *Draft EIS,* vol. 4, 1170.

160. Ibid., vol. 3, exhibit 56. See also, "Saving Alaska," *Not Man Apart,* September 1971.

161. Klein, the leader of the Cooperative Wildlife Research Unit at the University of Alaska, interview with the author, 4 November 1984, Fairbanks.

162. Hearings, *Draft EIS,* vol. 4, Supplemental Testimony, exhibit 12, 1. Evangeline Atwood organized the Cook Inlet Historical Society in 1955 and was the author of a book about the Matanuska colonization project in Alaska during the New Deal; *We Shall Be Remembered* (1966). She was born in Sitka of Scandinavian missionary parents and trained as a social caseworker in Chicago. She returned to Alaska in 1935 with her husband, Robert, the editor-publisher of the *Anchorage Daily Times.*

163. Hearings, *Draft EIS,* vol. 4, 3.

164. Ibid., 117.

165. Ibid., 2, 5.

166. William H. Wilson, "Alaska's Past, Alaska's Future: The Uses of Historical Interpretation," *Alaska Review* 4 (Summer 1970): 1–12.

167. Hearings, *Draft EIS,* vol. 4, 2–3.

168. His elder brother, Stanley, had also been a territorial legislator and was a member of the Alaska Statehood Committee in the mid-1950s.

169. Hearings, *Draft EIS,* vol. 2, 676–77.

170. Ibid., vol. 5, exhibit 103. His father, who ran the Loman Reindeer Company on the Seward Peninsula, had been the so-called Reindeer King of Nome.

171. Ibid., vol. 1, Washington, D.C., 41.

172. Ibid. The best-known exponent of nature's ability to heal itself is the eminent French microbiologist, René Dubos. See *The Wooing of Earth: New Perspectives on Man's Use of Nature* (New York: Scribner's, 1980), especially "The Resilience of Nature," 31–49. However, all the examples of recovery he cites are taken from temperate zones.

173. Hearings, *Draft EIS,* vol. 5, exhibit 101, 5.

174. For these and previous details, see Duane Koenig, "Ghost Railway in Alaska: The Story of the Tanana Valley Railroad," *Pacific Northwest Quarterly* 45 (January 1954): 8–12.

175. Melody Webb, *The Last Frontier: A History of the Yukon Basin of Canada and Alaska* (Albuquerque: University of New Mexico Press, 1985), 217.

176. Gates, *History of Public Land Law,* 535.

177. Hearings, *Draft EIS,* vol. 10, exhibit 145. Patton had played a leading role in the development of Alaska's post-war tourist industry. Having worked on the construction of the Alaska Highway, he was the first to operate tour buses along the road after 1949. For a biographical portrait, see *Journal of the Alaska Visitors Association* 24 (January 1987): 1, 4.

178. David Wharton, *The Alaska Gold Rush* (Bloomington: Indiana University Press, 1972), 268–69; Webb, *The Last Frontier,* 308.

179. S. B. Young, ed., *The Environment of the Yukon-Charley Rivers Area, Alaska,* Contributions from the Center for Northern Studies, no. 9, Center for Northern Studies, Wolcott, Vermont, June 1976, 47.

180. Department of the Interior, *Trans-Alaska Pipeline, Final Environmental Impact Statement,* 20 March 1972, Washington, D.C., vol. 2, "Environmental Setting of the Proposed Trans-Alaska Pipeline," 136. Fire was often associated

with the deforestation, which both had long-term consequences for range use patterns. Moose flourished in the second growth deciduous forest following fire and tree cutting, whereas caribou winter habitat was destroyed and population declined. David Klein, written communication, January 1989.

181. Hearings, *Draft EIS,* vol. 4, exhibit 101.

182. Ibid., vol. 3, 676.

183. Ibid., vol. 4, exhibit 12, 4–5.

184. Ibid., exhibit 19.

185. "Canol Pipeline: A History of the Norman Wells to Whitehorse Oil Pipeline," *Beaufort* 1 (May 1982): 6–7.

186. Hearings, *Draft EIS,* vol. 7, exhibit 51, 3–4. Dalton, vol. 5, exhibit 101, makes a similar point.

187. Ibid., exhibit 51, 4.

188. Petroleum Directorate, U.S. Army, Alaska, *The Haines Pipeline,* Seattle, 17 February 1971, 1.

189. Hearings, *Draft EIS,* vol. 1, 139.

190. Petroleum Directorate, *The Haines Pipeline,* 2–3. See also, Colonel Frederic Johnson, "Ecology and the Pipeline: Lessons Learned from the Haines-Fairbanks Line," U.S. Army, Alaska, undated draft. Included as part of the hearings record *(Draft EIS)* by Julian Rice.

191. Petroleum Directorate, *The Haines Pipeline,* 2; Moxness, "The Long Pipe," 19. A subsequent tuboscope inspection of the entire line in 1970 revealed that the whole southern section was in a generally deteriorated condition. Out of a total of thirteen spill incidents recorded between 1956 and 1970, four were caused by bullet holes. The Haines to Tok Junction section of the line was "inactivated" in the fall of 1971. The Tok to Wainwright section remained open, though scarcely used, as it was cheaper to transport fuel via the Alaska Railroad and Alaska Highway.

192. Dean to the BLM, 2 March 1971, hearings, *Draft EIS,* vol. 8, Supplemental Testimony. After the Trans-Alaska Pipeline had been authorized, Friends of the Earth did try to capitalize on the Dezadeash incident. See Harvey Manning, *Cry Crisis! Rehearsal in Alaska* (San Francisco: Friends of the Earth, 1974), 133–35.

193. Hearings, *Draft EIS,* vol. 1, prepared statement, 1.

194. Ibid., vol. 1, 265–66; vol. 2, 640.

195. Thomas Kimball, executive director, NWF, to Carl R. Arnold, American Petroleum Institute, 6 October 1967, Bureau of Mines; Oil Pollution, E. L. Bartlett Papers, UAF.

196. For the views of David Klein, leader of the University of Alaska's Cooperative Wildlife Research Unit, see Tom Brown, *Oil on Ice,* 101. See also Jerry Hout, "Oil and Moose on the Kenai," *Alaska Conservation Society News Bulletin* 4 (April 1963): 4–5; Robert Weeden, "Kenai National Moose Range: A Resource Management Microcosm," ibid., 10 (Spring 1969): 9.

197. See David Klein, "Reaction of Reindeer to Obstructions and Disturbances," *Science* 173 (30 July 1971), 393–98; "Reaction of Caribou and Reindeer to Obstructions—A Reassessment," 2d International Reindeer/Caribou Symposium, 1979, *Proceedings,* edited by Reimers, E.; Gaare, E.; and Skjenneberg; S., Roros, Norway, 1979, 519–27.

198. A. Starker Leopold and Frank Fraser Darling, *Wildlife in Alaska: An Ecological Reconnaissance* (New York: Conservation Foundation, New York Zoological Society and The Ronald Press, 1953), 56–59.

CHAPTER 8. THE TRANS-ALASKA PIPELINE CONTROVERSY: 2, 1971–1974

1. *Congressional Record,* 63d Congress, 2d Session, 51 (6 December 1913), 362.

2. *Congressional Quarterly Almanac* 25 (1969): 679, 793–94; 26 (1970): 333.

3. Ibid., 27 (1971): 640.

4. Ibid., 26 (1970): 1201.

5. Ibid., 1226.

6. *Solidarity* 32 (July 1969): 3; 32 (October 1969): 8.

7. Ibid., 33 (April 1970): 4.

8. Ibid., 33 (May 1970): 9.

9. His biographers describe him as "an ardent conservationist, a facet of his life not generally known." Jean Gould and Lorena Hickok, *Walter P. Reuther: Labor's Rugged Individualist* (New York: Dodd, Mead and Co., 1972), 362.

10. *Congressional Quarterly Almanac* 29 (1973): 605.

11. Southwestern advocates of Glen Canyon Dam and Reservoir (Lake Powell) were simultaneously involved in a similar initiative that sought to shift the suit filed by conservationists against the flooding of Rainbow Bridge National Monument from Washington, D.C., to Utah.

12. *New York Times,* 22 July 1971.

13. Interview with Tom Brown, *Anchorage Daily News,* September 1970, as reprinted in *Not Man Apart* 1 (January 1971): 4. The veteran conservationist, Ira N. Gabrielson, now president of the U.S. National Appeal, World Wildlife Fund, agreed that the danger of spills in Prince William Sound was a potentially greater threat than the pipeline itself. Gabrielson believed a pipeline across Canada was ecologically superior, though he accepted the inevitability of arctic oil development. For his view, see *Animals* 14 (1972): 115.

14. In 1969 the major employers in Valdez were the State Highways Department and a major state mental health institution.

15. James T. Payne, *"Our Way of Life is Threatened and Nobody Seems to Give a Damn:" The Cordova District Fisheries Union and the Trans-Alaska Pipeline,* Alaska Humanities Forum, Anchorage, October 1985, 24–27.

16. Kenneth Simpson, hearings, *Trans-Alaska Pipeline, Draft Environmental Impact Statement,* Supplemental Testimony, vol. 7, exhibit 56, 2.

17. Ibid., vol. 1, 10–11.

18. *The Thunder Mug,* 4 May 1971, 2. The bill died in committee.

19. Richard Corrigan, "Environment report: Fishing Town Joins Legal Fight to Stop Trans-Alaska Pipeline Project," *National Journal* 3 (3 July 1971): 1400. See *CDFU versus Rogers C. B. Morton and Clifford Hardin,* Civil Action 861-71, Alaska (carton 19), Friends of the Earth Papers, Bancroft Library, University of California, Berkeley. (Hereafter, FOE Papers, UCB.)

20. *The Thunder Mug,* 4 May 1971, 1.

21. *Fairbanks Daily News-Miner,* 30 April 1971.

22. *Anchorage Daily Times,* 15 June 1971.

23. According to the *Alaska Almanac* of 1985 (26, quoting a 1909 publication), "Cheechako" is a Native term derived from a combination of the Chinook word *"chee,"* meaning "new," and the Nootka Indian word *"chako,"* meaning "come" or "approach."

24. *Cordova Times,* 4, 11 March; *Anchorage Times,* 16 June 1971; *Fairbanks Daily News-Miner,* 25 June.

25. Goeres to Evans, 2 April 1971, M. Brock Evans Papers, University of Washington Libraries, Seattle. (Hereafter referred to as Evans Papers.)

26. Wright to Marshall, 21 January 1972, George Marshall Papers, Bancroft Library, UCB.

27. Ken Roemhildt, superintendent, Point Chekalis Packers (Cordova), hearings, *Draft EIS,* Anchorage, 1971. Supplemental Testimony, vol. 7, exhibit 56(b), 3.

28. Ross Mullins, *Anchorage Daily News,* 8 April 1971.

29. Corrigan, "Fishing Town Joins Legal Fight," 1402.

30. The *Thunder Mug,* 18 August 1971, 4.

31. The *Anchorage Daily Times* called the relationship between them "a torrid romance" (22 September 1973). James T. Payne agrees that the alliance was, at best, "a marriage of convenience." Letter to the author, 23 September 1985. The relationship between the fishermen and the other groups fighting the pipeline needs to be studied.

32. Corrigan, "Fishing Town Joins Legal Fight," 1400.

33. Alderson to Kowalsky, 2 June 1971, Oil and Gas File, box 42, Fairbanks Environmental Center Collection, UAF. (Hereafter referred to as FECC.)

34. Llewellyn R. Johnson, 1 May 1972, affidavit sent to Kowalsky, FECC.

35. J. E. Murphy, 10 June 1971, FECC.

36. Mueller to Kowalsky, 10 May 1971, FECC.

37. Ott to Kowalsky, 8 June 1972, FECC.

38. *Anchorage Daily Times,* 25 April 1974.

39. *Draft EIS,* 110.

40. Bob Maguire to Kowalsky, 7 June 1972, affidavit, box 42, FECC.

41. Bane to Kowalsky, 1 April 1972, box 42, FECC.

42. Strunka to the director of the BLM, box 42, 6 March 1971, FECC.

43. The task force took its name from the section of NEPA that contained the impact statement requirement.

44. *Anchorage Daily News,* 24 June 1971.

45. See *HRDC v. Morton,* 458 F2d 827, 2 ELR 20029, Washington, D.C. Circuit, 1972; Peter Hoffman, "Evolving Judicial Standards under the National Environmental Policy Act," *Yale Law Journal* 81 (1972): 1604–5; Henry Myers, "Federal Decision-Making and the Trans-Alaska Pipeline," *Ecology Law Quarterly* (1974): 936. The NRDC is an umbrella organization that included most members of APIC.

46. Title to the land (and subsurface rights) is held by twelve regional corporations in which all Native Alaskans born before 19 December 1971 own shares. In 1991 these stocks can be sold on the open market.

47. *Fairbanks Daily News-Miner,* 9 June 1972. This is not to say that all Alaskan Natives supported ANCSA, were satisfied with its provisions, and approved of the pipeline project hereafter. The Arctic Slope Native Association opposed ANCSA because it believed that the amount of land allocated to the Inupiat (4 million acres, based on population figures) was insufficient for their needs. Some Natives opposed the project because it would disrupt their existing lifestyles. The final environment impact statement declared that construction and operation of the pipeline "could entail significant impact on the resource utilization patterns of the Alaska natives. The extent of this impact is largely a matter of speculation." (*Final EIS,* vol. 4, 416). In 1973, the Doyon Native corporation, which covers much of the interior of Alaska and contained 200 miles of the proposed pipeline within its boundaries, announced its opposition to the project.

48. Department of the Interior, *Final Environmental Impact Statement, Proposed Trans-Alaska Pipeline*, Washington, D.C., 20 March 1972, vol. 1, "Introduction and Summary," 128. (Hereafter referred to as *Final EIS.*)

49. Ibid., 122–23.

50. Ibid., 191. See also vol. 3 ("Environmental Setting Between Port Valdez And West Coast Ports"), 20.

51. Ibid., 160, 174.

52. Ibid., vol. 2 ("Environmental Setting"), 218.

53. Ibid., vol. 1, 250.

54. Ibid., vol. 2, 216–17.

55. Ibid., 217.

56. Ibid., "Values of Wilderness," 1. See also vol. 4, 313–19.

57. *Anchorage Daily Times*, 20 March 1972.

58. Natural gas has half the density of crude oil. This bulkiness means that is is unprofitable to ship natural gas by tanker.

59. See Bureau of Domestic Commerce, Department of Commerce, "Economic Effects of Opening the Oil Reserves of Alaska Through the Trans-Alaska Pipeline," *Final EIS*, vol. 7 (Supporting Analyses, II), M-4-2-39.

60. *Congressional Quarterly Almanac* 28 (1972): 565.

61. Department of the Interior Press Release, 20 March 1972, ibid.

62. Wilderness Society, Friends of the Earth, and the Environmental Defense Fund, *Comments on the Environmental Impact Statement for the Trans-Alaska Pipeline*, Center for Law and Social Policy, Washington, D.C., 4 May 1972; vol. 1, "Technical Comments"; vol. 2, "Terrestrial Impact"; vol. 3, "Marine Impact"; vol. 4, "Economics, National Security and Balancing of Alternatives."

63. Ibid., vol. 3, 4. Myers, "Federal Decision-Making and the Trans-Alaska Pipeline," emphasized that Interior was predisposed to approve TAPS (950, 955). For an evaluation that sees the impact statement as inadequate in terms of NEPA and insignificant as a contribution to Interior's decision-making, see Hoffman, "Evolving Judicial Standards under NEPA," 1613–39.

64. Secretary of Defense, "National Security Aspects of Oil Pipeline from Alaska," *Final EIS*, vol. 7 (Supporting Analyses, II), M-5-3-5.

65. Ibid., M-5-6.

66. Ibid., M-5-3-6.

67. Myers, "Federal Decision-Making and the Trans-Alaska Pipeline," 938–39.

68. Richard D. Nehring, "Future Developments of Arctic Oil and Gas: An Analysis of the Economic Implications of the Possibilities and Alternatives," Office of Economic Analysis, Department of the Interior, Washington, D.C., 10 May 1972.

69. *Congressional Quarterly Almanac* 28 (1972): 568.

70. Ibid., 29 (1973): 565.

71. David Brower, "Conservationists Blast Morton Decision on Pipeline," news release, FOE, Washington, D.C., 11 May 1972.

72. See *American Forestry* 76 (November 1970): 2; hearings, *Draft EIS*, 1971, vol. 1, 194.

73. The construction of a gas line would probably be similar to that of a hot oil pipeline. The Soviet Union had found it necessary to elevate its gas pipelines in permafrost regions.

74. U.S. Congress, Joint Economic Committee, hearings, *Natural Gas Regulation and the Trans-Alaska Pipeline*, 92d Congress, 2d session, 7–9, 22 June 1972 (Washington, D.C.: GPO, 1972), 344–45.

75. Ibid., 214–15. The basic premise of his book, which used evidence from the "counter" impact statement extensively, was the desirability of developing North Slope oil. See Charles J. Cicchetti, *Alaskan Oil: Alternative Routes and Markets* (Baltimore: Resources for the Future and Johns Hopkins University Press, 1972), 116–17.

76. Nehring, letter of resignation, ibid., 256.

77. Ibid., 254. TAPS proponents emphasized that Nehring, a 28-year old doctoral candidate in political science, had only been employed at Interior since September 1971; that this was his first job; and that he was not an economist.

78. Frederick Jackson Turner, "The Old West," 1908, in *The Frontier in American History* (Huntington, N.Y.: Robert E. Krieger, reprint, 1976), 67.

79. Turner, "The Problem of the West," 1896, ibid, 205, 212.

80. Frederic Logan Paxson, *History of the American Frontier: 1763–1893* (Boston: Houghton Mifflin, 1924), 43. John T. Juricek claims that Turner was the first to give the term frontier an abstract meaning in this sense. See "American Usage of the Word 'Frontier' From Colonial Times to Frederick Jackson Turner, *Proceedings of the American Philosophical Society* 110 (February 1966): 29.

81. *Fairbanks Daily News-Miner,* 25 July 1972.

82. Conservationists in the Seattle region rallied to fight this expansion of Puget Sound's petro-refining facilities. The best source of information for this story is the M. Brock Evans Papers (Alaska, Oil) at the Suzzallo Library, the University of Washington, Seattle. This collection contains the records of the Sierra Club's Pacific Northwest Office, of which Evans was in charge before moving to the club's office in Washington, D.C., in 1971.

83. For Anderson's testimonies, see U.S. Congress, Senate, Committee on Interior and Insular Affairs, Subcommittee on Conservation and Natural Resources, hearings, *Protecting America's Estuaries: Puget Sound and the Straits of Georgia and Juan de Fuca,* 92d Congress, 2d session, 10–11 December 1971 (Washington, D.C.: GPO, 1972), 450–81; *Natural Gas Regulation and the Trans-Alaska Pipeline,* 9 June 1972, 213.

84. Judge J. Skelly Wright, Majority Opinion, *Wilderness Society versus Morton,* 479 F.2d, D.C. Circ. 1973, 842, 848, 892, Alaska (carton 19), FOE Papers.

85. *All-Alaska Weekly,* 16 February; *Anchorage Daily News,* 24 February 1973. Vogler, who had been a loud critic of ANWR and a supporter of Rampart, would stand as the Alaska Independence Party (AIP) candidate for governor in 1974 and polled 4,770 votes. He remains Alaska's leading secessionist. For a colorful portrait, see John McPhee, *Coming Into The Country* (New York: Farrar, Straus and Giroux, 1977), 314–21.

86. Egan to Sharon Cissna, Alaska Center for the Environment, Anchorage, 24 May 1973, Oil and Gas File (Pipeline Letters), FECC.

87. An analysis of the pipeline mail for 1973 in the Gravel Papers, II. Correspondence, series 3, shows that the senator received 79 letters in support of the line and 82 in opposition. From the lower forty-eight, 35 were against the line and 41 were for it. A small majority of Alaskan correspondents were opposed (47-38).

88. State of the Union Address, 22 January 1970, *Congressional Quarterly Almanac* 26 (1970), appendix ("Presidential Messages and Statements"), 4-A.

89. *Congressional Quarterly Almanac* 29 (1973), appendix 12-A.

90. Ibid., appendix 52-A. In 1971 Nixon had submitted to Congress the first presidential energy message in American history.

91. Mike Gravel, "A Call To Action," News Release, 9 February 1971, VI, Public Relations, series 31, Press Releases 1969–80, Gravel Papers.

92. "Pipeline Needed for Alaskans to Control Alaska," *Fairbanks Daily News-Miner.* (Progress Edition), 15 March 1973.

93. Senate Interior Committee, hearings, *Rights-of-Way Across Federal Lands: Transportation of Alaska's North Slope Oil,* 93d Congress, 1st session, part 1, 9 March; part 2, 27 March, part 3, 2–3 May; part 4, 4 May 1973 (Washington, D.C.: GPO, 1973). (Hereafter, hearings, *Rights-of-Way.*)

94. *Congressional Record,* 65th Congress, 2d session, 56 (25 May 1918), H7096–98.

95. Hearings, *Rights-of-Way,* May 1973, 2.

96. Ibid., 598.

97. Gravel to Helen Long, 15 August 1969, Pipeline Correspondence, 1969, series 3, subseries 2, Gravel Papers.

98. Commencement Address, Western New England College, Springfield, Mass., 23 May 1971. Newsletter, no. 6, 1971, 12–13, VI, Public Relations, subseries 31, Gravel Papers.

99. Gravel to Roy Schaeffer, 30 May 1972, II, Correspondence, series 3, subseries 5, Gravel Papers.

100. Hearings, *Rights-of-Way,* May 1973, 56.

101. Bill Auld, "A Statement of Position and Policy on Industry and Ecology in Alaska," *Alaska Conservation Review* 14 (May 1973): 3. The *Alaska Conservation Review* was the former *Alaska Conservation Society News Bulletin.* The name changed in the fall of 1967.

102. Ibid.

103. Dan Osborne, letter to the *Fairbanks Daily News-Miner,* 31 January 1971. Osborne was chairman of the Club's Fairbanks chapter.

104. This information was received during personal interviews. Brower and McCloskey were interviewed by the author in San Francisco on 14 May and 4 June 1985 respectively.

105. Ganapole to Wayburn, 10 April 1970, Edgar Wayburn, Private Papers, San Francisco, Calif.

106. Terris Moore, "Alaska's North Slope and Our Environmentalists," *Appalachia,* June 1973. Copy without page or volume number in the Press Clippings File (1973) of the Gruening Papers. For another of Moore's references to the "new colonialists," see the *Anchorage Daily Times,* 1 May 1973.

107. *Anchorage Daily News,* 6 June 1973.

108. Hearings, *Rights-of-Way,* Wayne Smith, President, National Parks and Conservation Association; Stewart Brandborg, Wilderness Society, 249; Saunders C. Hillyer, Washington, D.C. representative, Cordova District Fisheries Union, also speaking on behalf of the United Fishermen of Alaska and the Western Division of the National Federation of Fishermen, 277; Bruce Driver, Washington, D.C., representative, Environmental Policy Center, 307; Brock Evans, Washington representative, Sierra Club, 305 (27 March).

109. Ibid., 306.

110. U.S. Congress, Senate, Committee on Interior and Insular Affairs, hearings, *The Nomination of Rogers C. B. Morton to be Secretary of the Interior,* 92d Congress, 1st session, 25–26 January 1971 (Washington, D.C.: GPO, 1971).

111. U.S. Congress, Joint Economic Committee, *The Economy, Energy and the Environment,* Washington, D.C., 11 September 1970. As quoted in the *Draft EIS,* 180.

112. The RfF report containing this conclusion was published on the heels of the hearings on the draft impact statement held in February 1971. See *Alaska Conservation Review* 12 (Spring 1971).

113. Hearings, *Natural Gas Regulation and the Trans-Alaska Pipeline,* 22 June 1972, 446.

114. *Los Angeles Times,* 3 September 1972.

115. See, for example, "East, Midwest v. West Coast," Cleveland (Ohio) *Plain Dealer,* 19 April 1973.

116. Hearings, *Rights-of-Way,* part 4, 3 May 1973, 3.

117. *Fairbanks Daily News-Miner,* 13 July 1973.

118. Alderson to Brower and Kowalsky, 20 November 1972, Oil and Gas File, box 42, FECC.

119. Alderson to FOE regional representatives, FECC

120. Evans to Ken G. Farquarson, Sierra Club of British Columbia, 30 May 1973, Alaska, box 8, Sierra Club Office Files (San Francisco), Bancroft Library, UCB.

121. *Congressional Record,* 93d Congress, 1st session, 119 (17 July 1973), S24323.

122. Lille D'Easum, "Should We Develop The North?" May 1973, Alaska, box 8, Sierra Club Office Files, UCB. (Hereafter referred to as SCOF.)

123. Bohlen to McCloskey, 18 June 1973, Alaska, box 8, SCOF; Bohlen to Evans, 11 July 1973, SCOF; British Columbia Environmental Council to Evans, 8 March 1973, Evans Papers.

124. Bohlen to McCloskey, 18 June 1973, SCOF.

125. McCloskey to Bohlen, 6 July 1973, SCOF.

126. Harvey Manning, "Which Way Out?" *Not Man Apart* 3 (July 1973): 8.

127. The 453-mile road was finished and opened to the public in 1979.

128. See FOE advertisement, "The Wrong Way Out at the Wrong Time," July 1973. Also, the testimony of Stewart Brandborg, director of the Wilderness Society, hearings, *Rights-of-Way,* March 1973, 326. For the position of the railroad interests, see David Morgan, "Oil by Rail," *Environment* 14 (October 1972): 30–32. Morgan was the editor of *Trains.*

129. Manning, "Which Way Out?" 9.

130. CIGGT, *Railway to the Arctic: A Study of the Operational and Economic Feasibility of a Railway to Move Arctic Slope Oil to Market,* Queen's University, Ontario, January 1972.

131. David Klein, "Reaction of Reindeer to Obstructions and Disturbances," *Science* 173 (30 July 1971): 393–98. Klein cites a mean annual loss of 2,200 in Sweden between 1955 and 1964.

132. Manning, "Which Way Out?" 8.

133. Ibid.

134. Ibid., 8. See also Harvey Manning, *Cry Crisis! Rehearsal in Alaska* (San Francisco: Friends of the Earth, 1974), 203.

135. Manning, "Which Way Out?" 8.

136. These figures were substantiated in the final pipeline impact statement. See "Comparative Ranking of Environmental Impacts of Alternative Pipeline and Tanker Routes and Related Terminal Ports," "Unavoidable Environmental Impacts," vol. 5, table 1B. The Prudhoe Bay to Valdez route rates "C" on a scale of "A" (least) to "G" (worst). A Trans-Canadian route inland from Prudhoe Bay to Edmonton rates "F."

137. Manning, "Which Way Out?" 8.

138. Gravel, 6th Public Relations World Congress, Geneva, April 1973. Gravel appeared at this conference as lead speaker on the energy crisis, part of his

tour to promote the Alyeska pipeline. As reported in the *Anchorage Daily Times,* 17 April 1973.

139. Brewer to Gravel, 10 May 1973; as inserted in the *Congressional Record,* 93d Congress, 1st session, 119 (29 May 1973): 17098–100.

140. Hearings, *Rights-of-Way,* 290.

141. Ibid., 403.

142. Ibid., 43–44.

143. Barry Kay, "Stopping the Pipeline is Northern Native Cause," *Oilweek* 23 (20 March 1972): 20–21.

144. *Congressional Quarterly Almanac* 29 (1973): 603. See also Manning, *Cry Crisis!* Appendix II, "The Systematic Deception About the Canadian Alternative," 301–7.

145. The previous occasion was in 1969 when his vote had defeated an amendment that proposed to limit antiballistic missile development.

146. *Congressional Quarterly Almanac* 29 (1973): 600.

147. Young had replaced Congressman Nick Begich, whose plane disappeared on 16 October 1972 during a flight from Anchorage to Juneau.

148. *Baltimore Sun,* 18 July 1973.

149. *Congressional Record,* 93d Congress, 1st session, 119 (2 August 1973): 27627.

150. *Congressional Quarterly Almanac* 29 (1973): 608–9.

151. *Congressional Record,* 119 (2 August 1973): 27649.

152. Ibid., 27665, 27664, 27672, 27694, 27695.

153. Under the pressure of energy concerns, there was something of a general backlash against environmentalism in the latter part of 1973. This could be seen in the push for more offshore drilling and the relaxation of air quality standards. See Robert Sommer, "Ecology and the energy shortage," *The Nation* 217 (10 December 1973): 615–16.

154. Evans to Michael McCloskey, 13 September 1973, SCOF.

155. Wayburn, memorandum ("The Trans-Alaska Pipeline Affair") to Judge Ray Sherwin, 17 September 1973, SCOF. These memoranda had been written in response to an allegation by Judge Ray Sherwin, Superior Court, Solano County, Fairfield, Calif., the club's president, that efforts to prevent the authorization of the pipeline, "one of the great challenges in our history," had been inadequate. Sherwin's allegations aroused indignation among officials like Evans and Wayburn who had been closely associated with the project since 1968. Evans thought that Sherwin's assessment was ill-considered, harsh, and misinformed. See Sherwin, "memorandum to the Board of Directors, Council of the Sierra Club, Chapter Executive Committees, and Selected Sierra Club Leaders," 20 August 1973. Sierra Club Papers, William E. Colby Library, Sierra Club National Headquarters, San Francisco.

156. Evans to McCloskey, 13 September 1973, SCOF.

157. Ibid.

158. Wayburn to Sherwin, 17 September 1973.

159. *Congressional Quarterly Almanac* 29 (1973), appendix, 73-A.

160. *Congressional Record,* 93d Congress, 1st session, 119 (12 November 1973): 36599. There is an interesting parallel with the Hetch Hetchy controversy here. In the aftermath of the earthquake and fire that destroyed San Francisco in 1906, the city's mayor, James D. Phelan, claimed that if his original application to build a reservoir in Hetch Hetchy Valley had been approved in 1903, there would have been enough water to fight the fire.

161. *Congressional Quarterly Almanac* 29 (1973), appendix, 91-A.

162. Ibid., appendix, 93-A.

163. Don Young, *Washington Report,* Winter 1973–4, 1. Fairbanks Environmental Center Files, box 44.

164. Memorandum re. possibilities for further legal challenges to the authorization of the Trans-Alaska Pipeline, 16 May 1973 (marked confidential), George Marshall Papers, UCB.

165. Confidential memorandum, John F. Dienelt and Joseph N. Onek, re. Pipeline Litigation: Recommendations for Future Action, 5 September 1973, 18, 20, George Marshall Papers.

166. Arctic Institute of North America, *Arctic Environmental Council, Report on the On-Site Visit to the Trans-Alaska Pipeline System, October 1974,* Washington, D.C., 1, Alaska: Oil, box 8 (Arctic Environmental Council), SCOF.

167. George Alderson to Robert Faylor, director of AINA, 11 December 1973, ibid.

168. Stewart Brandborg to Robert Faylor, 11 February 1974, ibid.

169. Brock Evans to Gil Zemansky, 12 October 1977, SCOF. Many conservationists, including members of the Sierra Club's Alaska branch, believed that the club had compromised itself by joining. Gerald R. Brookman, vice chair, Sierra Club, Alaska chapter, to Brock Evans, 12 September 1975, SCOF.

170. Weeden resigned shortly after joining the council when he assumed state office in the incoming administration of Governor Jay Hammond.

CHAPTER 9. THE PIPELINE PROJECT

1. *Fairbanks Daily News-Miner,* 26 September 1969.

2. Robert Sherrill, "Unsafe At Any Width: The Trans-Alaska Pipeline," *The Nation* 218 (11 June 1973): 745.

3. Brock Evans, Sierra Club Office Files, Alaska: Oil, box 8 (Arctic Environmental Council), Bancroft Library, University of California, Berkeley (UCB). (Hereafter referred to as SCOF.)

4. Evans to Fisher, 8 November 1974, SCOF.

5. Arctic Institute of North America (AINA), *Arctic Environmental Council Report on the On-Site Visit to the Trans-Alaska Pipeline System, October 1974,* Washington, D.C., 1974, 23, SCOF.

6. Ibid, SCOF. For similar comments by Evans, see, "Two Roads North," *Sierra* 59 (November-December 1974): 19. Evans was interpreting "the North" loosely. Jack London never visited the Brooks Range or the North Slope.

7. Harvey Manning, *Cry Crisis! Rehearsal in Alaska* (San Francisco: Friends of the Earth, 1974), introduction.

8. Robert W. Weller, "Civilization Threatens 10 Alaskans," *International Herald Tribune,* 14 October 1974, 14. See also Spokane (Wash.) *Spokesman-Review,* 7 October 1974.

9. "Alaska Pipeline Road: A Tourist Mecca," *New York Times,* 29 September 1974. For conservationist fears, see *Not Man Apart* 4 (December 1974): 4–5.

10. A study of the impact on the socioeconomic fabric of Fairbanks by a community research group, the Fairbanks Impact Information Center, had been completed by its director by the time the pipeline project was finished. See Mim Dixon, *What Happened to Fairbanks? The Effects of the Trans-Alaska Oil Pipeline on the Community of Fairbanks, Alaska,* Social Impact Assessment

Series (Boulder, Colo.: Westview Press, 1978). See also, Peter Gruenstein and John Hanrahan, *Lost Frontier: The Marketing of Alaska* (New York: W. W. Norton, 1977), 167–76.

11. For example, Richard Fineberg, "X-Ray Faking—A Latter Day Alaskan Mellerdrama," *All-Alaska Weekly*, 24 October 1975, 1, 8–10. Fineberg had been a crew member aboard the ship that sailed into the Aleutian test zone to protest against the Cannikin nuclear blast of 1971.

12. For details of contemporary press allegations, see David A. Remley, *Crooked Road: A History of the Alaska Highway* (New York: McGraw-Hill Book Co., 1976), 242–43.

13. *Boston Globe*, 19 November 1975.

14. David Lamb, "Alaska Town Weathers the Pipeline Rush," *Los Angeles Times*, 20 May 1975. See also, Ed Fortier, "Alaska Pays The Pipers," *The National Observer*, 20 September 1975; Winthrop Griffith, "Blood, toil, tears and oil," *New York Times Magazine*, 27 July 1975; "Rush for Riches," *Time* 105 (2 June 1975); Jack Anderson, "A Pipeline Through the Last Frontier," *Washington Post*, 9 May 1976; Michael Roberts, "Dissolute Alaska: The Coming of the Pipe," *New Republic* 173 (1 November 1975): 17–21; Bryan Hodgson, "The Pipeline: Alaska's Troubled Colossus," *National Geographic* 150 (November 1976): 684–717; Ron Rau, "The Taming of Alaska," *National Wildlife*, November 1976, 19–23.

15. Advertisement, *Alaska Industry* 8 (June 1976): 96.

16. Frederic Logan Paxson, *When the West is Gone* (New York: Henry Holt and Co., 1930), 96, 115.

17. As quoted in William J. Tobin, "Economic Impact Beyond Expectations," *Alyeska Reports*, 1980, 10.

18. Peter De May, Alyeska Course, Tanana Valley Community College (TVCC), Fairbanks, 21 October 1975, Oral History Collection, UAF, Tape U75-10.

19. AINA, *Arctic Environmental Council, Report on the On-Site Visit to the Trans-Alaska Pipeline System*, October 1974, 7.

20. Evans to Bob Faylor, 31 December 1975, SCOF.

21. AINA, *Arctic Environmental Council, Second Report on the On-Site Visit to the Trans-Alaska Pipeline System*, May 1975, Washington, D.C., 2. Evans was always a critical member of the council. He protested against all its reports following the four on-site visits between 1974 and 1977, which were invariably praiseworthy. Robert B. Weeden had been replaced by David Klein as the sole Alaskan member. The former Project Chariot researcher, Norman J. Wilimovsky, Institute of Animal Resource Ecology, University of British Columbia, was also a member. Klein resigned on 29 June 1976, citing the council's ineffectiveness due to its financial dependence on Alyeska. See Klein's letter of that date to Fred Armstrong, the chairman of the council, SCOF.

22. E. F. Rice, "The Shakedown Years," *Alyeska Reports*, Special Issue, September 1980, 4. *Alyeska Reports* was Alyeska's public information journal.

23. Alyeska Pipeline Service Company (ALPS), "Conventional Burial," Data Sheet B-1, Alyeska Pipeline Service Company, Anchorage, July 1977; "Cathodic Protection," Data Sheet C-5, April 1976.

24. ALPS, "Heat Pipes," Data Sheet C-2, November 1976.

25. ALPS, "Insulation (Supports)," Data Sheet C-3, November 1976; "Insulation (Pipe)," Data Sheet C-1, November 1976.

26. ALPS, "Special Burial Designs," Data Sheet B-2, Anchorage, July 1977.

27. ALPS, "Zigzag Configuration," Data Sheet No. 8, February 1976.

28. ALPS, "Anchor Supports," Data Sheet C-6, January 1979.
29. A special issue of *Alyeska Reports* (vol. 2, no. 4, September 1976) was devoted to the welding controversy and the defense of Alyeska's welding standards.
30. ALPS, "Gate and Check Valves," Data Sheet, August 1975.
31. ALPS, "Drainage Structures," Data Sheet No. 9, December 1975.
32. ALPS, "Work Pad Insulation," Data Sheet D-3, Alyeska Pipeline Service Company, Anchorage, July 1977.
33. ALPS, "Revegetation," Data Sheet D-5, Anchorage, February 1979.
34. ALPS, "Pump Station Facilities," Data Sheet No. 14, Alyeska, Anchorage, March 1976; "Topping Plant," Data Sheet E-3, August 1976.
35. ALPS, "Ballast Treatment Facility," Data Sheet F/1, October 1976.
36. ALPS, *Facts; Trans-Alaska Pipeline,* January 1985, 34, 27. This figure included contractors' employees as well as Alyeska personnel.
37. *Alaska Industry* 6 (April 1974): 32.
38. Leo Marx, *The Machine in the Garden: Technology and the Pastoral Ideal in America* (New York: Oxford University Press, 1967), 214; *Congressional Record,* 93d Congress, 1st session, 119 (16 July 1973), 24081.
39. Atlantic Richfield, *Alaska Industry* 6 (March 1974): 9.
40. Paul Thomas, "Engineers: Designing for the North," *Alaska Industry* 8 (April 1976): 86. See also the federal stipulations for the construction of the pipeline, "Aesthetics," ibid., 37, regarding paint color for permanent structures.
41. Peter J. Brennan, "The Trans-Alaska Pipeline: Start-up and Operation," *Alyeska Reports,* reprint, January 1977, 1.
42. Ibid., 4.
43. Lawrence J. Allen, *The Trans-Alaska Pipeline: Vol. 1, The Beginning* (Seattle: Scribe Publishing Co., 1975), 29–30.
44. Kristina Lindbergh and Barry L. Provorse, *The Trans-Alaska Pipeline: Vol. 3, Emerging Alaska* (Seattle: Scribe Publishing Co., 1977), 66. John Gardey, *The Sophisticated Wilderness* (London: Wilton House Gentry, 1976) also reports this example (166).
45. Lindbergh and Provorse, *The Trans-Alaska Pipeline,* ibid.
46. Ibid., 69.
47. Roscow, *800 Miles to Valdez: The Building of the Alaska Pipeline* (Englewood Cliffs, N.J.: Prentice-Hall Inc., 1977), 98.
48. Ibid., 183–84.
49. See David Norton, "Pipeline Surveillance from the Inside," *Alaska Conservation Review* 16 (Summer-Fall 1975), 4–7. Norton, an Alaska Department of Fish and Game biologist, was in charge of the state half of the JSFWAT.
50. Zemansky represented FOE's Seattle branch at the 27th Alaska Science Conference in 1976, where he gave a critical paper based on his findings: Gil M. Zemansky, "Environmental Non-Compliance and the Public Interest During Construction of the Trans-Alaska Pipeline," *Science in Alaska,* vol. 2 (College: Alaska Division, AAAS, 1976), 26–37. For another report by Zemansky see, "Symbolic Environmental Monitoring on the Trans-Alaska Oil Pipeline," Corvallis, Oregon, June 1977. Zemansky's doctoral dissertation was based on this subject: "Water Quality Regulation During Construction of the Trans-Alaska Oil Pipeline System" (Ph.D. diss., University of Washington, Seattle, 1983). Many of the state and federal documents upon which he based his findings were obtained by lawsuit and the Freedom of Information Act.
51. Evans, Field Notes, 3, October 1975, SCOF. Sierra Club Papers, UCB. Michael McCloskey, the Sierra Club's former executive director, has claimed

that by mitigating the pipeline project's potential destructiveness, conservationists had substantially achieved their goal. Interview with the author, San Francisco, 5 June 1985.

52. Brower, memorandum, 4 January 1977, carton 20 (Alaska), Friends of the Earth Papers, Bancroft Library, UCB.

53. For representative comment, see Richard James, "Alaska Pipeline Opens on Time, but Short-cuts Scar the Environment," *Wall Street Journal*, 20 June 1977.

54. Michael Storper, Laura Baker, and Mary Lou Seaver, "Too Much, Too Soon (Too Bad)," *Not Man Apart* 7 (February 1977): 1–4; *Washington Post*, 28 April 1976.

55. William R. Wood, *Not From Stone*, University of Alaska Foundation, College, Alaska, 1983, 43.

56. Roscow, *800 Miles to Valdez*, 184.

57. See David Klein, "The Alaska Oil Pipeline in Retrospect," *Transactions of the 44th North American Wildlife and Natural Resources Conference*, 1979, (Washington, D.C.: Wildlife Management Institute, 1979) 241; "Reaction of Caribou and Reindeer to Obstructions—A Reassessment," *Proceedings of the 2nd International Reindeer/Caribou Symposium*, 1980, 523.

58. Stephanie Pain, "Alaska Lays its Wildlife on the Line," *New Scientist* 114 (30 April 1987): 52; R. T. Shideler, "Impacts of human developments and land use on caribou: A literature review." Vol. II. Impacts of oil and gas development on the Central Arctic herd. Technical Report No. 86-3, Alaska Department of Fish and Game, Juneau, 1986. Severe predation by insects can lead to deterioration in body condition, weight loss, and, perhaps, increased winter mortality. For a description of these possible effects, see U.S. Fish and Wildlife Service, Department of the Interior, *Arctic National Wildlife Refuge, Alaska, Coastal Plain Resource Assessment: Report and Recommendation to the Congress of the United States and Final Legislative Environmental Impact Statement*, (Washington, D.C.: GPO, 1987), vol. 1, 122. (Hereafter, *ANWR, Report and Recommendation*.)

59. Klein, "The Alaska Oil Pipeline in Retrospect," 240–41; Thomas A. Morehouse, "Fish, Wildlife, Pipeline: Some perspective on protection," *The Northern Engineer* 16 (Summer 1984): 18–26. Morehouse's article was based on a report prepared for the USFWS. In 1977 the USFWS contracted with the University of Alaska (Anchorage), Institute of Social and Economic Research for an investigation of the effectiveness of fish and wildlife surveillance during the construction of the pipeline. According to Morehouse, many environmental considerations were ignored in the interests of speedy construction and the overseeing bodies were reluctant to enforce stipulations when they conflicted with Alyeska's timetables.

60. David Klein, as quoted in Pain, "Alaska Lays its Wildlife on the Line," 53. See also James R. Udall, "Polar Opposites," *Sierra* 72 (September-October 1987): 45. *ANWR, Report and Recommendation*, vol. 1, also notes the large decrease in the number of wolves in the mid-1970s, and how this has coincided with the increased size of the Central Arctic herd (121).

61. The authors of the final environmental impact statement had warned that this might occur. See *Trans-Alaska Pipeline, Final Environmental Impact Statement*, 1972, vol. 1, 120. They were also sceptical about the ability of native grasses to eventually reclaim the reseeded areas because of the new environmental conditions established in the construction zone. Ibid., vol. 1, 116; vol. 4, 103. The recent report referred to is U.S. Fish and Wildlife Service, Fairbanks,

Fish and Wildlife Enhancement Office, "Comparison of actual and predicted impacts of the Trans-Alaska Pipeline System and Prudhoe Bay oilfields on the North Slope of Alaska," Fairbanks, December 1987, 48.

62. Klein, "The Alaska Oil Pipeline in Retrospect," 243. The authors of the section on large mammals in the final environmental impact statement for the pipeline had warned that this could happen. See vol. 4, 150–51. For the recent assessment, see USFWS, "Comparison of actual and predicted impacts," 38.

63. The final environmental impact statement for the pipeline project had anticipated this problem. See vol. 1, 113, 121. This particular matter has not yet been properly studied. However, it seems that, to date, the effect of atmospheric pollutants on lichens has been "negligible," thanks to the use of low sulphur fuels and the production of low sulphur crude at Prudhoe. USFW Service, "Comparison of actual and predicted impacts," 37, 43.

64. Any oil that reaches the Beaufort Sea in summer is likely to be confined along the ice-free coastal belt. Due to the relative absence of tidal action and waves, the oil will probably remain concentrated. It will be extremely difficult to clean up any oil that finds its way under the ice pack. Given the low temperatures of the Arctic, the natural processes of oil evaporation, ageing, and biodegradation will be very slow.

65. Pain, "Alaska Lays its Wildlife on the Line," 51–55.

66. *Anchorage Daily News,* 21, 26, 28 March, 4, 5 April, 5 May, 11, 12, 13, 15, 21 June.

67. Ibid., 23 July 1985. For details see James T. Payne, *"Our Way of Life is Threatened and Nobody Seems to Give a Damn:" The Cordova District Fisheries Union and the Trans-Alaska Pipeline,* Anchorage, October 1985, 169–80.

68. *Anchorage Daily News,* 4 April 1985.

CHAPTER 10. AFTER THE PIPELINE

1. Kowalsky, memorandum to Brower and FOE Legal Committee, 18 November 1976, carton 19 (Alaska), Friends of the Earth (FOE) Papers, Bancroft Library, University of California, Berkeley. (Hereafter referred to as FOE Papers.)

2. Rice to Pam Rich, FOE, Washington, D.C., 10 November 1976, carton 22 (Alaska), FOE Papers. The Alaska Conservation Society disbanded in the late 1970s. Many of its members joined Trustees for Alaska.

3. This should be distinguished from the Devil Canyon proposal, that Alaskan conservationists supported in the 1960s as an alternative to Rampart Dam. Susitna, further upstream, bore little resemblance to Devil Canyon. Susitna, costing $1.5 billion, would have been one of the biggest Corps of Engineers projects in American history, flooding 50,500 acres. For a variety of reasons, Susitna has never progressed beyond advocacy.

4. Rice to Rich, 10 November 1976, carton 22 (Alaska), FOE Papers.

5. Kowalsky to Michael McCloskey, 24 October 1973, Oil and Gas File, box 46, Fairbanks Environmental Center Collection, UAF.

6. *Philadelphia Inquirer,* 28 November 1976.

7. This extract from Robert Page's testimony at Yellowknife on 27 April 1987 is quoted in Martin O'Malley, *The Past and Future Land: An account of the Berger Inquiry into the Mackenzie Valley pipeline* (Toronto: Peter Martin Associates, 1976), 62.

8. The Berger Commission, *The Berger Report: Northern frontier, Northern homeland: The Report of the Mackenzie Valley Pipeline Inquiry*, (Ottawa: Department of Indian Affairs and Northern Development, Canadian Government Printing Office, 9 May 1977). The report was distilled from hearings in 45 communities. The full record consisted of 281 volumes and included 40,791 pages of testimony. The final report was 245 pages. Extracts were published in *Living Wilderness* 41 (April/June 1977). The following references are taken from these extracts.

9. *The Berger Report*, "Environmental Attitudes and Environmental Values," *Living Wilderness*, 6.

10. See Richard K. Nelson, *Make Prayers to the Raven: A Native View of the Koyukon Forest* (Chicago: University of Chicago Press, 1983), 245–46; "Cultural Values of the Land," in *Native Livelihood and Dependence, A Study of Land Use Values Through Time*, National Petroleum Reserve in Alaska, Field Study 1, Department of the Interior, 105(c) Land Use Study, Anchorage, June 1979, 27–36.

11. Nelson, *Make Prayers to the Raven*, 242.

12. *Alaska Business and Industry* (formerly *Alaska Industry*) 15 (April 1983): 42, 60–61.

13. Moreover, like oil from the North Slope, arctic gas requires presidential approval to allow exportation.

14. See prehearing press conference, 9 September 1974, as reported in the *Alaska Conservation Review* 17 (1976): 4.

15. Tina Stornorov, ibid. 18 (Spring 1976): 5.

16. Peter J. Brennan, "The Trans-Alaska Pipeline: Start-up and Operation," *Alyeska Reports*, January 1977.

17. Jay Hammond, statewide radio address, 18 January 1978, Haul Road File, Alaska Conservation Society (ACS) Papers, UAF. Jay Hammond, who ran on a "controlled development" ticket, narrowly defeated the incumbent Democrat William Egan for the governorship in 1974. Hammond had defeated Walter Hickel and Keith Miller in the contest for the Republican nomination.

18. David Cline to Jay Hammond, 5 January 1979, Haul Road File, ACS Papers. Cline was the National Audubon Society's Alaska representative.

19. For a review of the conflicting views on the road, see "Two Throughways to the Arctic," *Time* 113 (14 May 1979): 62–63. The second road referred to in the title was Canada's Dempster Highway.

20. Evans, memorandum to Judge Ray Sherwin, 3, Sierra Club Papers, William E. Colby Library, Sierra Club, San Francisco. See also Wayburn's memorandum to Sherwin, Sierra Club Papers. Michael McCloskey has denied that the fight against the Alaska Pipeline assumed the dimensions of a holy cause for the Sierra Club, whose ultimate strategic objective in Alaska was the "d(2)" settlement. Interview with author, San Francisco, 4 June 1985.

21. For an account of the political struggle, see Roderick Nash, *Wilderness and the American Mind* (New Haven: Yale University Press, 3d ed., rev. 1982), 292–315. ANILCA established a new category of federal protection; the national preserve. Sport hunting and trapping are permitted in preserves and so are "traditional" forms of mechanized access (airplane, snowmobile).

22. U.S. Congress, House, Committee on Interior and Insular Affairs, hearings, *Inclusion of Alaska lands in National Parks, Forests and Wildlife Refuges, and Wild and Scenic River Systems*, 95th Congress, 1st session, 14 May 1977 (Washington, D.C.: GPO, 1977), 2.

23. *Congressional Record*, 96th Congress, 2nd session, 126 (19 August 1980): S11202.

24. One session of the hearings was held in the Native village of Kaktovik (population circa 200), which is located on one of the Barrier Islands, just inside the northern boundary of ANWR. (The townsite itself is not managed as part of the refuge.) The others were in Anchorage and Washington, D.C. Many Alaskan conservationists believed there should have been hearings in Fairbanks as well.

25. James Eason, the department's director, as quoted in *Tundra Times*, 26 January 1987.

26. Ibid.

27. See *ANWR, Report and Recommendation*, vol. 2, 1.

28. *Tundra Times*, 23 February 1987.

29 *ANWR, Report and Recommendation*, vol. 1, introduction by the Secretary of the Interior, Donald P. Hodel, 1; conclusion, 185, 188, 190.

30. Donald P. Hodel, *ANWR, Report and Recommendation*, 181–82.

31. See David Lamb, "Boom or Bust—Alaska Is Unique Land; Spending Spree Ends," *Los Angeles Times*, 6 August 1987. Lamb quotes figures that estimate that a total of 15,000 people had already left the state in 1987.

32. ANWR, *Report and Recommendation*, 181, 186. The Kuparuk River field is located forty miles west of Prudhoe Bay and contains an estimated 1.3–1.5 billion barrels of recoverable oil. Oil flows to Pump Station 1 through an elevated, 24 inch pipe with a minimum of 5 feet of clearance for caribou. There are also gravel ramps to facilitate animal crossings.

33. Ibid., 190.

34. Ibid., vol. 1, 2, 187; vol. 2, O-7, O-9, O-28, O-38, O-66, O-68, O-69, O-88, O-121, O-129, O-135, O-295.

35. *Tundra Times*, 23 February 1987.

36. Alaska Oil and Gas Association, *ANWR, Report and Recommendation*, vol. 2, O-26, O-29.

37. Ibid., 5.

38. U.S. Fish and Wildlife Service, Fairbanks, Fish and Wildlife Enhancement Office, "Comparison of actual and predicted impacts of the Trans-Alaska Pipeline System and the Prudhoe Bay Oilfields on the North Slope of Alaska," Fairbanks, Alaska, December 1987, 51.

39. Ibid., 9, 42, 51.

40. Ibid., 17, 19–20, 22, 28, 43–44.

41. Ibid., 21, 23, 45.

42. Ibid., 31–35, 48.

43. Ibid., 9, 39–40, 47–48.

44. Ibid., 48.

45. Ibid., 36, 49.

46. *ANWR, Report and Recommendation*, vol. 2, David Cline, Anchorage Audubon Society, O–189; Ginny Wood, P-110–11; Celia Hunter, P-34–36.

47. Huge quantities of water, scarce in the 1002 area, are required during exploration for drilling purposes. Large amounts of gravel are needed to construct drilling pads, air strips, and access roads. Ibid, vol. 1, 84.

48. Ibid., vol. 1, 144–45, 153, 164; vol. 2, 37, 144–45, 152, 164. There are some "intrusions" in the wilderness, in the shape of abandoned DEWline airstrips and "deactivated" (1963) DEWline stations at Camden Bay and Beaufort Lagoon. The DEWline station at Barter Island, half a mile from Kaktovik, lies outside the wildlife refuge.

49. *ANWR, Report and Recommendation*, vol. 2, Appendix, Public Comments and Responses, O-300.

50. Ibid., O-419.

51. Michael L. Fischer, *Sierra* 72 (July-August 1987): 6. See also, ibid., (September-October): 41–48.

52. *ANWR, Report and Recommendation*, vol. 2, F-1–27. The ANWR coastal plain is an important staging area for snow geese, which are sensitive to disturbance by aircraft (ibid., vol. 1, 133). Polar bears maintain denning sites along the coast.

53. Ibid., vol. 2, S-19, S-32–34.

54. Joseph J. Tritt, ibid., S-17, S-41.

55. Walter J. Hickel, "Dramatic Development Ahead," *Alaska Industry* 6 (March 1974): 40. In this article, Hickel expressed relief that Alaska was finally moving forward in terms of economic development and described Alaska as a global storehouse of oil, coal, timber, and protein (krill).

56. "Pipeline Monument (Dedicated to the Men and Women Who Built the trans Alaska pipeline)," Anchorage, 1984.

57. Alyeska, *Facts: Trans-Alaska Pipeline*, Anchorage, January 1985, 34.

58. TAPS, *Draft EIS*, 1971, 10.

59. Morgan B. Sherwood, *Big Game in Alaska: A History of Wildlife and People* (New Haven: Yale University Press, 1981), 1, 76–77, 152.

60. Avery Craven, "Frederick Jackson Turner, Historian," in O. Lawrence Burnette, ed. *Wisconsin Witness to Frederick Jackson Turner: A Collection of Essays on the Historian and the Thesis* (Madison: State Historical Society of Wisconsin, 1961), 104.

61. Turner, "The West and American Ideals," in *The Frontier in American History* (Huntington, N.Y.: Robert E. Krieger Publishing Co., reprint, 1976), 293.

62. Morgan B. Sherwood, "The End of American Wilderness," *Environmental Review* 9 (Fall 1985): 200–201.

63. Patricia Nelson Limerick, *The Legacy of Conquest: The Unbroken Past of the American West* (New York: W. W. Norton, 1987), 24.

64. Ibid., 26.

65. Ibid., 145.

66. Ibid., 99, 100.

67. See Henry Nash Smith, *Virgin Land* (Cambridge, Mass.: Harvard University Press, 1971), 258–59; David M. Potter, *People of Plenty: Economic Abundance and the American Character* (Chicago: University of Chicago Press, 1954), 148–49.

68. Potter, *People of Plenty*, 160–61.

69. Ibid., 163–64.

70. James C. Malin, "Mobility and History: Reflections on the Agricultural Policies of the United States in Relation to a Mechanized World," *Agricultural History* 17 (1943): 178.

71. Ray Allen Billington, *America's Frontier Heritage* (New York: Holt, Rinehart and Winston, 1966), 25.

72. James C. Malin, "On the Nature of the History of Geographical Area, With Special Reference to the Western United States," in *History and Ecology: Studies of the Grassland*, Robert P. Swierenga, ed. (Lincoln: University of Nebraska Press, 1984), 130–31. Malin felt that Turner's approach to the history of the American frontier was far too narrow in time and place. He pioneered a bold and expansive approach to the study of the frontier in history as opposed to the study simply of the Anglo-American displacement of Native Americans between the seventeenth and nineteenth centuries. Malin believed that 10,000

years of history in the Americas was the appropriate timespan and his approach concentrated on the conflict over resources between indigenes and invaders.

73. Howard R. Lamar and Leonard Thompson, eds., *The Frontier in History: North America and Southern Africa Compared* (New Haven: Yale University Press, 1981), 7.

74. Ibid., Epilogue, 315.

75. Ibid., 311.

Select Bibliography

1. ARCHIVAL COLLECTIONS

I.I UNIVERSITY OF ALASKA ARCHIVES, FAIRBANKS (UAF)

Alaska Conservation Society Papers
E. L. Bartlett Papers
John A. Clark Papers
Anthony J. Dimond Papers
Fairbanks Chamber of Commerce Collection
Fairbanks Environmental Center Collection
Don C. Foote Papers
Mike Gravel Papers
Ernest Gruening Papers
Howard W. Pollock Papers
Ralph J. Rivers Papers

I.2 BANCROFT LIBRARY, UNIVERSITY OF CALIFORNIA, BERKELEY

Friends of the Earth (FOE) Papers
George Marshall Papers
Robert Marshall Papers
Sierra Club Office Files (San Francisco and Washington, D.C.)

I.3 WILLIAM E. COLBY LIBRARY, SIERRA CLUB, SAN FRANCISCO, CALIF.

Sierra Club Papers

I.4 PACIFIC NORTHWEST COLLECTION, MANUSCRIPTS DEPARTMENT, SUZZALLO LIBRARY, UNIVERSITY OF WASHINGTON LIBRARIES, SEATTLE

M. Brock Evans Papers

PRIVATE COLLECTION

Edgar Wayburn Papers, San Francisco

2. Interviews

David R. Brower, 14 May 1985, San Francisco.
Celia Hunter, 15 September 1985, Fairbanks. Interview conducted with William S. Schneider, curator of oral history, University of Alaska, Fairbanks.
David R. Klein, 4 November 1984, Fairbanks.
Michael J. McCloskey, 4 June 1985, San Francisco.
William R. Wood, 4 September 1984, Fairbanks.

3. Oral Sources

3.1 TAPE RECORDINGS (ORAL HISTORY COLLECTION, UAF)

Rampart Dam Conference. McKinley Park Hotel, Mount McKinley National Park, 7 September 1963. Recorded by Tom Snapp.
"Yukon Power for America Versus Conservation Society on Rampart Dam." Television Debate, Fairbanks, 17 April 1965.
Woodrow Johansen, The Construction of the Ice Road (Hickel Highway). Speech, American Institute of Mining, Metallurgical and Petroleum Engineers, 26 May 1969, Fairbanks.
Frank Moolin, Lecture, Alyeska Course. Tanana Valley Community College, Fairbanks, Fall 1975. Oral History Collection, Tape U75-10.

3.2 TRANSCRIBED TAPES

Howard Rock, interview by Levi Lott (transcribed by Ronald K. Inouye, October 1985). Oral History Collection, UAF.
George W. Rogers, interview conducted and transcribed by Ronald K. Inouye, April 1986. Oral History Collection, UAF.

4. Government Publications

4.1 UNITED STATES

4.1.1 U.S. GOVERNMENT PRINTING OFFICE (GPO), WASHINGTON, D.C. (UNLESS SPECIFIED)

Alaskan International Highway Commission. *Report of the International Highway Commission to the President.* 76th Cong., 3d sess., House document no. 711, April 1940.
Army Corps of Engineers. *Yukon and Kuskokwim River Basins, Harbors and Rivers in Alaska Survey Report.* Interim Report no. 7. Anchorage, 1 December 1959 (revised 1 June 1962).
―――. *Notice of Completion of Report on the Rampart Canyon Project, Yukon*

River Basin. Alaska North Pacific Division, Portland, Ore., 25 June 1971.

Atomic Energy Commission (AEC). Committee on Environmental Studies for Project Chariot, Plowshare Program, *Bioenvironmental Features of the Ogotoruk Creek Area, Cape Thompson, Alaska: A First Summary by the Committee on Environmental Studies for Project Chariot.* TID-12439. Division of Technical Information, Springfield, Va., December 1960.

————. *Bioenvironmental Features of the Ogotoruk Creek Area, Cape Thompson, Alaska. (Including Predictions and Assessments of the Effects of Nuclear Detonations Proposed for an Excavation Experiment), A Second Summary Report.* TID-17226. Edited by Weichold, Bette. Division of Technical Information, Springfield, Va., October 1962.

————. *Environment of the Cape Thompson Region, Alaska.* PNE-481. Edited by Wolfe, John N., and Wilimovsky, Norman J. Division of Technical Information, Springfield, Va., 1966.

Energy Research and Development Administration. Fuller, Glen, and Merritt, Melvin L. *The Environment of Amchitka Island, Alaska.* Department of Military Applications, Washington, D.C., 1977.

4.1.2 U.S. CONGRESS, HOUSE OF REPRESENTATIVES (BY DATE)

National Resources Committee. *Alaska: Its Resources and Development.* House Document no. 485, 75th Cong., 3d sess., 1938.

Committee on Roads, Hearings, *H.R. 3095 (Proposed Highway to Alaska).* 77th Cong., 2d sess., 1942.

The Alaska Highway—An Interim Report, (pursuant to House Resolution 255, Authorizing the Committee on Roads, as a whole or by Subcommittees, to investigate the federal road system and for other purposes). House Report no. 1705, 1946.

Subcommittee on Territories and Insular Possessions of the Committee on Public Lands. Hearings, *Alaska Statehood, (House Resolution 93).* 80th Cong., 1st sess., 1948.

Hackett, William H. Subcommittee on Territories and Insular Possessions. *Alaska's Vanishing Frontier: A Progress Report.* 1951.

Committee on Interior and Insular Affairs. Hearings, *Alaska Statehood and Elective Governorship, S.50 and S.224,* 83d Cong., 1st sess., 1953.

Subcommittee on Territories and Insular Possessions. Hearings, *Alaska Statehood, S.49 and S.35,* 85th Cong., 1st sess., 1957.

Committee on Interior and Insular Affairs. Hearings, *Inclusion of Alaska Lands in National Parks, Forests and Wildlife Refuges, and Wild and Scenic River Systems.* 95th Cong., 1st sess., 1977.

U.S. Congress, Joint Economic Committee. Hearings, *Natural Gas Regulation and the Trans-Alaska Pipeline.* 92d Cong., 2d sess., 1972.

4.1.3 U.S. CONGRESS, SENATE (BY DATE)

Subcommittee on Territories and Island Possessions. Committee on Interior and Insular Affairs. Hearings, *Establishment of Mount McKinley National Park.* 64th Cong., 1st sess., 1916.

Committee on Foreign Relations. Hearings, *Subcommittee on Senate Resolution 253 (A Resolution Providing for an Inquiry into the Location of the Alaska Highway on the So-Called C or Prairie Route)*. 77th Cong., 2d sess., 1942.

Special Senate Committee to Investigate the National Defense Program. Hearings, *Investigation of the National Defense Program, Part 22; Canol Project*. 78th Cong., 1st sess., 1944.

Committee on Interior and Insular Affairs. Hearings, *Alaska Statehood*, 81st Cong., 2d sess., 1950.

Subcommittee on Irrigation and Reclamation, Committee on Interior and Insular Affairs. Hearings, *Eklutna Project, Alaska, S. 3222, A Bill to Authorize Modification and Reconstruction*. 86th Cong., 2d sess., 1960.

Subcommittee on Merchant Marine and Fisheries, Committee on Interstate and Foreign Commerce. Hearings, *S. 1899, A Bill to Authorize the Establishment of the Arctic Wildlife Range, Alaska*. 86th Cong., 1st sess., 1960.

Subcommittee on Irrigation and Reclamation, Committee on Interior and Insular Affairs. Hearings, *Hydroelectric Power Requirements and Resources in Alaska*. 86th Cong., 2d sess.,1961.

Committee on Public Works. *The Market for Rampart Power, Yukon River, Alaska*. 87th Cong., 2d sess., 23 April 1962.

Committee on Interior and Insular Affairs. Hearings, *The Nomination of Governor Walter J. Hickel, of Alaska, to be Secretary of the Interior*. 79th Cong., 1st sess., 1969.

Hearings, *Federal Oversight of Oil Development of Public Lands in Alaska*, 91st Cong., 1st sess., 1969.

Hearings, *The Status of the Proposed Trans-Alaska Pipeline*. 91st Cong., 1st sess., 1969.

Hearings, *The Nomination of Rogers C. B. Morton to be Secretary of the Interior*. 92d Cong., 1st sess., 1971.

Subcommittee on Conservation and Natural Resources, Committee on Interior and Insular Affairs. Hearings, *Protecting America's Estuaries: Puget Sound and the Straits of Georgia and Juan De Fuca*. 92d Cong. 1st sess., 1972.

Committee on Interior and Insular Affairs. Hearings, *Trans-Alaska Pipeline, Draft Environmental Impact Statement*. 92d Cong., 1st sess., 1971.

Hearings, *Rights-of-Way Across Federal Lands: Transportations of Alaska's North Slope Oil*. 93d Cong., 1st sess., 1973.

4.1.4 U.S. DEPARTMENT OF AGRICULTURE

Murie, Olaus J. *The Alaska-Yukon Caribou*. Department of Biological Survey. North American Fauna Series, no. 54. June 1935.

Lutz, Harold J. *Ecological Effects of Forest Fires in the Interior of Alaska*. Technical Bulletin 1133, July 1956.

4.1.5 U.S. DEPARTMENT OF THE INTERIOR

Rampart Project, Alaska; Market for Power and Effect of Project on Natural Resources. Field Study Report 998, 3 vols., Juneau, Alaska, 9 February 1965.

Northern Alaska: How to Realize its Promise and Potential, Report of the NORTH Commission to the Governor of Alaska. 24 June 1970.

Environmental Statement, Yukon River-North Slope Road. 20 March 1970.

Federal Task Force on Alaskan Oil Development. *Preliminary Report of the Federal Task Force on Alaskan Oil Development.* 15 September 1969.

Draft Environmental Impact Statement, for the Trans-Alaska Pipeline, Section 102(2)c of the National Environmental Policy Act of 1969. 15 January 1971.

Special Interagency Task Force. Federal Task Force on Alaskan Oil Development, *Final Environmental Impact Statement, Proposed Trans-Alaska Pipeline.* 9 vols, March 1972.

Bureau of Land Management. Marion Clawson. *Alaska: Information relative to the disposal and leasing of Public Lands in Alaska.* Information Bulletin, no. 2. August 1948.

Bureau of Reclamation. *Alaska: A Reconnaissance Report on the Potential Development of Water Resources in the Territory of Alaska for Irrigation, Power Production and Other Beneficial Uses.* (2 May 1949), House Document 197, 82d Cong., 2d sess., January 1952.

Devil Canyon Project, Alaska, Report to the Secretary of Interior. 6 March 1961.

Fish and Wildlife Service (USFWS). Dufresne, Frank. *Mammals and Birds of Alaska.* Circular no. 3, 1942.

Rhode, Clarence J. *Alaska's Fish and Wildlife.* Circular 17. 1953.

Rietze, Harry L., Barnaby, J. T., and Watson, Gordon. *A Report to the Secretary of the Interior, Rampart Canyon, Dam and Reservoir Project Committee.* April 1964.

Arctic National Wildlife Range, Alaska, Coastal Plain Resource Assessment: Report and Recommendation to the Cong. of the United States and Final Legislative Environmental Impact Statement. 2 vols., April 1987.

National Park Service. Stanton, William J. *Alaska Recreation Survey.* (2 vols.), 1953.

Carnes, W., ed. *A Preliminary Survey of the Kongakut-Firth River Area, Alaska-Canada.* 1954.

4.1.6 UNITED STATES GEOLOGICAL SURVEY (USGS)

Ferrians, Oscar J., Kachadoorian, Reuben, and Greene, Gordon W. *Permafrost and Related Engineering Problems of Alaska.* Professional Paper 678, 1969.

Kachadoorian, R., Campbell, R. H., Sainsbury C. L., and Scholl, D. W. *Geology of the Ogotoruk Creek Area, Northwestern Alaska,* (Report TEM-976, October 1958), Menlo Park, Calif., 14 October 1959.

Lachenbruch, Arthur H. *Some Estimates of the Thermal Effects of a Heated Pipeline in Permafrost.* Alaska Branch, Menlo Park, Calif., December 1970.

Péwé, T. L., Hopkins, D. M., and Lachenbruch, A. H. *Engineering Geology Bearing on Harbor Site Selection Along the Northwest Coast of Alaska from Nome to Point Barrow.* (Atomic Energy Commission Report TEI-668. 1958). Menlo Park, Calif., 14 October 1959.

Reed, John C. *Exploration of Naval Petroleum Reserve No. 4 and Adjacent*

Areas of Northern Alaska, 1944–53, Part 1, History of the Exploration. USGS Professional Paper 301, 1958.

Smith, Philip S., Mertie, J. B., and Foran, W. T. *Summary of Recent Surveys in Northern Alaska.* USGS Bulletin 783-E, 1926.

Smith, Philip S. and Mertie, J. B. *Geology and Mineral Resources of Northwestern Alaska.* USGS Bulletin 815, 1930.

4.2 CANADA

British Columbia-Yukon-Alaska Highway Committee. *Report on the Proposed Highway Through British Columbia and the Yukon Territory to Alaska.* (June 1941) Ottawa: King's Printer, 1942.

Government of British Columbia. *The Way Out.* Victoria, B.C.: Queen's Printer, 1973.

The Berger Commission. *The Berger Report: Northern Frontier, Northern Homeland.* Department of Indian Affairs and Northern Development. Ottawa: Canadian Government Printing Office, 9 May 1975.

5. Books

Allen, Lawrence J. *The Trans-Alaska Pipeline.* Vol. 1: *The Beginning.* Vol. 2: *South to Valdez.* Seattle: Scribe Publishing Co., 1975 and 1976.

Allin, Craig W. *The Politics of Wilderness Preservation.* Westport, Conn.: Greenwood Press, 1982.

Andrasko, Kenneth, and Marcus Halevi. *Alaska Crude: Visions of the Last Frontier.* Boston: Little, Brown & Co., 1977.

Andrews, Clarence Leroy. *The Story of Alaska.* 1st ed. Seattle: Lowman & Hanford Co., 1931.

———. *The Story of Alaska.* 2d ed. Caldwell, Idaho: Caxton Printers, 1938.

Athearn, Robert G. *The Mythic West in Twentieth-Century America.* Lawrence: University Press of Kansas, 1986.

Atwood, Evangeline. *Anchorage: All-American City.* Portland, Oreg.: Binfords and Mort, 1957.

———. *Frontier Politics: Alaska's James Wickersham.* Portland, Oreg.: Binfords and Mort, 1979.

Atwood, Fred N. *The Alaska-Yukon Gold Book.* Seattle: Sourdough Stampede Association, Inc., 1930.

Badè, William Frederic, ed. *The Life and Letters of John Muir.* Vol. 2. Boston: Houghton Mifflin, 1924.

Bailes, Kendall, E., ed. *Environmental History: Critical Issues in Comparative Perspective.* Lanham, Md.: University Press of America, 1985.

Balch, Thomas Willing. *The Alaska Frontier.* Philadelphia: Allen, Lane and Scott, 1903.

Bancroft, Hubert Howe. *History of Alaska, 1730–1885.* 1886. Reprint. New York: Antiquarian Press, 1960.

Bartlett, Richard A. *The New Country: A Social History of the American Frontier, 1776–1890.* New York: Oxford University Press, 1974.

Bates, J. Leonard. *The Origins of Teapot Dome: Progressives, Parties, and Petroleum, 1909–21.* Urbana: University of Illinois Press, 1963.

Beach, Rex. *The Iron Trail.* New York: Harpers, 1913.

Berry, Mary Clay. *The Alaska Pipeline: The Politics of Oil and Native Land Claims.* Bloomington: Indiana University Press, 1975.

Berton, Pierre. *The Klondike Fever: The Life and Death of the Last Great Gold Rush.* New York: Alfred A. Knopf, 1959.

Billington, Ray Allen. *America's Frontier Heritage.* New York: Holt, Rinehart and Winston, 1966.

———. *Westward Expansion: A History of the American Frontier.* 1949. 3d ed. New York: Macmillan Co., 1967.

———. *Frederick Jackson Turner: Historian, Scholar, Teacher.* New York: Oxford University Press, 1973.

———. *America's Frontier Culture: Three Essays.* College Station, Tex.: Texas A & M University Press, 1977.

———. *Land of Savagery, Land of Promise: The European Image of the American Frontier in the Nineteenth Century.* New York: W. W. Norton, 1981.

Bockstoce, John R. *Whales, Ice and Men: The History of Whaling in the Western Arctic.* Seattle: University of Washington Press, 1986.

Bohn, Dave. *Glacier Bay: The Land and the Silence.* San Francisco: Sierra Club, 1967.

Bone, Scott C. *Alaska, Its Past, Present, and Future.* Juneau, Alaska: Governor's Office: 1925.

Boorstin, Daniel J. *The Americans: The National Experience.* New York: Random House, 1965.

———. *The Americans: The Democratic Experience.* New York: Random House, 1973.

Bright, Elizabeth Parks. *Alaska, Treasure Trove of Tomorrow.* New York: Banner Book, Exposition Press, 1956.

Brooks, Alfred Hulse. *Blazing Alaska's Trails.* College: University of Alaska and Arctic Institute of North America, 1953.

Brooks, Paul. *The Pursuit of Wilderness.* Boston: Houghton Mifflin, 1971.

Brower, Kenneth. *Earth and the Great Weather: The Brooks Range.* San Francisco: Friends of the Earth, 1971.

Brown, Ralph H. *Historical Geography of the United States.* New York: Harcourt, Brace & Co., 1948.

Brown, Tom. *Oil on Ice: Alaskan Wilderness at the Crossroads.* Edited by Richard Pollak. San Francisco: Sierra Club Battlebook, 1971.

Brown, William E. *This Last Treasure: Alaska National Parklands.* Anchorage: Alaska Natural History Association, 1982.

Burnette, O. Lawrence, Jr., ed. *Wisconsin Witness to Frederick Jackson Turner: A Collection of Essays on the Historian and the Thesis.* Madison: State Historical Society of Wisconsin, 1961.

Burroughs, John. *Harriman Alaska Expedition.* Vol. 1: *Narrative of the Expedition.* New York: Doubleday, Page & Co., 1902.

Candler, Isaac. *A Summary View of America.* London: T. Cadwell, 1824.

Chasan, Daniel Jack. *Klondike '70: The Alaska Oil Boom.* New York: Frederick A. Praeger, 1971.

Cicchetti, Charles J. *Alaskan Oil: Alternative Routes and Markets.* Baltimore: Resources for the Future and Johns Hopkins University Press, 1972.

Clark, John G., ed. *The Frontier Challenge: Responses to the Trans-Mississippi West.* Lawrence: University Press of Kansas, 1971.

Coates, Kenneth S., ed. *The Alaska Highway: Papers of the 40th Anniversary Symposium.* Vancouver: University of British Columbia Press, 1985.

Coates, Kenneth S., and William R. Morrison, *Land of the Midnight Sun: A History of the Yukon.* Edmonton, Alberta: Hurtig, 1988.

Coe, Douglas. *Road to Alaska: The Story of the Alaska Highway.* New York: Julian Messner, 1943.

Colby, Merle. *A Guide to Alaska: Last American Frontier.* American Guide Series. Federal Writers' Project. New York: Macmillan Co., 1939.

Commoner, Barry. *Science and Survival.* London: Victor Gollancz Ltd., 1966.

———. *The Closing Circle: Confronting the Environmental Crisis.* London: Jonathan Cape, 1972.

Cooley, Richard A. *Alaska: A Challenge in Conservation.* Madison: University of Wisconsin Press, 1967.

Crisler, Lois. *Arctic Wild.* New York: Harper and Brothers, 1958.

Crosby, Alfred W. *The Columbian Exchange: Biological and Cultural Consequences of 1492.* Westport, Conn.: Greenwood Press, 1972.

———. *Ecological Imperialism: The Biological Expansion of Europe, 900–1900.* Studies in Environment and History. Cambridge: Cambridge University Press, 1986.

Davis, Neil. *Energy/Alaska.* Fairbanks: University of Alaska Press, 1988.

Dall, William H. *Alaska and its Resources.* Boston: Lee and Shepard, 1870.

Dixon, Mim. *What Happened to Fairbanks? The Effects of the Trans-Alaska Oil Pipeline on the Community of Fairbanks, Alaska.* Social Impact Assessment Series. Boulder, Colo.: Westview Press, 1978.

Douglas, William O. *My Wilderness: The Pacific West.* New York: Doubleday, 1960.

Downs, Roger M., and David Stea. *Maps in Mind: Reflections on Cognitive Mapping.* Harper and Row Series in Geography. New York: Harper and Row, 1977.

Dubos, René. *The Wooing of Earth: New Perspectives on Man's Use of Nature.* New York: Scribner's, 1980.

Dufresne, Frank. *Alaska's Animals and Fishes.* New York: A. S. Barnes & Co., 1946.

———. *My Way Was North: An Alaskan Autobiography.* New York: Holt, Rinehart and Winston, 1966.

Dunlap, Thomas R. *DDT: Scientists, Citizens, and Public Policy.* Princeton: Princeton University Press, 1981.

East, William Gordon. *The Geography Behind History.* 1935. 2d ed. London: Discussion Books, 1965.

Elliott, Henry Wood. *A Report on the Condition of Affairs In the Territory of Alaska.* Washington, D.C.: GPO, 1875.

———. *An Arctic Province: Alaska and the Seal Islands.* London: Sampson Low, Marston, Searle and Rivington, 1886.

Engler, Robert. *The Politics of Oil: A Study of Private Power and Democratic Directions*. Chicago: University of Chicago Press, 1961.

Fischer, Victor. *Alaska's Constitutional Convention*. National Municipal League, State Constitutional Convention Series, No. 9. College, Alaska: Institute of Social, Economic and Government Research, 1975.

Fitch, Edwin M. *The Alaska Railroad*. New York: Frederick A. Praeger, 1967.

Fox, Stephen R. *John Muir and his Legacy: The American Conservation Movement, 1890–1975*. Boston: Little, Brown and Company, 1981.

Frederick, Robert A. *Frontier Alaska: A Study In Historical Interpretation and Opportunity*. Proceedings of the Conference on Alaskan History, Alaska Methodist University, Anchorage, June 1967. *Alaska Review* (Special Issue), Anchorage: 1968.

Gallagher, Hugh C. *Etok: A Story of Eskimo Power*. New York: G. P. Putnam, 1974.

Gates, Paul W. *History of Public Land Law Development*. Washington, D.C.: GPO, November 1968.

Gibson, James R. *Imperial Russia in Frontier America, The Changing Geography of Supply of Russian America, 1784–1867*. Andrew H. Clark Series in the Historical Geography of North America. New York: Oxford University Press, 1976.

Gilman, William. *Our Hidden Front*. New York: Reynal and Hitchcock, Inc., 1944.

Godsell, Philip H. *The Romance of the Alaska Highway*. Toronto: Ryerson Press, 1946.

Goetzmann, William H., and Kay Sloan. *Looking Far North: The Harriman Expedition to Alaska, 1899*. New York: Viking Press, 1982.

Gould, Jean, and Lorena Hickok. *Walter P. Reuther: Labor's Rugged Individualist*. New York: Dodd, Mead & Co., 1972.

Gould, Peter, and Rodney White. *Mental Maps*. Pelican Geography and Environmental Studies. Harmondsworth: Penguin, 1974.

Griffin, Harold. *Alaska and the Canadian Northwest; Our New Frontier*. New York: W. W. Norton, 1944.

Grinnell, George Bird. *Harriman Alaska Expedition*. Vol. 2: *History, Geography and Resources*. New York: Doubleday, Page & Co., 1902.

———. *A Brief History of the Boone and Crockett Club*. New York: Forest and Stream, 1910.

Gruening, Ernest. *The State of Alaska*. New York: Random House, 1954.

———. *The Battle for Alaska Statehood*. College: University of Alaska Press, 1967.

———. *Many Battles: The Autobiography of Ernest Gruening*. New York: Liveright, 1973.

Hallock, Charles. *Our New Alaska (or The Seward Purchase Vindicated)*. New York: Forest and Stream Publishing Co., 1886.

Hanrahan, John, and Peter Gruenstein. *Lost Frontier: The Marketing of Alaska*. New York: W. W. Norton, 1977.

Harrison, E. S. *Alaska Almanac, 1909*. The Arctic Club. Seattle, 1909.

Harrison, Gordon Scott, ed. *Alaska Public Policy: Current Problems and Issues*.

College, Alaska: Institute of Social, Economic and Government Research, 1971.

Hays, Samuel P. *Conservation and the Gospel of Efficiency: The Progressive Conservation Movement, 1890–1920.* Cambridge: Harvard University Press, 1959.

——. *Beauty, Health, and Permanence: Environmental Politics in the United States, 1955–85.* Studies in Environment and History. Cambridge: Cambridge University Press, 1987.

Hellenthal, John. *The Alaskan Melodrama.* New York: Liveright, 1936.

Helmericks, Harmon, and Constance Helmericks. *We Live in the Arctic.* London: Hodder and Stoughton, 1949.

Hewlett, Richard G., and Francis Duncan. *Atomic Shield: 1947–52.* Vol. 2: *A History of the United States Atomic Energy Commission.* University Park: Pennsylvania State University Press, 1969.

Hickel, Walter J. *Who Owns America?* Englewood Cliffs, N.J.: Prentice-Hall, 1971.

Hilscher, Herb, and Miriam Hilscher. *Alaska, USA.* Boston: Little, Brown, 1959.

Hofstadter, Richard, and Seymour Martin Lipset, eds. *Turner and the Sociology of the Frontier.* New York: Basic Books, 1968.

Hollon, W. Eugene. *The Great American Desert; Then and Now.* New York: Oxford University Press, 1966.

Hornaday, William T. *Our Vanishing Wild Life: Its Extermination and Preservation.* New York: New York Zoological Society, 1913.

Hulley, Clarence C. *Alaska: Past and Present.* Portland, Oreg.: Binfords and Mort, 1970.

Hunt, William R. *Alaska: A Bicentennial History.* New York: W. W. Norton, 1976.

——. *Stef: A Biography of Vilhjalmur Stefansson; Canadian Arctic Explorer.* Vancouver: University of British Columbia Press, 1986.

Jensen, Ronald J. *The Alaska Purchase and Russian-American Relations.* Seattle: University of Washington Press, 1975.

Johann, A. E. *Pelzjäger, Prärien und Präsidenten; Fahrten und Erlebnisse Zwischen New York und Alaska.* Berlin: Deutscher Verlag, 1937.

Lamar, Howard R., and Leonard Thompson, eds. *The Frontier in History: North America and Southern Africa Compared.* New Haven: Yale University Press, 1981.

Lambert, Derek. *Great Land.* London: Bantam, 1978.

Leighly, John, ed. *Land and Life: A Selection from the Writings of Carl Ortwin Sauer.* Berkeley and Los Angeles: University of California Press, 1967.

Leopold, Aldo. *A Sand County Almanac.* New York: Oxford University Press, 1949.

Leopold, A. Starker, and Frank Fraser Darling. *Wildlife in Alaska: An Ecological Reconnaissance.* New York: Ronald Press for the Conservation Foundation and the New York Zoological Society, 1953.

Limerick, Patricia Nelson. *The Legacy of Conquest: The Unbroken Past of the American West.* New York: W. W. Norton, 1987.

Lindbergh, Kristina, and Barry L. Provorse. *The Trans-Alaska Pipeline.* Vol. 3: *Emerging Alaska.* Seattle: Scribe Publishing Co., 1977.

Lopez, Barry. *Arctic Dreams: Imagination and Desire in a Northern Landscape.* New York: Bantam, 1987.

Lowenthal, David, and Martyn J. Bowden, eds. *Geographies of the Mind: Essays in Historical Geosophy in Honor of John Kirtland Wright.* American Geographical Society. New York: Oxford University Press, 1976.

Malin, James C. *The Grassland of North America: Prolegomena to Its History.* Lawrence, Kans.: author, 1947.

————. *History and Ecology: Studies of the Grassland.* Edited by Robert P. Swierenga. Lincoln: University of Nebraska Press, 1984.

Machlin, Milt. *Pipeline.* New York: Pyramid Books, 1976.

Malone, Michael P., ed. *Historians and the American West.* Lincoln: University of Nebraska Press, 1983.

Manning, Harvey. *Cry Crisis! Rehearsal in Alaska (A Case Study Of What Government By Oil Did To Alaska And Does To The Earth).* San Francisco: Friends of the Earth, 1974.

Marshall, Robert. *Arctic Wilderness.* Edited by George Marshall. Berkeley and Los Angeles: University of California Press, 1956.

————. *Alaska Wilderness: Exploring the Central Brooks Range.* Edited by George Marshall. Berkeley and Los Angeles: University of California Press, 1970.

————. *Arctic Village.* London: Jarrolds, 1934.

McCloskey, Maxine E., ed. *Wilderness: The Edge of Knowledge.* San Francisco: Sierra Club, 1970.

McGeary, Nelson. *Gifford Pinchot: Forester-Politician.* Princeton: Princeton University Press, 1960.

Mcphee, John. *Coming Into The Country.* New York: Farrar, Straus and Giroux, 1976.

————. *Encounters with the Archdruid; Narratives about a Conservationist and Three of His Natural Enemies.* New York: Farrar, Straus and Giroux, 1971.

Mead, Robert Douglas. *Journeys Down the Line: Building the Trans-Alaska Pipeline.* Garden City, N.Y.: Doubleday & Co., 1978.

Meinig, Donald W., ed. *The Interpretation of Ordinary Landscapes: Geographical Essays.* New York: Oxford University Press, 1979.

Menzies, Don. *The Alaska Highway: A Saga of the North.* Edmonton, Alberta: Stuart Douglas, 1943.

Miller, Orlando W. *The Frontier in Alaska and the Matanuska Colony.* Yale Western Americana Series, No. 26. New Haven: Yale University Press, 1975.

Milton, John P. *Nameless Valleys, Shining Mountains: The Record of an Expedition into the Vanishing Wilderness of Alaska's Brooks Range.* New York: Walker & Co., 1970.

Mitchell, William L. *The Opening of Alaska.* Edited by Lyman L. Woodman. Anchorage, Alaska: Cook Inlet Historical Society, 1982.

Moog, Vianna. *Bandierantes and Pioneers.* New York: George Braziller, 1964.

Morehouse, Thomas A., ed. *Alaskan Resources Development: Issues of the 1980s.* Boulder, Colo.: Westview Press, 1984.

Morenus, Richard. *Distant Early Warning, The Miracle of America's First Line of Defense.* New York: Rand McNally, 1957.

Morgan, Arthur E. *Dams And Other Disasters.* Boston: Porter Sargent, 1971.

Morgan, Lael. *Art and Eskimo Power: The Life and Times of Alaskan Howard Rock.* Fairbanks, Alaska: Epicenter Press, 1988.

Muir, John. *Travels in Alaska.* Boston: Houghton Mifflin, 1915.

———. *Our National Parks.* 1901. Reprint. Madison: University of Wisconsin Press, 1981.

———. *The Cruise of the Corwin: Journal of the Arctic Expedition of 1881 in Search of De Long and the Jeanette.* Edited by William Frederic Badè. Boston: Houghton Mifflin, 1917.

———. *Steep Trails.* Edited by William Frederic Badè. Boston: N. S. Berg, 1918.

Murie, Adolph. *A Naturalist in Alaska.* Devin-Adair American Naturalists Series. New York: Devin-Adair Co., 1961.

Murie, Margaret E. *Two In The Far North.* New York: Alfred A. Knopf, 1957.

Murie, Olaus J. *Journeys to the Far North.* Palo Alto, Calif.: Wilderness Society and Crown Publishers, 1973.

Nash, Roderick. *Wilderness and the American Mind.* 1967. 3d ed. rev. New Haven: Yale University Press, 1981.

Naske, Claus-M. *An Interpretative History of Alaska Statehood.* Anchorage: Alaska Northwest Publishing Co., 1973.

———. *E. L. "Bob" Bartlett of Alaska: A Life in Politics.* College, Alaska: University of Alaska Press, 1979.

———. *Paving Alaska's Trails: The Work of the Alaska Road Commission.* Lanham, Md.: University Press of America, 1986.

Naske, Claus-M., and Herman E. Slotnick. *Alaska: A History of the 49th State.* Grand Rapids, Mich.: William E. Eerdmans, 1979.

Nelson, Richard K. *Make Prayers to the Raven: A Native View of the Koyukon Forest.* Chicago: University of Chicago Press, 1983.

Nichols, Jeannette Paddock. *Alaska: A History of its Administration, Exploitation, and Industrial Development During Its First Half Century Under the Rule of the United States.* Cleveland: Arthur H. Clark, 1924.

Nichols, Roger L., ed. *American Frontier and Western Issues: A Historiographical Review.* Contributions in American History 118. Westport, Conn.: Greenwood Press, 1986.

Nielson, Johnathan M. *Armed Forces on a Northern Frontier: The Military in Alaska's History, 1867–1987.* Contributions in Military Studies, no. 74. Westport, Conn.: Greenwood Press, 1988.

Noggle, Burl. *Teapot Dome: Oil and Politics in the 1920s.* Baton Rouge: Louisiana State University Press, 1962.

O'Malley, Martin. *The Past and Future Land: An Account of the Berger Inquiry into the Mackenzie Valley Pipeline.* Toronto: Peter Martin Associates Ltd., 1976.

Paolino, Ernest N. *The Foundations of the American Empire: William H. Seward and United States Foreign Policy.* Ithaca: Cornell University Press, 1973.

Patty, Ernest. *North Country Challenge.* New York: David McKay Co., 1969.

Paxson, Frederic Logan, *The Last American Frontier.* New York: Macmillan Co., 1910.

———. *History of the American Frontier: 1763–1893.* Boston: Houghton Mifflin, 1924.

————. *When the West is Gone.* New York: Henry Holt & Co., 1930.

Penick, James, Jr. *Progressive Politics and Conservation: The Ballinger-Pinchot Affair.* Chicago: University of Chicago Press, 1968.

Pinchot, Gifford. *Breaking New Ground.* New York: Harcourt, Brace & Co., 1947.

Pomeroy, Earl. *The Pacific Slope: A History of California, Oregon, Washington, Idaho, Utah and Nevada.* New York: Alfred A. Knopf, 1966.

Pyne, Stephen, J. *Fire in America: A Cultural History of Wildland and Rural Fire.* Princeton: Princeton University Press, 1982.

Reid, Virginia Hancock. *The Purchase of Alaska: Contemporary Opinion.* Long Beach, Calif.: author, 1939.

Reiger, John F. *American Sportsmen and the Origins of Conservation.* New York: Winchester Press, 1975.

Remley, David A. *Crooked Road: A History of the Alaska Highway.* American Trail Series. New York: McGraw-Hill Book Co., 1976.

Rogers, George W. *The Future of Alaska: The Economic Consequences of Statehood.* Baltimore: Resources for the Future and Johns Hopkins University Press, 1962.

————. ed. *Change in Alaska: People, Petroleum, and Politics.* College: University of Alaska Press, 1970.

Roscow, James P. *800 Miles to Valdez: The Building of the Alaska Pipeline.* Englewood Cliffs, N.J.: Prentice-Hall, Inc., 1977.

Rosenman, Samuel I., ed. *The Public Papers and Addresses of FDR.* Vol. 1: *The Genesis of the New Deal, 1928–32.* New York: Random House, 1938.

Runte, Alfred. *National Parks: The American Experience.* Lincoln: University of Nebraska Press, 1979. 2d ed. rev., 1987.

Savage, William W., Jr., and Stephen I. Thompson, eds. *The Frontier: Comparative Studies.* Vol. 2. Norman: University of Oklahoma Press, 1979.

Service, Robert W. *Collected Verse.* Vol. 1. London: Ernest Benn, 1930.

Seton, Ernest Thompson. *The Arctic Prairies.* London: Constable, 1915.

Seward, Oliver Risley. *William H. Seward's Travels Around The World.* New York: D. Appleton and Co., 1873.

Sheldon, Charles. *The Wilderness of Denali: Explorations of a Hunter-Naturalist in Northern Alaska.* New York: Charles Scribner's Sons, 1930.

Shepard, Paul, and Daniel McKinley. *The Subversive Science: Essays Toward an Ecology of Man.* Boston: Houghton Mifflin, 1969.

Sherwood, Morgan B. *The Exploration of Alaska: 1865–1900.* Yale Western Americana, no. 7. New Haven: Yale University Press, 1965.

————. *Big Game in Alaska: A History of Wildlife and People.* Yale Western Americana, no. 33. New Haven: Yale University Press, 1981.

Shiels, Archie W. *The Purchase of Alaska.* University of Alaska and Alaska Centennial Commission. College: University of Alaska Press, 1967.

Simonson, Harold P. *The Closed Frontier: Studies in American Literary Tragedy.* New York: Holt, Rinehart and Winston, 1970.

Slotkin, Richard. *The Fatal Environment: The Myth of the Frontier in the Age of Industrialization, 1800–1890.* New York: Atheneum, 1985.

Smith, Henry Nash. *Virgin Land: The American West as Symbol and Myth.* 1951. 2d ed. Cambridge: Harvard University Press, 1971.

Starr, S. Frederick. *Russia's American Colony.* Kennan Institute for Advanced Russian Studies. Durham, N.C.: Duke University Press, 1987.

Stefansson, Vilhjalmur. *The Northward Course of Empire.* New York: Macmillan Co., 1924.

————. *Northwest to Fortune: The Search of Western Man for a Commercially Practical Route to the Far East.* London: George Allen and Unwin, 1960.

Stephenson, William B., Jr. *The Land of Tomorrow.* New York: George H. Doran Co., 1919.

Sundborg, George. *Opportunity in Alaska.* New York: Macmillan Co., 1945.

————. *Hail Columbia: The Thirty-Year Struggle for Grand Coulee Dam.* New York: Macmillan Co., 1954.

Tompkins, Stuart Ramsay. *Alaska: Promyshlennik and Sourdough.* Norman: University of Oklahoma Press, 1945.

Treadgold, Donald W. *The Great Siberian Migration.* Princeton: Princeton University Press, 1957.

Trefethen, James B. *Crusade for Wildlife: Highlights in Conservation Progress.* Harrisburg, Pa.: Stackpole Co., 1961.

Turner, Frederick Jackson. *The Frontier in American History.* 1920. Reprint. Huntington, N.Y.: Robert E. Krieger Publishing Co., 1976.

Underwood, John J. *Alaska: An Empire in the Making.* New York: Dodd, Mead & Co., 1913.

Wade, Richard C. *The Urban Frontier: The Rise of Western Cities, 1790–1830.* Cambridge: Harvard University Press, 1959.

Watson, J. Wreford, and Timothy O'Riordan, eds. *The American Environment: Perceptions and Policies.* Chichester, U.K.: John Wiley & Sons, 1976.

Webb, Melody. *The Last Frontier: A History of the Yukon Basin of Canada and Alaska.* Albuquerque: University of New Mexico Press, 1985.

Webb, Walter Prescott. *The Great Frontier.* Cambridge, Mass.: Houghton Mifflin, 1951.

Wharton, David. *The Alaska Gold Rush.* Bloomington: Indiana University Press, 1972.

Wilson, William H. *Railroad in the Clouds: The Alaska Railroad in the Age of Steam, 1914–45.* Boulder, Colo.: Pruett Publishing Co., 1977.

Wood, William R. *Not From Stone.* College: University of Alaska Foundation, 1983.

Wooley, Wesley T. *Alternatives to Anarchy: American Supranationalism Since World War Two.* Bloomington: Indiana University Press, 1988.

Worster, Donald. *Nature's Economy: A History of Ecological Ideas.* 1977. Reprint. Studies in Environment and History. Cambridge: Cambridge University Press, 1985.

Wright, Billie. *Four Seasons North: A Journal of Life in the Alaskan Wilderness.* New York: Harper & Row, 1973.

Wyman, Walker D., and Clifton B. Kroeber, eds. *The Frontier in Perspective.* Madison: University of Wisconsin Press, 1957.

Young, S. Hall. *Alaska Days With John Muir.* New York: Fleming H. Revell Co., 1915.

Zaslow, Morris. *The Northward Expansion of Canada, 1914–67.* The Canadian Centenary Series, no. 17. Toronto: McClelland and Stewart, 1988.

6. OTHER PUBLICATIONS

Arctic Environmental Council. *Report on the On-Site Visit to the Trans-Alaska Pipeline System.* Washington, D.C.: Arctic Institute of North America (AINA), October 1974.

———. *Second Report on the On-Site Visit to the Trans-Alaska Pipeline System.* Washington, D.C.: AINA, May 1975.

Atlantic Pipeline Company, et al. *Transcontinental Pipeline Project: Transportation of Alaskan Crude Oil, Economic Feasibility Study: Seattle to Chicago, Prudhoe Bay to Chicago, Chicago to East Coast.* Houston, Texas, 31 September 1968.

Bechtel Summary Report. *Technological Feasibility and Cost Study: Mackenzie Valley Pipeline Research, Ltd.,* Houston, January 1970.

Billington, Ray Allen. "Cowboys, Indians, and the Land of Promise: The World Image of the American Frontier." Proceedings, 14th World International Congress of the Historical Sciences, New York, 1976: 60–79.

Billington, R. A., and Albert Camarillo. *The American Southwest: Image and Reality.* William Andrews Clark Memorial Library, University of California, Los Angeles, 16 April 1979.

Boreal Institute. *Alaska Highway 1942–82, Sources of Information in the Yukon Archives: A Partial Listing of Highway Literature.* Edmonton, Alberta, June 1982.

Bowen, Brent R. *Defense Spending in Alaska.* Institute of Social, Economic and Government Research, University of Alaska, College, vol. 8 (July 1971).

Bowman, Isaiah. *The Pioneer Fringe.* Special Publication no. 13. American Geographical Society. New York, 1931.

Buckley, John. *Wildlife in the Economy of Alaska.* Alaska Cooperative Wildlife Research Unit, Biological Papers of the University of Alaska, no. 1. College, 1955.

Cameron, Raymond D., and Kenneth R. Whitten. *First Interim Report of the Effects of the Trans-Alaska Pipeline on Caribou Movements.* Joint State-Federal Fish and Wildlife Advisory Team, Anchorage, 1976.

Canadian Institute of Guided Ground Transportation (CIGGT). *Railway to the Arctic: A Study of the Operational and Economic Feasibility of a Railway to Move Arctic Slope Oil to Market.* Queen's University, Ontario, January 1972.

Center for Law and Social Policy. *Comments on the Environmental Impact Statement for the Trans-Alaska Pipeline.* Washington, D.C., May 1972.

Cole, Terrence, ed. "Wheels On Ice." *Alaska Journal* 15 (Winter 1985): 27–47.

Development and Resources Corporation. Edited by Clapp, Gordon, R. *The Market for Rampart Power: Working Papers pertaining to the report of the U.S. Army Corps of Engineers, Alaska District.* New York, 1962.

First National Bank of Seattle, Economic Research Department. *Alaska: Frontier for Industry.* Seattle, April 1954.

Fischer, Victor. "Soviet Northern Development." Proceedings, 27th Alaska Science Conference, Alaska Division, American Association for the Advancement of Science, College, 4–7 August 1976. *Science in Alaska,* 1976: 1–11.

Flanders, H. F., "Contributions of the DEWline in Arctic Engineering." Proceedings, 8th Alaska Science Conference, Anchorage, 10–13 September 1957. Edited by Albert W. Johnson. *Science in Alaska*, 1958: 53–66.

Georgeson, C. C. "The Possibilities of Agricultural Settlement in Alaska." *Pioneer Settlement: Cooperative Studies*, American Geographical Society, Special Publication No. 14. New York, 1932: 50–60.

Institute of Business, Economic and Government Studies. "The Petroleum Industry in Alaska." *Monthly Review of Alaska Business and Economic Conditions* 1 (August 1964). College, Alaska.

———. "Alaska's Economy in 1967." *Monthly Review of Alaska Business and Economic Conditions* 5 (1967).

Lutz, Harold J. *Aboriginal Man and White Man as Historical Causes of Fires in the Boreal Forest, With Particular Reference to Alaska.* Yale School of Forestry, New Haven. Bulletin No. 65, 1959.

Nash, Roderick. "Tourism, Parks and the Wilderness Idea in the History of Alaska." *Alaska in Perspective* 4 (1981).

Naske, Claus-M. *Alaska Road Commission Historical Narrative: Final report.* State of Alaska, Department of Transportation and Public Facilities. Fairbanks, June 1983.

Naske, Claus-M., and William R. Hunt. *The Politics of Hydroelectric Power in Alaska: Rampart and Devil Canyon—A Case Study.* Institute of Water Resources, University of Alaska. Fairbanks, 1978.

National Research Council, National Academy of Sciences, Division of Biology and Agriculture, Rampart Canyon Dam and Reservoir Project Committee. *A Report to the Secretary of the Interior.* Washington, D.C., 1 November 1965.

Nelson, Richard K. "Cultural Values of the Land." *Native Livelihood and Dependence, A Study of Land Use Values Through Time.* National Petroleum Reserve in Alaska, Field Study 1. U.S. Department of the Interior, 105(c) Land Use Study. Anchorage, June 1979, 27–36.

Paxson, Frederic Logan. "The Pacific Railroads and the Disappearance of the Frontier in America." *Annual Report of the American Historical Association.* Washington, D.C., 1907: 105–18.

Payne, James T. *"Our Way of Life is Threatened and Nobody Seems to Give a Damn": The Cordova District Fisheries Union and the Trans-Alaska Pipeline.* Anchorage, October 1985.

Spencer, David L., Claus-M. Naske, and John Carnahan. *National Wildlife Refuges of Alaska; An Historical Perspective.* Prepared for the U.S. Fish and Wildlife Service, Arctic Environmental Information and Data Center. Anchorage, January 1979.

Spurr, Stephen H., ed. *Rampart Dam and the Economic Development of Alaska.* University of Michigan, Ann Arbor, March 1966.

St. Louis Committee for Nuclear Information (CNI). "Project Chariot: A Complete Report on the Probable Gains and Risks of the AEC's Plowshare Project in Alaska." *Nuclear Information* 3 (3 June 1961). St. Louis, Missouri.

Sundborg, George. *Statehood for Alaska: The Issues Involved and the Facts About the Issues.* Anchorage, July 1946.

Trefethen, James B., ed. Transactions, 28th North American Wildlife and Natural Resources Conference, Detroit, 4–6 March 1963. Washington, D.C.: Wildlife Management Institute, 1963.

——. Transactions, 29th North American Wildlife and Natural Resources Conference, Las Vegas, Nev., 9–11 March 1964. Washington, D.C.: Wildlife Management Institute, 1964.

——. Transactions, 30th North American Wildlife and Natural Resources Conference, Mexico City, March 1965. Washington, D.C.: Wildlife Management Institute, 1965.

——. Transactions, 31st North American Wildlife and Natural Resources Conference, Pittsburgh, Pa., 14–16 March 1966. Washington, D.C.: Wildlife Management Institute, 1966.

U.S. Army Corps of Engineers. Proceedings, *Conferences of the Rampart Economic Advisory Board.* Anchorage, 27 May 1961; 10–12 January 1962; 30–31 March 1962.

Woodman, Lyman L. *The Army's Role in the Building of Alaska.* U.S. Army, Alaska, Pamphlet 360-5. Anchorage, 1 April 1969.

Yukon Power for America. *Addresses Presented at Rampart Dam Conference,* McKinley Hotel, Mount McKinley National Park, Alaska, 7–9 September 1963. Fairbanks, 1963.

——. *The Rampart Canyon Project.* Fairbanks, 3 September 1964.

Zemansky, Gil. "Environmental Non-Compliance and the Public Interest During Construction of the Trans-Alaska Pipeline." Proceedings, 27th Alaska Science Conference. Fairbanks, 4–7 August 1976. *Science in Alaska* 2 (1976): 26–37.

——. *Symbolic Environmental Monitoring on the Trans-Alaska Oil Pipeline.* Oregon State University, Corvallis, June 1977.

7. ARTICLES AND PAPERS

Bagnall, V. B. "Operation DEW line." *Journal of the Franklin Institute* 259 (June 1955): 481–90.

Bailey, Thomas A. "Why the United States Purchased Alaska." *Pacific Historical Review* 3 (1934): 39–94.

——. "The North Pacific Sealing Convention of 1911." *Pacific Historical Review* 4 (1935): 1–14.

Baldwin, George. "Conservative Faddists Arrest Progress and Seek to Supplant Self-Government with Bureaucracy." *Alaska-Yukon Magazine* 13 (February 1912): 44–46.

Barry, A. "Now That You Own Alaska, Friends, What Are You Going to Do with It?" *Esquire* 72 (April 1969): 119–25.

Bates, J. Leonard. "Fulfilling American Democracy: The Conservation Movement, 1907–1921." *Mississippi Valley Historical Review* 44 (June 1957): 29–57.

Beaman, Charles C. "Our New Northwest." *Harper's* 35 (June 1867): 170–85.

Belous, Robert. "Unsolved Problems of Alaska's North Slope." *National Parks and Conservation Magazine* 44, Special Issue: Alaska (November 1970): 16–17, 20–1.

Black, I. G. "USSR Oil Leak: A Warning To Alaska's Pipeliners?" *Iron Age* 208 (July 1971): 41.

Boyle, Robert H. "America Down The Drain." *Sports Illustrated* 21 (16 November 1964): 78–90.

Brooks, Alfred Hulse. "The Value of Alaska." *Geographical Review* 15 (January 1925): 25–50.

———. "The Future of Alaska." *Annals of the Association of American Geographers* 15 (December 1925): 163–79.

Brooks, Paul. "Alaska: The Last Frontier." *Atlantic Monthly* 210 (September 1962): 73–78.

———. "The Plot to Drown Alaska." *Atlantic Monthly* 215 (May 1965): 53–60.

Brooks, Paul, and Joseph Foote. "The Disturbing Story of Project Chariot." *Harper's Weekly* 224 (April 1962): 60–67.

Buske, Frank E. "Rex Beach: A Frustrated Goldseeker Lobbies to 'Win the Wilderness over to Order.'" *Alaska Journal* 10 (Autumn 1980): 37–42.

Cameron, Raymond D., and Kenneth R. Whitten. "Caribou and Petroleum Development." *Alaska Journal* 12 (Spring 1982): 28–32.

"Canada's First Arctic Coast Mega-Project." *Beaufort* 2 (August 1982): 6–9.

"Canol Pipeline: A History of the Norman Wells to Whitehorse Oil Pipeline." *Beaufort* 1 (May 1982): 6–9.

Carter, Luther J. "North Slope: Oil Rush—Alaska May or May Not Become a Kuwait of the Arctic But the Oil Men's Arrival Is Changing a Wilderness." *Science* 166 (3 October 1969): 85–92.

Collins, George L., and Lowell Sumner. "The Northeast Arctic: The Last Great Wilderness." *Sierra Club Bulletin* 38 (October 1953): 13–26.

Cooke, Alan. "The Rampart Dam Proposal for Yukon River." *Polar Record* 12 (1964): 277–80.

Cooley, Richard. "State Land Policy in Alaska: Progress and Prospects." *Natural Resources Journal* 4 (January 1965): 455–67.

Corrigan, Richard. "Environment Report: Fishing Town Joins Legal Fight to Stop Trans-Alaska Pipeline Project." *National Journal* 3 (3 July 1971): 1400–1404.

Daley, Patrick, and Beverly James. "An Authentic Voice in the Technocratic Wilderness: Alaskan Natives and the *Tundra Times*." *Journal of Communication* 36 (Summer 1986): 10–30.

Dannenbaum, Jed. "John Muir and Alaska." *Alaska Journal* 2 (Autumn 1972): 14–21.

De Novo, John. "Petroleum and the United States Navy." *Mississippi Valley Historical Review* 41 (March 1955): 647–56.

Devall, Bill. "The Deep Ecology Movement." *Natural Resources Journal* 20 (April 1980): 299–322.

Drew, Elizabeth, B. "Dam Outrage: The Story of the Army Engineers." *Atlantic Monthly* 225 (April 1970): 51–62.

Dufresne, Frank. "What of Tomorrow?" *Alaska Sportsman* 3 (April 1937): 9–10.

———. "Rampart Roulette." *Field and Stream* 69 (August 1964): 10–13, 77.

East, Ben. "Is it TAPS for Wild Alaska?" *Outdoor Life* 145 (May 1970): 43–46, 86, 88–89.

Elliott, Henry W. "Ten Years' Acquaintance with Alaska." *Harper's* 55 (November 1877): 801–16.

———. "The Loot and Ruin of the Fur-Seal Herd of Alaska." *North American Review* 185 (June 1907): 426–36.

Evans, Brock. "Two Roads North." *Sierra Club Bulletin* 59 (November-December 1974): 19.

Evans, Charles D. "Environmental Effects of Oil Development in the Cook Inlet Area." 21st Alaska Science Conference, Alaska Branch, American Association for the Advancement of Science (AAAS). Edited by Eleanor Viereck. *Science in Alaska*, 1970: 213–21.

Fitzpatrick, Brian, "The Big Man's Frontier and Australian Farming." *Agricultural History* 21 (January 1947): 8–12.

Fleming, Donald. "Roots of the New Conservation Movement." *Perspectives in American History* 6 (1972): 7–91.

Foote, Don C. "Conservation and Arctic Alaska Eskimos." *Alaska Conservation Society News Bulletin* 3 (April 1962): 1–3.

Fortier, Edward J. "Alaska Discovers 'Intruder' Threatening Wildlife." *National Observer* 26 February 1968.

Francis, Karl E. "Outpost Agriculture, The Case of Alaska." *Geographical Review* 57 (October 1967): 496–505.

Friggins, Paul. "The Great Alaska Oil Rush." *Reader's Digest* 95 (July 1969): 66–70.

———. "The Great Alaska Pipeline Controversy." 100 (November 1972): 125–29.

Frome, Michael. "Hickel and the Arctic." *Field and Stream* 74 (November 1969): 12–49.

Gabrielson, Ira N. "Alaska Wildlife." *Audubon* 45 (November-December 1943): 329–35.

Gay, James Thomas. "Henry Wood Elliott, Crusading Conservationist." *Alaska Journal* 3 (Autumn 1973): 211–16.

Gilman, William. "Colonists for Alaska." *Yale Review* 34 (Summer 1945): 666–82.

Grahame, Arthur. "A-Test Alaska Threat?" *Outdoor Life* 127 (January 1961): 137–39.

Gruening, Ernest. "Go North, Young Man." *Reader's Digest* 44 (January 1944): 53–57.

———. "Alaska: Progress and Problems." *Scientific Monthly* 77 (July 1953): 3–12.

———. "The Plot to Strangle Alaska." *Atlantic Monthly* 216 (July 1965): 56–59.

Haas, Stanley B. "Marine Transport—The Northwest Passage." *Science in Alaska*, 1970: 76–81.

Hellenthal, John. "Why Not Go Modern On Conservation?" *Alaska Life: The Territorial Magazine* 4 (January 1941): 3, 13, 17–19, 27–28, 30.

Hinckley, Ted C. "Alaska and the Emergence of America's Conservation Consciousness." *Prairie Scout* 2 (1975): 79–111.

———. "The Inside Passage: A Popular Golden Age Tour." *Pacific Northwest Quarterly* 56 (1965): 67–74.

Hodgson, Bryan. "The Pipeline: Alaska's Troubled Colossus." *National Geographic* 150 (November 1976): 684–717.

Hoffman, Peter M. "Evolving Judicial Standards under the National Environmental Policy Act and the Challenge of the Alaska Pipeline." *Yale Law Journal* 81 (1972): 1592–1639.

Huber, Louis R. "What Price Fontier? Alaska Has Inherited The Role Of Pioneer!

What Keeps Her From Fulfilling It?" *Alaska Life* 2 (February 1941): 4–5, 15.

Jacobs, Wilbur R. "The Great Despoliation: Environmental Themes in American Frontier History." *Pacific Historical Review* 67 (February 1978): 1–26.

Johannsen, Robert W. "James C. Malin: An Appreciation." *Kansas Historical Quarterly* 38 (Winter 1972): 457–66.

Juricek, John T. "American Usage of the Word 'Frontier' From Colonial Times to Frederick Jackson Turner." *Proceedings of the American Philosophical Society* 110 (February 1966): 10–34.

Karamanski, Theodore J. "The Canol Project: A Poorly-Planned Pipeline." *Alaska Journal* 9 (Autumn 1979): 17–21.

Kay, Barry. "Stopping the Pipeline is Northern Native Cause." *Oilweek* 23 (20 March 1972): 20–21.

Keithahn, Edward L. "Alaska Ice, Inc." *Pacific Northwest Quarterly* 36 (April 1945): 121–31.

Kennedy, Michael S. "Belmore Browne and Alaska." *Alaska Journal* 3 (Spring 1973): 96–104.

Klein, David R. "The Alaska Oil Pipeline in Retrospect." *Transactions of the 44th North American Wildlife and Natural Resources Conference*, Wildlife Management Institute, Washington, D.C. (1979), 235–46.

———. "Reaction of Caribou and Reindeer to Obstructions—A Reassessment." In Reimers, E., E. Gaare, and S. Skjenneberg, eds. *Proceedings of the 2d International Reindeer/Caribou Symposium*, (Roros, Norway, 1979). Trondheim, Norway, 1980: 519–27.

Koenig, Duane. "Ghost Railway in Alaska: The Story of the Tanana Valley Railroad." *Pacific Northwest Quarterly* 45 (January 1954): 8–12.

Kent, P. E. "Entry into Alaska." *BP Shield* January 1970: 5–9.

Knox, Robert. "Will The TAPS Pipeline Really Go?" *Alaska Industry* 2 (June 1970): 26–28.

Lapham, Lewis. "Alaska: Politicians and Natives, Money and Oil." *Harper's* 240 (May 1970): 85–102.

Long, John Sherman. "Webb's Frontier and Alaska." *Southwest Review* 56 (Autumn 1971): 301–9.

Malin, James C. "Mobility and History: Reflections on the Agricultural Policies of the United States in Relation to a Mechanized World." *Agricultural History* 17 (1943): 177–91.

———. "Space and History: Reflections on the Closed Space Doctrines of Turner and Mackinder and the Challenge of Those Ideas by the Air Age, Part 1." *Agricultural History* 18 (April 1944): 65–74; "Part 2" (July 1944): 107–26.

Marshall, George. "Robert Marshall as a Writer: Bibliography, 1901–39." *Living Wilderness* 16 (Autumn 1951): 1–23.

Marshall, Robert. "Should We Settle Alaska?" *New Republic* 102 (8 January 1940): 49–50.

McAlister, Bruce W., and Wayne V. Burt. "Predicted Water Temperatures for the Rampart Reservoir on the Yukon River." Proceedings, 14th Alaska Science Conference, Anchorage, 27–30 August 1963. *Science in Alaska*. Edited by George Dahlgren, 1964: 249–65.

McCloskey, Michael. "The Wilderness Act of 1964: Its Background and Meaning." *Oregon Law Review* 45 (June 1966): 288–321.

———. "Wilderness Movement at the Crossroads." *Pacific Historical Review* 5 (August 1972): 346–63.

McDonald, Donald. "Highway . . . Hell!" *Alaska Life* 2 (September 1942): 3–12.

Means, Richard L. "The New Conservation." *Natural History* 78 (August-September 1969): 16–25.

Meinig, Donald W. "Americans Wests: Preface to a Geographical Interpretation." *Annals of the Association of American Geographers* 62 (June 1972): 159–84.

———. "The Continuous Shaping of America: A Prospectus for Geographers and Historians." *American Historical Review* 83 (December 1978): 1186–1205.

Mikesell, Marvin W. "Comparative Studies in Frontier History." *Annals of the Association of American Geographers* 50 (1960): 64–74.

Milton, John P. "Arctic Walk." *Natural History* 78 (May 1969): 45–52.

———. "The Web of Wildness." *Living Wilderness* (Special Alaska Issue) 35 (Winter 1971–72): 15–19.

Mood, Fulmer. "Notes on the History of the Word *Frontier*." *Agricultural History* 22 (1948): 78–83.

———. "The Concept of the Frontier, 1871–1898: Comments On A Select List of Source Documents." *Agricultural History* 19 (January 1945): 24–30.

Moore, Terris. "Alaska's First American Century: The View Ahead." *Polar Record* 14 (1968): 3–15.

Moran, Casey. "Land to Loot." *Collier's* (Special Alaska Issue), 6 August 1910.

———. "The Truth About Alaska." *Alaska-Yukon Magazine* 11 (January 1911): 53–60.

Morehouse, Thomas A. "Fish, Wildlife, Pipeline: Some Perspective on Protection." *The Northern Engineer* 16 (Summer 1984): 18–26.

Morgan, David D. "Oil by Rail." *Environment* 14 (October 1972): 30–32.

Moxness, Ron. "The Long Pipe." *Environment* 12 (October 1970): 12–23, 36.

Murie, Olaus J. "Aircraft and Wilderness." *Living Wilderness* 12 (Autumn 1947): 2–6.

———. "Planning for Alaska's Big Game." Inaugural Alaska Science Conference, National Academy of Sciences, National Research Council, Washington, D.C., 9–11 November 1950. Selected Papers of the Alaska Science Conference, edited by Henry B. Collins. *Science in Alaska*. Washington, D.C.: Arctic Institute of North America, Special Publication no. 1, June 1952: 258–68.

———. "Alaska with Olaus J. Murie." *Living Wilderness* 21 (Fall-Winter 1956–57): 28–30.

———. "Wilderness Philosophy, Science, and the Arctic National Wildlife Range." Proceedings, 12th Alaska Science Conference, College, 28 August–1 September 1961. Edited by George Dahlgren. *Science in Alaska*, May 1962: 58–69.

Myers, Henry R. "Federal Decision-Making and the Trans-Alaska Pipeline." *Ecology Law Quarterly* 4 (1975): 915–61.

Nash, Roderick. "Do Rocks Have Rights?" *Center Magazine* 10 (November-December 1977): 2–12.

————. "Ideal and Reality in 'Ultimate' Wilderness: Aviation and Gates of the Arctic National Park." *Orion* (Spring 1983): 5–13.

Naske, Claus-M. "The Alcan: Its Impact on Alaska." *The Northern Engineer* 8 (Spring 1976): 12–18.

Nelson, Arnold and Helen. "The Bubble of Oil at Katalla." *Alaska Journal* (A 1981 Collection): 19–27.

Nelson, Edward W. "Notes on the Wild Fowl and Game Animals of Alaska." *National Geographic* 9 (4 April 1898): 132.

Neuberger, Richard L. "How Much Conservation?" *Saturday Evening Post* 212 (15 June 1940): 12–13, 89–90, 92, 94–96.

Norgaard, Richard. "Petroleum Development in Alaska: Prospects and Conflicts." *Natural Resources Journal* 12 (1972): 83–107.

Norton, David W. "Pipeline Surveillance from the Inside." *Alaska Conservation Review* 16 (Summer-Fall, 1975): 4–7.

Norwood, Gus. "Rampart Dam in Perspective." *Alaska Construction* (July-August 1966): 18–20.

Ostrander, J. Y. "What Conservation Does to Alaska." *Alaska-Yukon Magazine* 11 (January 1911): 45–7.

Pain, Stephanie. "Alaska Lays its Wildlife on the Line." *New Scientist* 114 (30 April 1987): 51–5.

Panitch, Mark. "Alaska's Pipeline Road: New Conflicts Loom." *Science* 189 (4 July 1975): 30–32.

Pekkanen, John. "Ecology: a Cause Becomes a Mass Movement." *Life* 68 (January 1970): 22–32.

"'Phooey—on the Highway.' 'By An Alaskan Businessman.'" *Alaska Life* 4 (July 1941): 3, 14–15, 30.

"Project Chariot—The Long Look." *Sierra Club Bulletin* 46 (May 1961): 5–9, 12–13.

Pollak, Richard. "Are We about to Plunder Alaska?" *Current* 129 (May 1971): 39–46.

Pruitt, William O. "Animal Ecology and the Arctic National Wildlife Range." *Science in Alaska,* 1962: 44–51.

————. "High Radiation in Eskimos." *Audubon* 65 (September-October 1963): (1–3, 6.)

Quigley, Carroll. "Our Ecological Crisis." *Current History* 59 (1970): 3–12, 49.

Rainey, Froelich. "Alaskan Highway an Engineering Epic." *National Geographic* 83 (February 1943): 143–68.

Rau, Ron. "The Taming of Alaska." *National Wildlife* 14 (November 1976): 19–23.

Richardson, Harold W. "Alcan—America's Glory Road: Part 1: Strategy and Location." *Engineering News-Record* 129 (17 December 1942): 81–96.

————. "Alcan—America's Glory Road: Part 2: Supply, Equipment, Camps." *Engineering News-Record* 129 (31 December 1942): 35–42.

————. "Alcan—America's Glory Road: Part 3: Construction Tactics." *Engineering News-Record* 130 (14 January 1943): 131–38.

Riggs, Anna E. "E. W. Nelson, Unpaid Collector." *Alaska Journal* 10 (Winter 1980): 91.

Roberts, Michael. "Dissolute Alaska: The Coming of the Pipe." *New Republic* 173 (1 November 1976): 17–21.

"Rush for Riches on the Great Pipeline." *Time* (European Edition) 105 (2 June 1975): 26–32.

Roby, D. P., R. D. Cameron, K. R. Whitten, and Smith, W. T. "Caribou and the Trans-Alaska Pipeline: A Summary of Current Knowledge." Proceedings, 27th Alaska Science Conference, Fairbanks, 4–7 August 1976. *Science in Alaska*, 1976: 38–43.

Ryan, Pat. "The Earth As Seen From Alaska: Hickel In Earth Day Observances." *Sports Illustrated* 32 (4 May 1970): 26–28, 31.

Saltonstal, Richard. "The Education of Wally Hickel." *Time* 94 (1 August 1969): 13.

Sanders, Ralph. "Nuclear Dynamite: A New Dimension in Foreign Policy." *Orbis—A Quarterly Journal of World Affairs* 4 (Fall 1960): 307–22.

Sanford, David. "Hickel's Pickle." *New Republic* 160 (18 January 1969): 11–12.

Sauer, Carl O. "Foreword to Historical Geography." *Annals of the Association of American Geographers* 31 (1941): 1–24.

———. "Theme of Plant and Animal Destruction in Economic History." *Journal of Farm Economics* 20 (1938): 765–75.

Schoenfeld, Clay S. "The Ecology of the New Conservation." *Proceedings, 35th North American Wildlife Conference, Chicago, 1970.* Washington, D.C.: Wildlife Management Institute, 1970: 351–67.

Shalkop, Robert L. "Henry W. Elliott, Fighter for the Fur Seals." *Alaska Journal* 13 (Winter 1983): 4–12.

Schueler, R. L. "Ecology: the New Religion." *America* 122 (21 March 1970): 292–95.

Sherrill, Robert. "Unsafe at Any Width; The Trans-Alaska Pipeline." *The Nation* 218 (11 June 1973): 745–52.

Sherwood, Morgan B. "The End of American Wilderness." *Environmental Review* 9 (Fall 1985): 197–209.

Slotnick, Herman E. "The Ballinger-Pinchot Affair in Alaska." *Journal of the West* 10 (April 1971): 337–47.

Smith, Anthony Wayne. "Alaskan Prospect." *National Parks Magazine* 43 (September 1969): 2.

Smith, Richard G. "The Kenai National Moose Range." *Living Wilderness* 30 (Winter 1966–67): 24–32.

Sommer, Robert. "Ecology and the Energy Shortage." *The Nation* 217 (10 December 1973): 615–16.

Starnes, Richard. "The Rampart We Watch." *Field and Stream* 68 (August 1963): 12, 64–66.

———. "Can the Arctic be Saved?" *Field and Stream* 75 (July 1970): 16–17, 82.

Stefansson, Vilhjalmur. "Routes to Alaska." *Foreign Affairs* 19 (July 1941): 861–69.

Stone, Christopher D. "Should Trees Have Standing?—Toward Legal Rights for Natural Objects." *Southern California Law Review* 45 (1972): 450–501.

Storper, Michael. Laura Baker, and Mary Lou Seaver. "Too Much, Too Soon (Too Bad):" *Not Man Apart* 7 (February 1977): 1–4.

Sumner, Lowell. "Alaska's Biological Wealth: Why Let History Repeat Itself?" Proceedings, 2d Alaska Science Conference, Mount McKinley Park, 4–8 September 1951. *Science in Alaska,* 1951: 337–39.

———. "Arctic Wilderness." *Living Wilderness* 18 (Winter 1953–54): 5–16.

———. "Your Stake in Alaska's Wildlife and Wilderness." *Sierra Club Bulletin* 41 (December 1956): 54–71.

Sundborg, George. "The 'Biggest Dam' on the Mighty Yukon." *Rural Electrification* 17 (July 1959): 15–17.

"TAPS Not Tundra's First Line—Not By 45 Years." *Oil and Gas Journal* 68 (10 August 1970): 104.

Thomas, Paul. "Engineers: Designing for the North." *Alaska Industry* 8 (April 1976): 50, 85–86, 98.

Treadgold, Donald W. "Russian Expansion in the Light of Turner's Study of the American Frontier." *Agricultural History* 26 (October 1952): 147–51.

Tribe, Lawrence H. "Ways Not to Think About Plastic Trees: New Foundations for Environmental Law." *Yale Law Journal* 83 (June 1974): 1315–48.

Tucker, William. "Is Nature Too Good For Us?" *Harper's* 264 (March 1982): 27–35.

Udall, James R. "Polar Opposites." *Sierra* 72 (September-October 1987): 41–48.

Underwood, John J. "Population for Alaska Awaits Transportation Facilities." *Alaska-Yukon Magazine* 13 (February 1912): 20–27.

Vevier, Charles. "The Collins Overland Line and American Continentalism." *Pacific Historical Review* 28 (1959): 237–53.

Walker, Francis A. "The Indian Question." *The North American Review* 116 (April 1873): 329–88.

Watson, J. Wreford. "Image Geography: The Myth of America in the American Scene." *Advancement of Science* 27 (September 1970): 71–80.

Webb, Walter Prescott. "Geographical-Historical Concepts in American History." *Annals of the Association of American Geographers* 50 (June 1960): 85–95.

Weeden, Robert B. "Conservation Society and Kilowatts." *Alaska Conservation Society News Bulletin* 2 (May 1961): 3–5.

———. "Kenai National Moose Range: A Resource Management Microcosm." *Alaska Conservation Society News Bulletin* 10 (Spring 1970): 8–9.

Weeden, Robert, and David R. Klein. "Wildlife and Oil: A Survey of Critical Issues in Alaska." *Polar Record* 15 (1971): 479–95.

Weisberg, Barry. "The Ecology of Oil: Raping Alaska." *Ramparts* 8 (January 1970): 25–33.

Welch, Richard E. "American Public Opinion and the Purchase of Russian America." *American Slavic and East European Review* 17 (1958): 481–94.

White, Lynn, Jr. "The Historical Roots of Our Ecological Crisis." *Science* 155 (10 March 1967): 1203–7.

Wilson, William H. "The Alaska Railroad and the Agricultural Frontier." *Agricultural History* 52 (April 1978): 263–79.

———. "The Alaska Engineering Commission and a New Agricultural Frontier." 42 (October 1968): 339–50.

———. "Alaska's Past, Alaska's Future." *Alaska Review* 4 (Spring-Summer 1970): 1–11.

Wood, Virginia Hill. "Rampart—Foolish Dam." *Living Wilderness* 29 (Spring 1965): 4–6.

Woodman, Lyman L. "Building the Alaska Highway: A Saga of the Northland." *The Northern Engineer* 8 (Summer 1976): 11–28.

Wright, John K. "Terra Incognitae: The Place of Imagination in Geography." *Annals of the Association of American Geographers* 37 (1947): 1–15.

Wright, Samuel. "A Letter from the Arctic." *Living Wilderness* 33 (Spring 1969): 5–6.

———. "Last Frontier." *American West* 7 (November 1970): 32–37.

Zemansky, Gil. "Promise and Practice: Water Quality Regulation On The Alaska Pipeline." *The Northwest Environmental Journal* 1 (Spring 1985): 35–64.

8. DISSERTATIONS

Buske, Frank E. "The Wilderness, The Frontier and the Literature of Alaska to 1914: John Muir, Jack London and Rex Beach." Ph.D. diss., University of California, Davis, 1976.

Coyne, John. "Alaska: Image of a Resource Frontier Region, 1867–1900." M.A. thesis, Birkbeck College, University of London, 1976.

Cuba, James Lee. "A Moveable Frontier: Frontier Images in Contemporary Alaska." Ph.D. diss., Yale University, 1981.

Evans, Gail Edith Hallett. "From Myth to Reality: Travel Experiences and Land-scape Perceptions in the Shadow of Mount McKinley, Alaska." M.A. thesis, University of California, Santa Barbara, 1987.

Glines, Carroll V. "Alaska's Press and the Battle for Statehood." M.A. thesis, American University, Washington, D.C., 1969.

Jody, Marilyn. "Alaska in the American Literary Imagination: A Literary History of Frontier Alaska." Ph.D. diss., University of Indiana, 1969.

Lacey, Michael J. "The Mysteries of Earth-Making Dissolve: A Study of Wash-ington's Intellectual Community and the Origins of American Environmen-talism in the Late Nineteenth Century." Ph.D. diss., George Washington University, 1979.

Wolfe, John Hilton. "Alaskan Literature: The Fiction of America's Last Wilder-ness." Ph.D. diss., Michigan State University, 1973.

Zemansky, Gil. "Water Quality Regulation during Construction of the Trans-Alaska Oil Pipeline." Ph.D. diss., University of Washington, 1983.

Index

Page numbers in boldface type refer to maps, charts, and illustrations.

433

Wilderness Society: campaign for ANWR, 96, 97, 99, 107; founding of, 79; Lois Crisler and, 80; and statehood, 88; and Kenai oil leasing, 94; and Project Chariot, 121, 124, 128; and the Rampart Dam, 144; and oil development, 169, 176, 185, 186, 189, 217, 218, 230, 238, 239
Wildlife: concern for, 42–44, 78–80, 315–16; exploitation of, 37–39; impact on TAPS, 263–64, 266–67, 313. *See also* Antelope; Bald eagle; Buffalo; Caribou; Ducks; Grizzly bear; Hunting; Marine mammals; Moose; Musk ox; Otter; Seals; Sportsmen; Wolf; Wolverine
Wildlife Management Institute, 87, 94, 142, 217
Wildlife Society, 142, 185
Wilson, William H., 209
Wiseman, 13, 57, 58, 166, 225, 252
Wolf, 170, 267
Wolf, David P., 190
Wolfe, John N., 120, 123, 125, 130
Wolff, Ernest, 104, 175, 201–2

Wolverine, 170
Wood, Ginny, 103, 116, 146
Wood, William R.: supports Project Chariot, 122, 125; supports Rampart Dam, 136, 140, 143, 155; supports TAPS, 203, 254, 265
Wood Buffalo Park, 105
Worthless lands thesis, 316, 355n.110
Wright, Samuel A., 166, 198, 201, 223

Yannacone, Victor J., 173
Yellowstone National Park, 42, 105, 106, 107
Yosemite National Park, 19, 107, 217
Young, Don, 148, 246, 249, 311
Yukon Flats, 134, 141, 142, 144, 145, 146, 147, 148, 150, 151, 153, 158, 159
Yukon Flats National Wildlife Refuge, 309
Yukon River, 134–35, 159, 209, 258

Zelnick, Robert, 192
Zemansky, Gil M., 264
Zero Population Growth, 218, 219

About the Author

Peter A. Coates was born in the United Kingdom at Southport, Lancashire. His first degree is from St. Andrews University, Scotland, and his doctorate is from Cambridge University. He is currently a lecturer in the Department of Historical Studies at Bristol University, England.